Materials Forming, Machining and Tribology

Series editor

J. Paulo Davim, Aveiro, Portugal

More information about this series at http://www.springer.com/series/11181

Prakriti Kumar Ghosh

Pulse Current Gas Metal Arc Welding

Characteristics, Control and Applications

Springer

Prakriti Kumar Ghosh
Department of Metallurgical and Materials Engineering
Indian Institute of Technology Roorkee
Roorkee, Uttarakhand
India

ISSN 2195-0911 ISSN 2195-092X (electronic)
Materials Forming, Machining and Tribology
ISBN 978-981-10-3556-2 ISBN 978-981-10-3557-9 (eBook)
DOI 10.1007/978-981-10-3557-9

Library of Congress Control Number: 2017932417

Printed on acid-free paper

This Springer imprint is published by Springer Nature
The registered company is Springer Nature Singapore Pte Ltd.
The registered company address is: 152 Beach Road, #21-01/04 Gateway East, Singapore 189721, Singapore

The book is dedicated to my beloved parents and in-laws

Preface

Pulse current gas metal arc welding (P-GMAW) is rapidly becoming a popular welding process in a large variety of industries engaged in weld fabrication of light to heavy sections of ferrous and nonferrous metals. In order to better understand the intricacies of P-GMAW, well-known welding researchers have registered their understandings from different points of view to explore its novelty in applications. Effective use of this welding process primarily involves understanding of the process parameters governing the behavior of metal transfer affecting thermal characteristics of the weld. The P-GMAW was initially well recognized as a unique process of hazard-free joining of relatively thin section with one drop transfer per pulse at a low heat input. The knowledge of controlling the characteristics of pulsed arc has lead to development of some innovative welding processes especially suitable to join sub-millimeter range thin sections of ferrous and nonferrous metals. But later on the enormous potential of P-GMAW process in precise control of weld characteristics could also be successfully used in weld fabrication of large variety of relatively heavy structures. This is primarily to join thick sections in different positions with better weld quality using multiple drop transfer per pulse at high deposition rate with low heat buildup in weld. However the implications of controlling the pulse parameters influencing these conditions under the multiple drop transfer per pulse are not yet systematically well understood. This book is exclusively dedicated to discuss the knowledge base developed to use the P-GMAW process for welding of practically every section size of ferrous and nonferrous materials at any position of engineering structure.

The multipurpose application of P-GMAW primarily requires the knowledge of designing and control of pulse form in order to manipulate energy input and its distribution in the process. To facilitate these kinds of welding operations with ease and comfort, by negotiating relatively larger number of pulse parameters, variety of user-friendly efficient power sources have been developed. They are dedicated to one knob control system of programmable power sources operating in non-synergic and synergic modes. However, regardless of all these facilities, the P-GMAW process is still handicapped to attract the welding fraternity for its large-scale use with unanimous choice and confidence. This is primarily because of deficiency of knowledge about the rather complicated influence of each pulse parameters on weld quality when they simultaneously interact during operation. Paucity of knowledge especially exists to control the relatively large number of interactive process parameters in case of multiple drop transfer per pulse.

The use of comparatively higher energy makes the process control more critical with respect to regulation of its thermal behavior. Author and his coworkers have tried to analyze these aspects from various perspective of basic and applied nature in their untiring efforts of more than three decades. Getting an insight of this subject became possible after development of a concept of summarized influence of simultaneously interactive pulse parameters of welding operation. It has opened a scope for in-depth critical understanding of the process and its performance in the light of a summarized influence of pulse parameters defined by a hypothetical dimensionless factor *phi*. A relatively smooth understanding of P-GMAW process as a function of this factor explores ample scope of its use in innovative applications with desired weld quality, which was otherwise difficult. The knowledge of P-GMAW has been systematically handled in order to present the subject from basic understanding of the process control to critical aspects of welding metallurgy and joint properties. The book also covers some advanced concept of the control of pulsed current for application of GMAW in joining of extra thin sections of ferrous and nonferrous metals by application of appropriate mode of metal transfer. The knowledge of largely smooth correlations among the *phi*, process control, weld characteristics, and joint properties has been emerged to explore the development of computer-aided P-GMAW process to produce desired weld quality. In the background of the basic understandings about the control of high deposition P-GMAW, this book as a monograph has been planned to serve the academicians, researchers, and practitioners of welding engineering.

The first compilation of the basic to application-oriented understanding of high deposition pulse current GMAW process in the nine chapters of this monograph may serve as first-hand comprehensive knowledge to practice advanced welding engineering in prospective applications. It may give confidence to the shop floor welding engineer to handle this otherwise highly critical welding process with confidence. It may inspire the researcher to think further for more versatile application of unique nature of pulse current in GMAW process to develop cutting edge welding technology. The author is aware of many other fundamental issues of the P-GMAW process which remains uncovered in this book and he also duly apologizes for non-intentional oversight of many potential contributions of various researchers to refer in this book. All such deficiencies of this presentation are kept in mind to establish the potentials of P-GMAW more fundamentally in another write-up planned for near future.

I would like to thankfully acknowledge the support of the Alexander von Humboldt Foundation to execute several collaborative advanced researches in this area at different institutes of Germany, and writing this book by compiling the unique understandings of P-GMAW. I am deeply indebted to the collaborative support, encouragement, and initial help provided by Prof. Dr.-Ing. Dr. h.c. Lutz Dorn of Technical University of Berlin in structuring the write-up of this book.

I also wish to thank the untiring company given by my wife Mita with lots of patience and tolerance all along the journey to be qualified, and develop the ability to write this book. At this point it always comes in memory the sweet and lively support of my son Aritra and daughter-in-law Gita that always encouraged me in completing this project. The relentless constructive arguments and active involvement of Aritra in execution of deliberations have brought perfection to this creation.

Roorkee, India
December 2016

Prakriti Kumar Ghosh

The original version of the book was inadvertently published without the preface in Frontmatter. The book has been updated with the change. The errratum to the book frontmatter is available at 10.1007/978-981-10-3557-9_10

Contents

About the Author

Dr. P.K. Ghosh, FIE, FIIW, MNASI graduated from University of Burdwan and received his Master's and Ph.D. in Metallurgical Engineering from IIT Roorkee (erstwhile University of Roorkee). He is Senior Professor in Department of Metallurgical & Materials Engineering, Govt. of India IPR Chair Professor and Deputy Director (also held the position of Director Officiating) of IIT Roorkee. He is a recipient of the Alexander von Humboldt Fellowship (Germany) and also several national awards including Binani Gold Medal and SAIL Gold Medal of Indian Institute of Metals. He has extensively travelled to many countries as Visiting Scientist and Professor for scientific and engineering research and education. His research interests lie in the area of welding and adhesive joining of materials, polymer-based nano filler composite, metal matrix composites and fatigue and fracture mechanics. His work includes more than 20 national and international sponsored research projects. He has guided a large number of Master's and Ph.D. theses, published more than 325 research papers, and filed five patents.

Nomenclature

Chapter 1

A_w	Cross-sectional area of the filler wire (m^2)
$V_{w(cc)}$	Wire feed rate for the conventional GMAW (ms^{-1})
ρ_w	Density of the filler wire (kg m^{-3})
Q_m	Heat per unit mass required for melting the filler wire (J.kg^{-1})
V_a	Anode fall voltage (V)
ζ	Work function of metal surface (V)
R_o	Resistivity of the filler wire (Ω m)
E_w	Electrode extension (m)
I	Welding current (A)
ψ	Effective melting potential at anode
A	Constant representing wire melting due to arc heating
B	Constant representing wire melting due to resistive heating
k	Boltzmann constant
T	Temperature (K)
e	Energy of one electron volt in (J)
R_s	Resistance of the electrode extension
Q^*	Power transferred from the molten anodic tip to solid part of the electrode
h	Depth of the cavity formed in weld pool by impact of droplet
d_t	Distance travelled by the centre of mass of the slug
R_c	Radius of the cylindrical cavity formed in weld pool by impact of droplet
T	Temperature of a point at a distance with respect to central axis of welding arc
T_0	Initial temperature of base plate
Q_p	Heat transferred per unit time by point heat source (J s^{-1})
K	Thermal conductivity of the base metal
q(x,y,z)	Heat flux at a point (x,y,z) within weld pool
a_h, b_h and c_h	Ellipsoidal heat source parameters

η_a	Heat transfer efficiency of welding arc
Q_{arc}	Arc heat input
Q_t	Total heat transferred to weld pool per unit time ($J\ s^{-1}$)
A	Thermal diffusivity of base metal
L_w	Half length of base plate
B_w	Half width of base plate
D	Thickness of base plate
ΔG^*	Free energy of formation of critical sized nucleus of solid
ΔG^*_{het}	Free energy of formation of critical sized nucleus in presence of inoculants
$f(\beta)$	Shape factor of nucleus

Chapter 2

(I_m)	Mean current
(I_p)	Peak current
(I_b)	Base current
I_e	Peak current in excess over the base current
I_{eff}	Effective current
(f)	Pulse frequency
(t_p)	Peak current duration
t_b	Base current duration
t_{pul}	Pulse cycle time
β	Minimum base current for stable arc
(V_w)	Wire feed rate
(V_d)	Droplet volume
D_n	Drop detachment parameter
V'_w	Estimated wire melting rate
V'_d	Estimated volume of a droplet
(λ_c)	Critical wavelength of cylindrical current conductor liquid column
δ	Electrode tapering coefficient
D	Diameter of droplet
μ_o	Permeability of free space
γ	Coefficient of surface tension of liquid filler metal
ρ_d	Density of liquid filler metal
N_D	Number of droplet transferred per pulse
R_w	Radius of filler wire
$V_{w(pc)}$	Overall wire burn off rate of filler wire
V_{wp}	Wire burn off rates during t_p
V_{wb}	Wire burn off rates during t_b
k_p	Pulse duty cycle

Chapter 3

ϕ	A Dimensionless factor defining summarised influence of pulse parameters
D_R	Diameter of arc root
D_P	Projected diameter of arc
L	Arc length
D	Diameter of droplet
M_t	Mass of filler metal transferred per pulse
A_s	Total surface area of molten metal transferred per pulse
γ	Surface tension
ρ	Density of filler wire
d	Diameter of filler wire
C_P	Specific heat of filler wire
T_m	Melting point of filler wire
μ_o	Permeability of free space
N_d	Number of metal drops transferred per pulse
V_i	Velocity of metal drop at the time of detachment from electrode
V_{de}	Velocity of a metal drop at the time of deposition in weld pool
V_t	Velocity of metal drop during its travel from the electrode tip to weld pool
τ	Flight time of metal drop from the electrode tip to weld pool
h	Distance travelled by a metal drop in between two consecutive frames of video-graph
g	Acceleration due to gravity
a_{cc}	Acceleration of droplet due to plasma aerodynamics drag force
V_{eff}	Effective velocity of plasma
ρ_g	Mass density of plasma in arc column
C_d	Drag coefficient,
ρ_d	Density of molten filler metal at its detachment temperature
R_e	Reynolds number
ν	Kinematic viscosity of ionized shielding gas
j_g	Current density of plasma in arc column
I_{eff}	Effective current
k_p	Pulse duty cycle
t_{pul}	Pulse cycle time period ($t_p + t_b$)
k_w	Wave number
k_c	Constant

Chapter 4

H_{AW}	Arc heat transfers to the weld pool
H_A	Heat consumed for melting of filler wire per unit time
H_{arc}	Arc heat generated by the energy input
V	Arc voltage

I_{eff}	Effective current
ψ	Effective melting potential at anode
η_a	Arc heat transfer efficiency
H_o	Heat generation at the tip of electrode per unit time
H_R	Heat generation by resistive heating per unit time
R_W	Electrical resistance of the filler wire extension
R_o	Electrical resistivity of filler wire at ambient temperature
A_w	Cross-sectional area of filler wire
H_i	Heat content per unit mass of molten metal at the time of detachment
H_w	Heat required per unit time for melting and producing a superheated drop of electrode
T_a	Ambient temperature
T_i	Temperature of molten metal at the time of detachment
V_w	Wire feed speed
ρ_w	Mass density of filler wire
j_{eff}	Effective current density at electrode tip
$C_{p(s)}$	Specific heat of the solid filler metal
$C_{p(l)}$	Specific heat of molten filler metal
T_m	Melting point of filler metal
L_h	Latent heat of fusion of filler metal
H_{cv}	Convective heat loss per unit mass of the filler metal during flight
H_r	Radiation heat loss per unit mass of the filler metal during flight
H_{de}	Heat content per unit mass of molten metal at the time of deposition
T_{de}	Temperature of droplet at the time of deposition
α_h	Heat transfer coefficient of droplets in flight
ε	Emissivity
σ	Stefan–Boltzmann constant
τ	Flight time of droplet
V_{max}	Maximum axial velocity of electro-magnetically induced plasma jet at peak current
j_g	Current density in plasma of arc column
V_{eff}	Effective velocity of plasma in arc column
F	Aerodynamic drag force on a droplet within arc column
C_d	Drag coefficient during transfer of droplet
R_e	Reynolds number
N_u	Nusselt number
λ_a	Thermal conductivity of gas shielding
C_p	Specific heat of argon plasma
η	Viscosity of ionized argon
H_{dp}	Heat content of filler metal transferred per pulse
H_f	Heat content of filler metal transferred unit time
H_{fpp}	Heat transfer by superheated filler metal during peak current
H_{fpb}	Heat transfer by superheated filler metal during base current

H_{dep}	Heat content per unit mass of filler metal at the time of deposition during peak current
H_{deb}	Heat content per unit mass of filler metal at the time of deposition during base current
M_{tp}	Mass of filler metal transferred during peak current
M_{tb}	Mass of filler metal transferred during base current
H	Total heat transferred per pulse to the weld pool
W	Work done in creating a paraboloidal cavity in weld pool
r_1	Focal length of paraboloidal cavity in weld pool
h	Depth of paraboloidal cavity in weld pool
p	Hydrostatic pressure
V_c	Volume of the paraboloidal cavity
r_c	Radius at projected surface of the cavity
ρ_m	Density of molten metal in weld pool
E	Kinetic energy of droplet during vertically downward transfer
M_d	Mass of a droplet
V_{de}	Velocity of a droplet at the time of vertically downward deposition on weld pool
T	Temperature at any point of a weld
R	Distance of any point with respect to central axis of welding arc
H_{tl}	Total heat loss in flight of molten metal transfers per pulse
D_T	Depth of thermocouple tip from its surface in fused weld pool
E_c	Conventional energy input per unit length
V	Welding speed
k_t	Thermal conductivity of base metal

Chapter 5

F_b	Fraction of base metal fusion per unit mass of weld deposition
D_{nb}	Dilution of weld bead
D_{nw}	Dilution of weld
P_n	Penetration or depth of fusion
W_b	Bead width
H_b	Bead height or reinforcement
θ	Weld toe angle
A_f	Area of plate fusion
A_d	Area of weld deposit
ρ_b	Density of base metal
ρ_d	Density of weld deposit
F_f	Form factor
D_r	Rate of deposition
k_p	Pulse duty cycle

Chapter 6

l_d	Arm length of primary dendrite
D_s	Arm spacing of primary dendrites
A_r	Aspect ratio of primary dendritic arm
C_R	Cooling rate
Δl	Longitudinal shrinkage
σ_{tr}	Transverse shrinkage stress
$\Delta_{tr(est)}$	Transverse shrinkage
E	Young's modulus
L_S	Straining length
L_C	Distance of the dial gauge tip from weld centre
λ_1	Linear thermal expansion with a change of temperature from T_0 to $((T_1-T_0)/2)$
λ_2	Linear thermal expansion with a change of temperature from T_0 to T_1
T_0	Initial temperature
T_1	Temperature above which the material is no longer elastic
A_w	Cross-sectional area of weld deposit
t_h	Plate thickness
b	Average breadth of the weld
K_0	Constant depending on thermal output of welding process
σ_{tr}	Transverse shrinkage stress
R_f	Shape factor of restraint intensity
μ_t	Transverse stiffness factor
α	Coefficient of thermal expansion
C_ρ	Specific heat per unit volume
q_w	Heat input
η_m	Fusion efficiency of base metal
η_t	Thermal efficiency of base metal
η_h	Arc efficiency
p	Correction factor
σ_b	Bending stress
M	Bending moment
I_S	Moment of inertia
b_w	Plate width
t_h	Plate thickness
F_d	Force generated at plate due to distortion
δ	Plate deflection due to distortion
k_f	Proportionality factor
w_{pl}	Width of plastic zone
(h)	Plate thickness
l	Weld length

Chapter 7

σ_{max}	Maximum stress
σ_u	Ultimate tensile strength
σ_y	Yield strength
e_f	Elongation at fracture
σ_m	Mean stress
P	Porosity content
Z_p	Zinc pick-up
D_f	Dendrite fraction
N	Fatigue cycle of loading
K	Stress intensity factor
K_C	Plane-stress critical stress intensity factor
K_{IC}	Plane-strain critical stress intensity factor
J_{IC}	Initiation fracture toughness
(J-R)	Fracture resistance curve
CTOD	Crack tip opening displacement
δ_m	CTOD fracture toughness
C-T	Compact tension
TPB	Three point bend
T_{mod}	Tearing modulus
E	Modulus of elasticity
R	Stress ratio
2a	Crack length
ν	Poisons ratio
ΔK	Stress intensity factor range under dynamic loading
ΔK_{th}	Threshold of stress intensity factor range
n	Material constant of Paris Law
C	Material constant of Paris Law
R_G	Ratio of bottom to top width of weld

Chapter 1
Introduction to Gas Metal Arc Welding Process

Abstract The operational characteristics of gas metal arc welding (GMAW) process influencing the behavior of weld with respect to its geometry, microstructure and defects has been described. It is presented in reference to the behavior of metal transfer, thermal characteristics of the process and thermal description of the weld. Three primary mode of metal transfer from electrode tip commonly known as short circuit, globular and spray transfer as a function of current and voltage of the GMAW process have been explained. Thermal behavior of the entire system which is basically responsible to govern the weld characteristics has been described in terms of the cause considered as thermal characteristics of the process and its effect as thermal behaviour of the weld. The thermal characteristics of the process are analytically stated in terms of melting of filler wire and thermal nature of metal transfer. The thermal description of weld is analytically discussed with respect to the effects of heat flow, fluid flow and metal transfer. The weld characteristics considered in terms of its geometry, microstructure and defects are primarily discussed in terms of the behavior of deposition and flow of liquid metal followed by the mechanism of its solidification.

Keywords GMAW · Operational characteristics · Metal transfer · Thermal characteristics of process · Thermal behaviour of weld · Heat flow · Fluid flow · Weld geometry · Microstructure · Weld defects

Gas metal arc welding (GMAW) process is largely being used in weld fabrication industries since more than half a century for semiautomatic and later on automatic welding of many metals and alloys used in engineering applications. It is a consumable electrode process where a relatively low voltage electric arc is established between filler wire electrode and base metal under shielding of inert or reactive gases or their mixture. It melts both the filler and base metals primarily by arc heating in order to deposit the superheated former one on the latter. Thus, the GMAW process primarily uses two types of consumables as solid or tubular flux cored filler wire of different diameters generally lying in the range of 0.8–2 mm and shielding gas of different compositions. Basic GMAW setup consists of a power

P.K. Ghosh, *Pulse Current Gas Metal Arc Welding*, Materials Forming, Machining and Tribology, DOI 10.1007/978-981-10-3557-9_1

source, wire feed unit and welding torch including a system of shielding gas supply. The various components of the GMA welding set up are schematically shown in (Fig. 1.1). In this process direct current (DC) with electrode positive is mostly used, which leads to a stable arc with smooth metal transfer, resulting in low spatter and good quality weld. The use of alternate current (AC) is not generally preferred as it results in arc extinction at every half cycle. The DC with electrode negative is not normally preferred as it may cause an unstable erratic arcing adversely affecting the non contaminated gas shield of arc environment. Both the transformer rectifier set and a motor generator power source with rated duty cycle of 60 and 100% for semi automatic and automatic modes respectively can be used.

In this process the wire feed is regulated simultaneously to control welding current by keeping a balance with its burn off rate to maintain constant arc length and a steady state of metal transfer from electrode tip to weld pool. Out of various drooping to flat characteristics of power source, the GMAW process operates by self-regulated arc length under constant potential tends to flat voltage (V)-current (I) characteristics (Fig. 1.2). Here a small variation in arc voltage considerably changes the welding current facilitating maintenance of a practically constant optimum arc length (l) required for a stable welding process. However, the arc length can be appreciably varied (Δl) with respect to its optimum size by changing the arc voltage. This is generally referred as self-regulation of arc length and which is an essential feature of GMAW process. However, the flat V–I characteristic of GMAW power source retains a mild drooping nature of 1–3 V per 100 A. Welding current and heat input are commonly used of the order of 60–500 A and 1–25 kJ/s

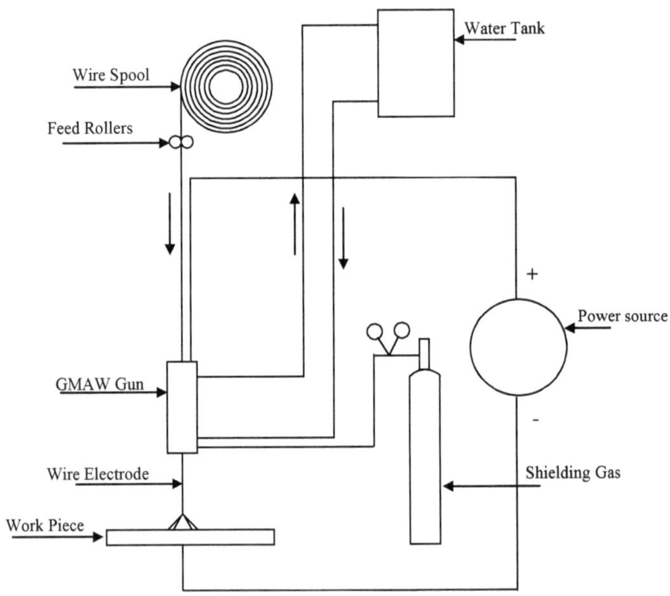

Fig. 1.1 Schematic diagram of GMAW process

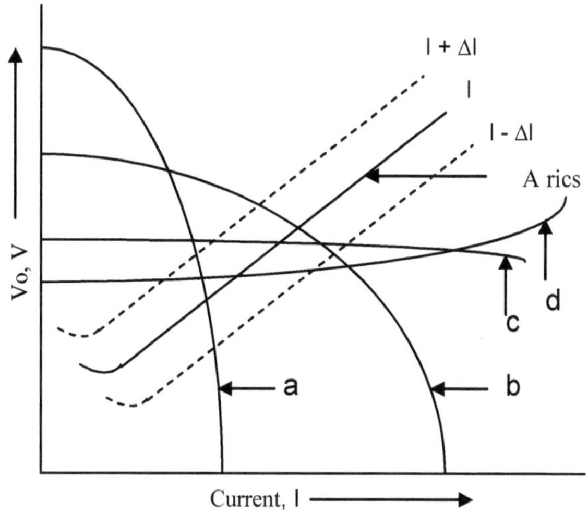

Fig. 1.2 Static volt-ampere characteristics of welding power source

respectively [1]. Welding current of the process is primarily dictated by type of weld metal and shielding gas used and also the required mode of metal transfer to weld pool such as by short circuiting (dip), free flight globular or forced projection (spray) process. However, besides the welding current and shielding gas the phenomenon of metal transfer in GMAW process also largely depends upon arc voltage (V) and size of filler. The spray mode of metal transfer offers optimum ease of operation and high quality weld but that is possible only at comparatively high welding current above transition level of producing globular to spray mode transfer of filler metal. It often results into excessive energy input and enhances the size and temperature of weld pool adversely affecting characteristics of the weld deposit and heat affected zone (HAZ). The regulation of thermal characteristics of the process is largely made by managing the behaviour of metal transfer primarily through a control on wire melting and thermal behaviour of the droplet. It is a versatile process and can be used in almost all kind of welding operations using various kinds of ferrous and nonferrous materials and their alloys. But the success of GMAW largely depends upon mode of metal transfer and manipulation of metal deposition in weld pool giving rise to optimum fusion of base metal with desired penetration [2].

1.1 Behaviour of Metal Transfer

The three distinct modes of short circuit, globular and spray transfer of molten metal from filler electrode tip to weld pool in GMAW depend upon welding current, arc voltage, electrode extension, size and composition of filler wire and type of shielding gas [3]. They maintain their own characteristic arc length affecting the penetration and shape of the weld pool [3–5]. In short circuiting or dip mode of metal transfer the electrode feed rate is generally kept slightly higher than its melting rate, so that the electrode tip eventually dips into the weld pool and extinguishes the arc [6]. Consequently heavy flow of current occurs resulting in an increased heating of the bridge formed between electrode tip and weld pool. A decreased viscosity and surface tension with increased electromagnetic and hydrodynamic forces result in the transfer of molten metal. As the metal transfer takes place, bridge breaks and the voltage tends to jump to the open circuit voltage (OCV) and thus, arc re-ignites. It is usually preferred for welding of thin sheets, heat sensitive materials and component with poor joint fit up as well as in root passes and positional welding [7, 8]. But, the process often requires a compromise with the weld quality due to lack of penetration along with excessive spatter and fast solidification of the weld [9, 10].

In globular mode the molten metal transfers in drops relatively larger than the diameter of the filler wire. It detaches primarily under the influence of gravitational force on the bulky drop when its weight grown sufficient to overcome the retaining force of surface tension causing an eventual necking to fall down. It occurs at a comparatively lower level of welding current with respect to the transition current of switch over from globular to spray mode of transfer and at a relatively higher arc voltage. Argon containing high percentage of CO_2 is usually employed as shielding gas. It is suitable for down hand welding and generally used in flat position to weld relatively thinner work piece. In globular mode of metal transfer if after detachment from electrode the drop does not travel directly towards the weld pool but repelled away from it under the action of certain forces, such as reverse plasma jet, then it is called repelled transfer. It occurs when concentration of current inside the drop (convergent current density) increases due to lower anode area in contact with the arc root which results in an upward electromagnetic force along with vaporization from the surface of the molten droplet. Thus, a thermal acceleration of vapour constituent in arc plasma takes place giving rise to an opposite force acting on the droplet resisting detachment [11, 12]. These forces act jointly with surface tension to counter the gravitational force and increase the size up to 4–6 times the diameter of the electrode [13]. Due to upward resultant force the drop oscillates around the electrode axis and when detached it may rotate and move through an uncontrolled trajectory [14].

An increase in welding current up to certain extent steadily increases the melting rate of electrode and gradually enhances the frequency of transfer of the drops with a reduction in their size through tapering of electrode. At a high value of current, a short column of molten metal streams off at the end of the electrode tip and small drops are formed as a result of breaking of this column and at a further higher value of current, asymmetric magnetic force becomes significant compared to the inertial force in the streaming column of the metal and the column spirals about the electrode axis. But, welding current beyond a critical level sharply influences reduction in size and increase of transfer frequency of the drops introducing a characteristic of spray transfer to them. This critical value of current is called "transition current". Increase of welding current, decreases gravitational and drag forces acting on droplets but significantly increases electromagnetic force. It plays a primary role on detachment [12] resulting in reduction of size and enhancement of transfer frequency of droplets. It prominently occurs under the argon rich shielding producing a series of droplets of size smaller than the filler wire diameter. It gives good arc stability, directionality of transfer, minimum spatter and lesser fume generation [6]. Thus, it produces comparatively cleaner welds with higher deposition rate than that of the globular transfer [15] and is ideal for flat and horizontal welds but, with relatively high heat input due to large welding current. In this process the droplets propelled axially from the tapered end of electrode and hence favour to employ more conveniently in positional welding. However, the use of high welding current increases the fluidity of molten metal which makes it difficult to control the weld pool in position welding. Nevertheless, the high conducting materials such as aluminium, which prevent the formation of too large weld pool, can be welded in all positions.

An increase of electrode extension from contact tube to work piece reduces the transition current [16] and thus affects mode of metal transfer by changing extent of ohmic heating. The electrode extension influences the arc length [17] and at a given arc voltage the wire feed rate should be increased with the increase in distance from contact tube to work piece [5]. The burn off rate is directly proportional to the electrode extension and resistivity of filler wire [18]. Within the operating range of process variables (5–10 mm arc length, 10–20 mm electrode extension, 5–20% CO_2 mixture in argon shielding) the specific burn-off rate of 1–1.2 mm diameter steel filler wire and V–I relationship is linear. Consequently, the generalized burn off relationship in that range of parameters may be represented by linear control equations and can be applied for synergic control [18].

1.2 Thermal Characteristics of GMAW Process

Due to its nature of consumable electrode the thermal characteristics of GMAW process very much depend upon the melting of filler wire and thermal nature of metal transfer affecting heat distribution in the weld. In this matter welding current, arc voltage, shielding gas and electrode extension of process simultaneously play some important role and finally governs the weld characteristics. The basic characteristics and primary role of these aspects in GMAW process are discussed below.

1.2.1 Melting of Filler Wire

The GMAW process successfully operates at the burn off rate equal to wire feed rate to maintain constant arc length by following energy balance per unit time expressed [19, 20] as.

$$A_w V_{w(cc)} \rho_w Q_m = \left\{ V_a + \zeta + \frac{3}{2}\frac{kT}{e} \right\} I + \frac{R_o E_w I^2}{A_w} \quad (1.1)$$

where, A_w is the cross sectional area of filler wire (m^2), $V_{w(cc)}$ is the wire feed rate (ms^{-1}) for the continuous current conventional GMA welding, ρ_w is the density ($kg\ m^{-3}$) of the filler wire, Q_m is heat per unit mass ($J\ kg^{-1}$) required for melting the filler wire, V_a is anode fall voltage (V), ζ is work function of metal surface (V), $\frac{3}{2} \cdot \frac{kT}{e}$ is thermal energy of electrons (V), R_o is resistivity (Ω m) of the filler wire, E_w is electrode extension (m) and I is welding current (A). For the wire feed rate of conventional GMAW the Eq. (1.1) may be re-written as,

$$V_{w(cc)} = \frac{\psi I}{A_w V_w Q_m} + \frac{R_o E_w I^2}{A_w^2 \rho_w Q_m} \quad (1.2)$$

$$\psi = \left\{ V_a + \zeta + \frac{3}{2} \cdot \frac{kT}{e} \right\} \quad (1.3)$$

where, ψ is the effective melting potential at anode. In consideration of $V_{w(cc)}$ as a function of two different components contributing to the wire melting the Eq. (1.2) may also be expressed as

$$V_{w(cc)} = A I + B E_w I^2 \quad (1.4)$$

where, A and B are the constants representing wire melting due to arc heating and resistive heating respectively expressed as

$$A = \frac{\psi}{A_w \rho_w Q_m} \quad (1.5)$$

$$B = \frac{R_o}{A_w^2 \rho_w Q_m} \quad (1.6)$$

The physical constants for the filler wire of different materials [19, 21–28] may be considered as given in Table 1.1.

1.2.2 Thermal Nature of Metal Transfer

The resistive heating of filler wire in general is not sufficient to melt the filler wire [29] and to influence the thermo fluid behaviour of the electrode tip and droplet affecting the thermal behaviour of metal transfer and consequently the weld quality [30]. During welding certain amount of heat flows from arc to outer surface of the liquid film at the electrode tip that acts as an anode. Thus, the resistive heating along

Table 1.1 Physical properties of different filler wires

Property	Mild steel	Stainless steel	Al–5Si alloy	Al–5 Mg alloy
Melting point, T_m, (K)	1750	1728	903	913
Specific heat, $c_{p(s)}$, (J kg^{-1} K^{-1})	686	500	1082	900
Specific heat, $c_{p(l)}$, (J kg^{-1} K^{-1})	855	760	1160	1180
Latent heat of fusion, L, (J kg^{-1})	2.76E5	2.84E5	3.89E5	3.97E5
Density of the solid metal, ρ_w, (kg m^{-3})	7870	7750	2690	2660
Density of the molten metal, ρ_d, (kg m^{-3})	6500	7507	2200	2335
Coefficient of surface tension, γ, (N m^{-1})	1.03	1.35	0.86	0.86
Resistivity at melting point, R_o, (Ω m)	8.2E–7	1.3E–6	1.3E–7	2.5E–7
Effective work function, ξ, (V)	5.9	5.3	6.66	6.66
Emissivity of molten droplet, ε	0.25	0.25	0.25	0.25

with heat flow through the anode surface governs melting rate of the filler wire. The remainder of heat generated at electrode surface affects the super heating of the droplet. The total power supplied to the solid filler wire is expressed as $\left(I^2R_s + I\,\psi\right)$, where R_s is the resistance of the electrode extension. The melting rate is then determined by the power supplied to the solid i.e., $\left(I^2R_s + Q^*\right)$, where Q* is the power transferred from the molten anodic tip to solid part of the electrode. The temperature of the droplet is determined by the mass flow rate of the filler wire $(A_w V_w \rho_w)$ and part of the power remain at the anode surface i.e., $(I\xi - Q^*)$, if the heat losses are ignored. The dependence of Q* on the mechanism of heat transfer varies with change in mode of metal transfer, which consequently affects the temperature distribution in the electrode extension and its resistivity. The resistance of electrode extension increases marginally with an initial increase of current, but it falls markedly around 250 A welding current (drop spray transfer) as it is marked in 1.2 mm diameter steel wire under shielding of argon +5% CO_2 [31]. During stream spray transfer, the resistance of electrode extension further increases with welding current. The electrical resistance thus depends upon the mode of metal transfer because each mode gives rise to different temperature distribution in the electrode extension [29].

During globular transfer, Q* varies linearly with melting rate by keeping the slope practically independent of welding current whereas, in the case of spray transfer which shows a non-linear correlation of Q* with the melting rate, it initially reduces steeply with the increase in melting rate and levels off at higher values of melting rate. Thus, the heat balance at the wire tip is closely related to the mode of metal transfer. In case of globular transfer, the heat transfer from the droplet to the solid electrode tip primarily happens due to convection within the drop caused by the Lorentz force if current diverges in the droplet [7, 29]. The liquid metal enters the droplet with certain initial velocity and is accelerated downwards by the electromagnetic force in the region around the axis and reflected from the inner surface of the droplet. A boundary layer is formed at the drop surface above the anode, and heat is transferred through it to the circulating liquid. The power Q* transferred to the solid part of the electrode and circulating velocity of the liquid metal within the droplet both are approximately proportional to the current therefore, droplet temperature is practically independent of current. However, the spray transfer is characterised by continuous flow of liquid film surrounding a tapered end of electrode filler wire down to an emerging liquid tail just below it and final release in the form of series of droplets. The heat evolved at the anode surface of the liquid film is partly transferred to the un-melted filler wire by conduction. It contributes to the heating and subsequent melting of the filler wire. The remaining heat is carried away by convection to the transferring metal and raises the temperature of the droplet. Hence the mean temperature of the droplet depends upon welding current and hence the modes of metal transfer [31]. In case of spray transfer, heat content of the droplet is found to be relatively less than that with globular transfer. Thus, the spray transfer endures comparatively less super heating of droplets resulting in higher melting rate than that of other transfer modes. In view of the above it is clearly understood that the welding consumables and welding parameters significantly affect the thermal behaviour of weld due to their considerable influence on thermal and transfer behaviours of weld deposit.

1.3 Thermal Description of Weld

Thermal description of gas metal arc (GMA) weld primarily depends upon welding parameters due to their significant influence on thermal nature of plasmatized arc environment and various modes of metal transfer. The thermal and transfer behaviours of filler metal affecting the heat content of the weld pool manipulates its solidification behaviour and dictates the characteristics of heat-affected zone (HAZ). In GMAW process the shielding gas protects the weld metal from the atmospheric contamination but it also affects the thermal characteristics of the weld up to certain extent based on its ionisation potential and flow rate. Considering the welding current for a stable arc system in a GMAW process the analysis of thermal behaviour of weld is primarily addressed to a great extent by considering the Rosenthal expressions of heat distributions.

The analysis of thermal distribution in arc weld by conduction process with the help of Fourier partial differential equation (PDE) was founded by Rosenthal [32]. The analytical process based upon moving coordinate system is proposed to estimate temperature distribution in the point and line heat sources applied to large variety of welding process. The solutions are assumed to be valid in quasi steady state condition and it is experimentally justified for a long weld. The quasi steady state condition is defined by a situation when it is missed to notice any change in temperature around a moving point heat source. Considering the partial differential equation of heat conduction, expressed by rectangular coordinates (x, y, z) with respect to a fixed origin in solid, it has been successfully applied by assuming a moving coordinate system in welding. Based on analysis of temperature distribution in fusion welding number of valid conclusions are made. But, at the outset it is marked that Rosenthal's solutions are largely valid at locations away from heat source but not so significantly at or near heat source in reference to weld pool. It may have possibly happened due to inappropriate assumption of point or line heat sources [33–36] in arc welding process. Several investigators tried to address the shortcomings of the Rosenthal's model by considering different configurations of heat source and heat flow pattern for its more universal application in different arc welding processes in case of both the weld pool and HAZ. In this regard the effect of fluid flow and nature of metal transfer has also been considered in certain cases. The solution to the problem was approached through the analytical solutions as well as by finite difference (FD) and finite element (FE) methods of analyses. Due to its relative simplicity most of the work in this area has been carried out in the field of gas tungsten arc welding (GTAW). However, some useful efforts have also been made in the area of GMAW.

1.3.1 Effect of Heat Flow

The model of heat flow by considering two-dimensional (2-D) moving rectangular heat source with uniform distribution of heat was developed as early as in the year 1952 for estimation of heat input and welding speed in arc welding process [37]. Latter on the heat flow in argon arc welding (without a filler wire) of different ferrous and non ferrous metals was reported [38] in the year 1955. The heat flow in welding of thick steel plates by considering the effects of boundary surfaces of bottom and edge of a plate was investigated [39] by assuming them adiabatic. Based on the Rosenthal's solution for point heat source analytical solution [40] for generalized plots of temperature distribution at surface as well as at different depths of any substrate was found useful for estimation of HAZ, cooling rate and residence time in between two temperatures at some specific cases of welding. The temperature solutions for the point heat source in semi infinite plate and semi infinite plate of finite thickness are described as follows.

Considering origin of a point heat source along central axis of welding arc [32] the temperature distribution at a semi-infinite plate in terms of rectangular coordinates (Fig. 1.3) x, y and z is expressed as

$$T - T_0 = \frac{Q_p}{2\pi k} e^{-\lambda v \xi} \cdot \frac{e^{-\lambda v R}}{R} \quad (1.7)$$

where, T is temperature of a point at a distance R with respect to central axis of welding arc, T_0 is initial temperature of base plate, Q_p is heat transfer per unit time

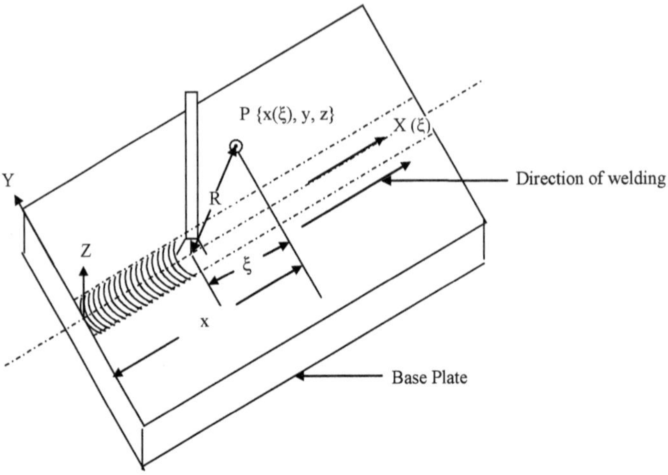

Fig. 1.3 Rectangular coordinates of a moving heat source

by point heat source ($J\ s^{-1}$), K is thermal conductivity of base metal and R is expressed as

$$R = \sqrt{\xi^2 + y^2 + z^2} \tag{1.8}$$

$$\xi = x - v\,t \tag{1.9}$$

In a semi-infinite plate of thickness, d, the temperature distribution for point heat source [32] is expressed by modifying the Eq. (1.7) for a limited range of z dimension as follows.

$$T - T_0 = \frac{Q_p}{2\,\pi\,k} e^{-\lambda\,v\,\xi} \left[\frac{e^{-\lambda\,v\,R}}{R} + \sum_{n=1}^{n=\infty} \left(\frac{e^{-\lambda\,v\,R_n}}{R_n} + \frac{e^{-\lambda\,v\,R'_n}}{R'_n} \right) \right] \tag{1.10}$$

where,

$$R_n = \sqrt{(2\,n\,d - z)^2 + \xi^2 + y^2} \tag{1.11}$$

and

$$R'_n = \sqrt{(2\,n\,d + z)^2 + \xi^2 + y^2} \tag{1.12}$$

With the help of finite difference method [41] to determine temperature distribution in 2-D plate using line heat source, the shape of weld pool has been correlated with welding variables by using melting temperature isotherms as boundary condition. In this method some experimental work is necessary to determine certain boundary conditions of the analysis. However, the details of the weld pool shape could not be correlated using the line source theory and therefore a concept of distributed heat source with Gaussian distribution of heat flux (W/m^2) was suggested. The steady state two-dimensional heat flow during welding of thin plates has also been described [42, 43] by a computer model using finite difference method and compared to those of Rosenthal's [32] analytical solutions as well as with the experimental results of GTAW in bead on plate of aluminium alloy.

The use of finite element method (FEM) has been found [44] effective in analysis of heat conduction in complex shape of solids with the boundary conditions of temperature and heat flux. An extensive study of temperatures in actual welds was conducted [45] by comparing the values computed from heat conduction equation with those obtained from actual welds. In cotemporary studies the FEM is extensively used to analyse various aspects of welding. A model of thermo mechanical analysis of welding process using FEM has been developed [46] to calculate the temperatures, stresses, and distortions resulting in a welding process. The FEM could significantly analyse [47] the transient heat conduction to investigate the effect of various welding parameters on thermal response characteristics affecting the weld bead shape and depth of penetration. In this analysis welding parameters were considered as the heat input along with its duration and distribution at the surface of weldments arising out of arcing. A semi empirical 2-D finite element heat transfer model has been found [48] effective to treat the molten interface as inner boundary to calculate quasi steady state temperature field and cooling rate in HAZ.

Various mathematical models of weld thermal behaviour of metals have been evaluated [49] in order to find quick solution or to formulate meaningful experimental procedure for quantitative thermal analysis of fusion welding. An extended Rosenthal model in multi polar expansion form was developed [50] to analyse phase change under thermal dipoles and heat circulation in molten weld pool. A modification of Rosenthal's model has been proposed [51] by considering 2-D Gaussian distributed heat source. The model assumes presence of no convective or radiative heat flow, prevailing constant average thermal properties, and existence of quasi steady state semi-infinite medium. In this analysis a solution for temperature of a semi-infinite body using constant distribution parameter has also been developed and its results are compared with their experimental values on various metals. It is practically considered as a turning point for more precise temperature prediction near heat source region. Some techniques of finite element analysis of temperature distribution in arc welding have been developed [52, 53] by using transient adaptive grid technique. They use relatively finer mesh around arc source, where temperature gradients are high and relatively coarser mesh in other places to increase computational efficiency of the analysis. An analytical solution for transient temperature distribution in fillet arc weld has also been developed [34] considering 2-D Gaussian heat source that uses different distribution parameters in x and y directions. Here the solution of temperature field in a plate of finite thickness is transformed to fillet-weld joint by conformal mapping technique. Although available solutions using 2-D Gaussian heat source can predict temperature at regions close to heat source but, their effectiveness is still limited due to shortcoming in appropriate consideration of the heat source itself. Thus it cannot correctly predict in depth direction. The limitations of 2-D heat source are to a large extent overcome by using more generalised heat source.

In view of the knowledge gathered so far on estimation of heat flow in welding, mathematical model has been developed based on a Gaussian distribution of power density [33] and proposed to consider a three-dimensional double ellipsoidal moving heat source. Finite element modelling was used to calculate temperature field in weld bead on plate deposition. Here it is found that the use of 3-D heat source can predict temperature in weld joints up to much deeper penetration compared to the 2-D model. Computed results of temperature distributions in submerged arc welds of thick work piece have been successfully compared with its experimental values [40]. The 3-D transient heat conduction [27, 54] and transient temperature distribution models for GTA welding has been developed by using FEM analysis along the welding heat source. The size of weld pool and HAZ predicted by using this model are satisfactorily compared with the experimental results of welding of medium carbon steel under various conditions. Transient temperature distribution in finite thick plate has also been studied [55] by analytical solution of 3-D heat conduction equation with convective boundary conditions at welding surface. The results are verified with the experimental values of GTA bead on plate weld of medium carbon steel. For determining three dimensional transient temperature variations during gas tungsten arc welding the heat diffusion equation has been solved [56] by using semi discrete technique employing unequally spaced grids. The analytical solution of heat conduction using quasi steady state instantaneous line heat source model and unsteady heat flow model for moving heat source has been developed [57, 58]. It predicts shape and cooling characteristics of HAZ and the results are compared with the experimental values. Transient three dimensional computer simulation model has also been studied [59, 60] for weld pool in autogenously and non-autogenously used GTAW and a generalised approach of numerical technique has been proposed [61] for predicting the thermal cycle of the weld.

The heat flux q(x, y, z) at a point (x, y, z) in moving coordinate system within a semi ellipsoidal heat source [33] is expressed as

$$q(x, y, z) = \frac{6\sqrt{3}\eta_a V I}{a_h b_h c_h \pi\sqrt{\pi}} \exp\left(-\frac{3\,x^2}{c_h^2} - \frac{3\,y^2}{a_h^2} - \frac{3\,z^2}{b_h^2}\right) \tag{1.13}$$

Whereas, the heat flux q(x, y, z) at a point (x, y, z) within the weld pool under double ellipsoidal heat source (Fig. 1.4) [33] is estimated by the following expressions representing two different semi ellipsoids. The heat flux at the points x, y and z within the first semi ellipsoid located in front of the welding arc is expressed as

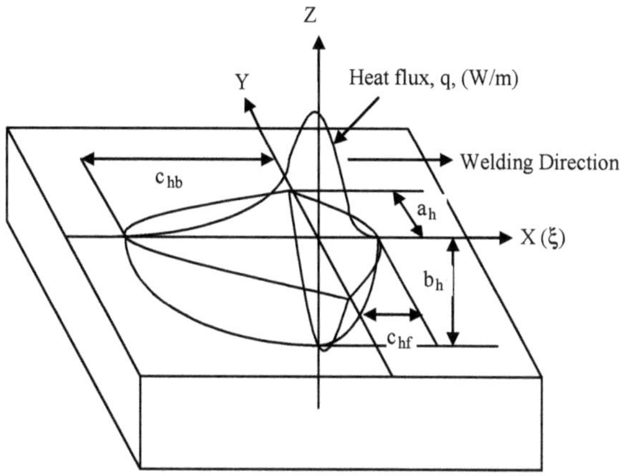

Fig. 1.4 Double ellipsoidal heat source

$$q_f(x, y, z) = \frac{6\sqrt{3}r_f Q_{arc}}{a_h b_h c_{hf} \pi \sqrt{\pi}} \exp\left(-\frac{3x^2}{c_{hf}^2} - \frac{3y^2}{a_h^2} - \frac{3z^2}{b_h^2}\right) \tag{1.14}$$

For the points x, y and z within the second semi ellipsoid, covering rear section of the arc, it is expressed as

$$q_b(x, y, z) = \frac{6\sqrt{3}r_b Q_{arc}}{a_h b_h c_{hb} \pi \sqrt{\pi}} \exp\left(-\frac{3x^2}{c_{hb}^2} - \frac{3y^2}{a_h^2} - \frac{3z^2}{b_h^2}\right) \tag{1.15}$$

where, a_h, b_h and c_h are the ellipsoidal heat source parameters defined by a location having minimum power density of 5% to that of the centre on surface of the ellipsoid and η_a is heat transfer efficiency of welding arc. The Q_{arc} is arc heat input expressed as

$$Q_{arc} = \eta_a VI \tag{1.16}$$

and, r_f and r_b are the proportion coefficients in front and behind the heat source respectively. They keep the relationship as $r_f + r_b = 2$. The continuity of heat source at x = 0 plane is maintained by introducing [33] another constraint for r_f and r_b as

$$\frac{r_f}{c_{hf}} = \frac{r_b}{c_{hb}} \tag{1.17}$$

$$r_f = 2\, c_{hf}/(c_{hf} + c_{hb}) \tag{1.18}$$

and

$$r_b = 2c_{hb}/(c_{hf} + c_{hb}) \tag{1.19}$$

Arc efficiency η_a and ellipsoidal parameters a_h, b_h, c_{hf} and c_{hb} are the five unknown parameters involved in the expression of double ellipsoidal heat source. The arc efficiency in GMA welding of aluminium is considered [28] as 85% and appropriate values of ellipsoidal axes parameters are selected [33] by measuring weld pool geometry. The a_h, b_h, c_{hf} and c_{hb} are length parameters of ellipsoidal axes of weld pool. They are described as the side, underneath, front and behind the arc respectively observed in reference to transverse direction of welding. The heat source parameter c_{hf} (in front of the heat source) and c_{hb} (behind the arc) may be considered [33, 36] as $c_{hf} = a_h$ and $c_{hb} = 2\ c_{hf}$. The parameters a_h and b_h are properly selected commensurate with the physical welding conditions. The selection is made [33, 36, 61] out of best fit values obtained from a comparison of measured width and penetration of actual weld pool with their corresponding values estimated on the basis of assumed values of the a_h and b_h.

The analytical solution [36] for the transient temperature field from time, $t' = 0$ to $t' = t$ at any point (x, y, z) of a semi ellipsoidal heat source distributed in a semi infinite plate is expressed as

$$T - T_0 = \frac{3\sqrt{3}\,Q_t}{\rho\, c\, \pi\sqrt{\pi}} \int_0^t \left[\frac{\partial t'}{\sqrt{(12\ a(t-t') + a_h^2)}\sqrt{(12\,a(t-t') + b_h^2)}\sqrt{(12\,a(t-t') + c_h^2)}} \cdot \exp\left(-\frac{3(x - v\,t')^2}{12\,a(t-t') + c_h^2} - \frac{3y^2}{12\,a(t-t') + a_h^2} - \frac{3z^2}{12\,a(t-t') + b_h^2} \right) \right] \tag{1.20}$$

The notations a, ρ and c are the thermal diffusivity, mass density and specific heat of base metal respectively. Similarly the analytical solution [36] for the transient temperature field from time, $t' = 0$ to $t' = t$ at any point (x, y, z) of a double ellipsoidal heat source distributed in a semi infinite plate is expressed as

$$T - T_0 = \frac{3\sqrt{3}Q_t}{\rho\, c\, \pi\sqrt{\pi}} \int_0^t \left[\frac{\partial t'}{\sqrt{(12\,a(t-t') + a_h^2)}.\sqrt{(12\,a(t-t') + b_h^2)}} \cdot \left(\frac{A'}{\sqrt{(12\,a(t-t') + c_{hf}^2)}} + \frac{B'}{\sqrt{(12\,a(t-t') + C_{hb}^2)}} \right) \right] \tag{1.21}$$

$$A' = r_f \exp\left(-\frac{3(x - v\,t')^2}{12\,a(t-t') + c_{hf}^2} - \frac{3y^2}{12\,a(t-t') + a_h^2} - \frac{3z^2}{12\,a(t-t') + b_h^2} \right) \tag{1.22}$$

and

$$B' = r_b \exp\left(-\frac{3(x - v\,t')^2}{12\,a(t - t') + c_{hb}^2} - \frac{3y^2}{12\,a(t - t') + a_h^2} - \frac{3z^2}{12\,a(t - t') + b_h^2}\right) \quad (1.23)$$

where, Q_t is the total heat transferred to weld pool per unit time (J s^{-1}) and a is thermal diffusivity of base metal.

The analytical solution [62] for the transient temperature field from time, $t' = 0$ to $t' = t$ at any point (x, y, z) of a semi ellipsoidal heat source distributed in a semi infinite plate of finite thickness is expressed as

$$T - T_0 = \frac{3\sqrt{3}.Q}{\rho.c.\pi\sqrt{\pi}} \int_0^t \left[\begin{array}{c} \left(\dfrac{E(L_w, x - vt', c_h).E(B, y, a_h).E(d, z, b_h)}{\sqrt{(12a(t - t') + a_h^2)}.\sqrt{(12a(t - t') + b_h^2)}\sqrt{(12a(t - t') + c_{hf}^2)}} \right) \\ .\exp\left(-\dfrac{3(x - vt')^2}{12a(t - t') + c_h^2} - \dfrac{3y^2}{12a(t - t') + a_h^2} - \dfrac{3z^2}{12a(t - t') + b_h^2}\right) \end{array} \right] dt' \quad (1.24)$$

However, the analytical solution [62] for the transient temperature field from time, $t' = 0$ to $t' = t$ at any point (x, y, z) of a double ellipsoidal heat source distributed in a semi infinite plate of finite thickness is expressed as

$$T - T_0 = \frac{3\sqrt{3}\,Q_t}{2\,\rho\,c\,\pi\sqrt{\pi}} \int_0^t \left[\begin{array}{c} \left(\dfrac{\left[\exp\left(-\dfrac{3y^2}{12\,a(t - t') + a_h^2} - \dfrac{3z^2}{12\,a(t - t') + b_h^2}\right).E(B_w,y,a_h).E(d,z,b_h)\right]}{\sqrt{(12a(t - t') + a_h^2)}\sqrt{(12\,a(t - t') + b_h^2)}} \right) \\ . \left[\begin{array}{c} \dfrac{r_f E(L_w, x - vt', c_{hf}).\exp\left(-\dfrac{3(x - vt')^2}{12\,a(t - t') + c_{hf}^2}\right)}{\sqrt{(12\,a(t - t') + c_{hf}^2)}} \\ + \dfrac{r_b E(L_w, x - vt', c_{hb}).\exp\left(-\dfrac{3(x - vt')^2}{12\,a(t - t') + c_{hb}^2}\right)}{\sqrt{(12\,a(t - t') + c_{hb}^2)}} \end{array} \right] \end{array} \right] dt' \quad (1.25)$$

where, L_w is half-length, B_w is half width and d is thickness of the base plate. $E(L_w, x - vt', c_{hf})$, $E(L_w, x - vt', c_{hb})$, $E(L_w, x - vt', c_h)$, $E(B, y, a_h)$, $E(d, z, b_h)$, r_f and r_b are expressed as

$$E(L_w, x - vt', c_{hf}) = \frac{\left[\begin{array}{l} \operatorname{erf}\left(\dfrac{(12a(t-t') + c_{hf}^2)(L_w - (x - vt')) - c_{hf}^2(x - vt')}{c_{hf}\sqrt{4a(t-t')}\sqrt{12a(t-t') + c_{hf}^2}}\right) \\ + \operatorname{erf}\left(\dfrac{(12a(t-t') + c_{hf}^2)(L_w + (x - vt')) - c_{hf}^2(x - vt')}{c_{hf}\sqrt{4a(t-t')}\sqrt{12a(t-t') + c_{hf}^2}}\right) \end{array}\right]}{\left[\operatorname{erf}\left(\dfrac{L_w - (x - vt')\sqrt{3}}{c_{hf}}\right) + \operatorname{erf}\left(\dfrac{L_w + (x - vt')\sqrt{3}}{c_{hf}}\right)\right]} \tag{1.26}$$

$$E(L_w, x - vt', c_{hb}) = \frac{\left[\begin{array}{l} \operatorname{erf}\left(\dfrac{(12a(t-t') - c_{hb}^2)(L_w - (x - vt')) - c_{hb}^2(x - vt')}{c_{hb}\sqrt{4a(t-t')}\sqrt{12a(t-t') + c_{hb}^2}}\right) \\ + \operatorname{erf}\left(\dfrac{(12a(t-t') + c_{hb}^2)(L_w + (x - vt')) - c_{hb}^2(x - vt')}{c_{hb}\sqrt{4a(t-t')}\sqrt{12a(t-t') + c_{hb}^2}}\right) \end{array}\right]}{\left[\operatorname{erf}\left(\dfrac{L_w - (x - vt')\sqrt{3}}{c_{hb}}\right) + \operatorname{erf}\left(\dfrac{L_w + (x - vt')\sqrt{3}}{c_{hb}}\right)\right]} \tag{1.27}$$

$$E(B, y, a_h) = \frac{\left[\begin{array}{l} \operatorname{erf}\left(\dfrac{(12a(t-t') + a_h^2)(B - y) - a_h^2 y}{a_h\sqrt{4a(t-t')}\sqrt{12a(t-t') + a_h^2}}\right) \\ + \operatorname{erf}\left(\dfrac{(12a(t-t') + a_h^2)(B + y) - a_h^2 y}{a_h\sqrt{4a(t-t')}\sqrt{12a(t-t') + a_h^2}}\right) \end{array}\right]}{\left[\operatorname{erf}\left(\dfrac{(B - y)\sqrt{3}}{a_h}\right) + \operatorname{erf}\left(\dfrac{(B + y)\sqrt{3}}{a_h}\right)\right]} \tag{1.28}$$

$$E(d, z, b_h) = \left[\begin{array}{l} \operatorname{erf}\left(\dfrac{(12a(t-t')d) + b_h^2.(d - z)}{b_h\sqrt{4a(t-t')}\sqrt{12a(t-t') + b_h^2}}\right) \\ + \operatorname{erf}\left(\dfrac{b_h z}{\sqrt{4a(t-t')}\sqrt{12a(t-t') + b_h^2}}\right) \end{array}\right] . \operatorname{erf}^{-1}\left(\frac{d\sqrt{3}}{b_h}\right) \tag{1.29}$$

$$r_f = \frac{2c_{hf}\left[\operatorname{erf}\left(\dfrac{(L_w - x)\sqrt{3}}{c_{hf}}\right) + \operatorname{erf}\left(\dfrac{(L_w + x)\sqrt{3}}{c_{hf}}\right)\right]}{\left[\begin{array}{l} c_{hf}\left[\operatorname{erf}\left(\dfrac{(L_w - x)\sqrt{3}}{c_{hf}}\right) + \operatorname{erf}\left(\dfrac{(L_w + x)\sqrt{3}}{c_{hf}}\right)\right] \\ + c_{hb}\left[\operatorname{erf}\left(\dfrac{(L_w - x)\sqrt{3}}{c_{hb}}\right) + \operatorname{erf}\left(\dfrac{(L_w + x)\sqrt{3}}{c_{hb}}\right)\right] \end{array}\right]} \tag{1.30}$$

$$r_b = \frac{2c_{hb}\left[\operatorname{erf}\left(\frac{(L_w - x)\sqrt{3}}{c_{hb}}\right) + \operatorname{erf}\left(\frac{(L_w + x)\sqrt{3}}{c_{hb}}\right)\right]}{\left[c_{hf}\left[\operatorname{erf}\left(\frac{(L_w - x)\sqrt{3}}{c_{hf}}\right) + \operatorname{erf}\left(\frac{(L_w + x)\sqrt{3}}{c_{hf}}\right)\right] + c_{hb}\left[\operatorname{erf}\left(\frac{(L_w - x)\sqrt{3}}{c_{hb}}\right) + \operatorname{erf}\left(\frac{(L_w + x)\sqrt{3}}{c_{hb}}\right)\right]\right]} \quad (1.31)$$

Analysis of transient temperature field in a semi-infinite conductive body subjected to a power density of double ellipsoidal moving heat source has been resolved [36, 62] and compared to the observations made in GMA weld bead on plate deposition on HT-780 plate. But, the model failed to predict shape of the weld pool in transverse cross section. Results of a three-dimensional transient finite element analysis of heat flow in arc welding using Gaussian distributed heat source are compared with the published values and found [63, 64] that the arc length significantly affects the temperature of weld pool and HAZ adjacent to base metal. Analytical solution for temperature distribution in a small arc weld has been developed [35] by considering arc beam as a moving plane heat source with a pseudo-Gaussian distribution of heat intensity. A neural network approach [65] for GMA butt joint welding has also been proposed in this area.

1.3.2 *Effect of Fluid Flow*

In a mathematical model analysis, considering the convection and temperature distribution in stationary arc weld pool driven by buoyancy, electromagnetic and surface tension forces, it is found [66] that the electromagnetic and surface tension forces dominate the fluid flow behaviour in weld. In some cases, these forces produce double circulation loops, which are indirectly confirmed by experimental measurement of segregation in the weld pool. A mathematical model [67] for the transient fluid flow and temperature field develops in GTA weld pool show that characteristic flux of heat and electric current acting on its free surface significantly affects both the fields of temperature and convective flow of liquid metal. The role of convective fluid flow in weld pool does not significantly define heat transfer process in absence of surface tensional effect under the condition of a broadly distributed heat flux. However, in case of a sharply focused heat flux and current density on free surface the convection may play an important role in determining shape of the weld pool. The surface tension gradient has been found to affect the surface velocities and consequently the weld pool shape. Convection of fluid flow and temperature distribution in moving weld pool driven by buoyancy, electromagnetic and surface tension forces during GTA welding have been mathematically modelled [68]. A comprehensive numerical model developed for turbulent thermo-fluids in stationary GTA welding [69] could predict the weld pool size and shape in AISI 304 stainless steel plates. A three-dimensional steady incompressible fluid flow and heat transfer during fusion welding has been investigated [70] with stationary and moving heat sources using non uniform, adjustable grids. Individual and summarized influence of different forces acting on fluid flow governing the heat transfer and geometry of weld pool have been studied and found that the effect of buoyancy force is insignificant. The Marngoni force increases the length and width but reduces depth of weld pool while electromagnetic force reduces the length and width and increases depth of weld pool. Summarized influence of all the forces on geometry of weld pool has been found to be insignificant. In an interesting effort [71] the evolution of temperature and velocity fields have been studied by using transient numerical model in GTA spot welding of AISI 1005 steel where the estimated cooling rates are found to vary significantly at the onset of solidification at different positions of weld pool.

A mathematical model has been proposed [72] to evaluate effect of the fields of electromagnetic force, fluid flow velocity and temperature in a weld pool of GMAW. Unsteady state model considers not only the interactions of buoyancy, electromagnetic and surface tension forces, but also the energy exchange between weld pool and molten filler metal droplets via spray transfer. It is found that the frequency of spray transfer is a dominant factor besides the Lorentz force in shaping the weld pool geometry while the buoyancy and surface tension effect plays a relatively minor role in GMA welds. In a 3-D model [73] of heat and fluid flow in GMA welding process it is found that momentum and heat content of droplets play an important role in shaping weld pool. An unsteady two-dimensional axis symmetric model [74] for investigating heat and fluid flow in weld pool determines weld bead geometry, velocity of fluid in weld pool and temperature profiles of GMA weld pool. It shows that the electromagnetic, surface tension forces and molten metal droplet influences weld pool geometry. The impact and kinetic energy of droplets at the time of touching weld pool surface and the force exerted by them during entering the weld pool introduces a new mode of distribution of arc heat flux on the deformed weld pool surface [75]. It determines the volume of weld pool having distribution of heat from the droplet.

1.3.3 Effect of Metal Transfer

Based on a hypothesis, that in GMAW the impact droplets transfer to weld pool at high velocity may form a cylindrical cavity in liquid weld pool, modelling has been made [19, 76, 77] to predict how droplets transfer heat at certain depth inside weld pool. The model [19] assumes that, the cavity formed by kinetic energy of molten metal droplet transferred to weld pool is having a radius equal to diameter of droplet. Between impingement of two successive droplets cavity tends to fill thus, they transfer their heat content at an effective depth of the cavity defined by $(h - d_t + R_c)$ where, h is depth of the cavity formed due to impact of the droplet and

d_t is distance travelled by the centre of mass of the slug and R_c is radius of the cylindrical cavity which was assumed [77] equivalent to diameter of the droplet.

At this stage it may be realised that in spite of significant progress in understanding the characteristics of welding processes and weld materials, several key problems and issues are yet to be addressed. Main difficulties of quantitative analysis of welding processes are complexity of considering physical aspects of fusion welding and paucity of relevant characteristic data of super heated weld deposit. Under these circumstances analytical modelling can provide significant understanding on the dynamics of welding processes and properties of weld metal, which could not be obtained otherwise. The simulation of thermal behaviour of arc welding has been carried out by considering two different approaches. In the first one only heat flow to weld pool has been considered while in other one heat and fluid flow both have been taken into account. In some investigations the mechanism of metal transfer has also been considered. But, in spite of several efforts made for solving temperature field in arc welding, no single direct solution is available for estimating temperature field and geometry of weld pool in GMAW process. However, available solutions [36, 62] for double ellipsoidal heat source may be modified to estimate temperature distribution and geometry of weld pool of GMAW process. Moreover, in view of the complexities of welding processes, theoretical estimation must be tested by well-designed experiments.

1.4 Weld Characteristics

From the knowledge as described above it may be realised that the quality of GMA weld largely depends upon selection of proper type and size of filler wire, appropriate use of shielding gas and right application of welding parameters [78]. In automatic applications of GMAW, the analytical and empirical correlations describing the interaction of process variables and their influence on average weld characteristics are necessary for development of process control algorithms. For all practical purposes the influence of various welding parameters on geometry and microstructure of weld is well known to a great extent.

1.4.1 Weld Geometry

Weld penetration in GMAW is primarily determined by the arc heat supplied to surface of weld pool and the transfer of heat by droplets pushed inside weld pool [79] up to a certain depth. Thus it primarily depends upon welding current and joint geometry. In general penetration is indirectly proportional to welding speed and directly proportional to electrode diameter. At a given welding current penetration decreases non-linearly with increase of electrode diameter, due to reduction in current density [80, 81]. Increasing root gap or included angle also leads to increased penetration. Based upon direct relation with the momentum of individual droplet impinging the molten weld pool a finger type of penetration may occur during spray transfer. Height of reinforcement of weld bead is directly proportional to welding current and indirectly varies with arc voltage and electrode diameter. Whereas bead width is directly proportional to arc voltage and electrode diameter, which affects the area of distribution of arc force and is indirectly proportional to the welding speed [81]. Area of weld bead directly varies with the welding current and arc voltage because of their significant influence on wire melting and shows an inverse proportionality to welding speed and electrode diameter due to reduction in deposition per unit area [81].

1.4.2 Microstructure

Due to significant possibilities of manipulation in thermal and mass transfer in weld by change in welding parameters the microstructure of GMA weld can be varied to a great extent which may be used in favour of improvement in weld quality. However, the control of microstructure of a given material requires basic understanding of its metallurgical transformations in the light of the thermal and mechanical behaviours of welding. The microstructures develop in a weld may be primarily classified as the zone of solidification at plane front of base metal along the fusion line followed by primary growth of columnar dendrite in the interior of weld and an equiaxed region results in certain cases at the weld centre. Weld metal microstructure is appreciably controlled by the basic structure of base metal and welding conditions because initial epitaxial growth occurs at partially melted grains in base metal [82–84]. In polycrystalline weld metal, amongst the randomly oriented grains of base metal, the most closely aligned grains of $\langle 100 \rangle$ direction favour heat flow direction and preferred grain selection process that promotes a columnar structure [85]. The effect of crystallographic growth and dendrite formation process on development of fusion zone microstructures has been studied [86–88] to predict stable dendrite growth direction as a function of shape of weld pool for various crystal orientations.

Defining weld metal microstructure especially with respect to its refinement is a complex function of chemical composition, heat content, thermal cycle and fine inclusion content [89–92]. The effect of different parameters on microstructure is understood by analyzing their effect on total heat content of weld pool and its thermal cycle. Increase in cooling rate refines the microstructure of weld metal as well as reduces width of HAZ. Increase in fine inclusion content also refines the microstructure by providing more nucleation sites. Finally the microstructure of arc weld deposit primarily depends upon solidification behaviour of weld metal through a nucleation and growth process of primary solids. During welding the solidification behaviour dictates solute redistribution in order to control the shape and size of grains of microstructure. Often the mechanism of weld pool solidification is derived from an extrapolation of the knowledge of cooling of cast ingots at a comparatively relatively lower thermal gradients and growth rates [93] than those exists in weld pool. In welding the solidification behaviour of molten pool is interpreted through both the temperature gradient with the effective sink and cooling rate that varies considerably across the fusion zone. Depending upon the type of welding process and location, the cooling rate in weld pool lies in the range from 10^2 to 10^7 °C/sec [94]. The critical parameters determining the microstructure of weld metal are the temperature gradient, undercooling and alloy constitution influencing the nucleation and growth rate of primary solids.

In conventional solidification, nucleation of primary solid may be considered to happen in homogeneous or heterogeneous manner. When the solid forms from liquid without any influence of foreign material, it may largely nucleate homogeneously. This process requires a large driving force or a large equivalent under cooling below freezing point. The free energy, ΔG*, of formation of solid nucleus of critical size is given by

$$\Delta G^* = \frac{16.\pi.\gamma_{SL}^3}{3.\Delta G_v^2} \tag{1.32}$$

where, γ_{SL} is solid liquid interfacial energy and ΔG_v is free energy change per unit volume of solidification. However, in presence of foreign solid matter such as oxides or other solid particles as externally added inoculants the nucleation of solid may be further facilitated. Here the free energy, ΔG^*_{het}, of formation of nucleus of critical size is relatively less than ΔG* by a factor f(β), expressed as

$$\Delta G^*_{het} = \Delta G^*.f(\beta) \tag{1.33}$$

where, f(β) is shape factor that depends upon wetting angle. For heterogeneous nucleation $f(\beta) \ll 1$, indicating a significant decrease in nucleation barrier when the solidification occurs spontaneously by epitaxial growth. In weld pool the factor f(β) approaches to zero because solidification proceeds from a pre existing solid substrate. In addition, a more classical case of heterogeneous nucleation may also arise if nucleation barrier is sufficiently reduced in absence of inoculants in welds. The dynamic methods of weld pool stirring and arc oscillations may fragment dendrites and displace them from the solidification front so that they can also act as nuclei for the growth of additional grains. However, an understanding of the principles of refinement in fusion zone microstructure needs to be further understood in reference to fluid dynamics of weld pool. In this case the computational fluid flow models need to be coupled with development of microstructure to ascertain the influence of convective fluid flow on dendrite fragmentation and its distribution contributing to refinement of microstructure of molten weld metal.

Temperature gradient at solid-liquid interface significantly determines the growth of solidifying substructure in fusion zone. In reference to solid-liquid interface, the thermal gradient exists in liquid with respect to that in solid component of weld pool is more critical in determining morphology of the interface. Evaluation of temperature gradient in liquid is complicated due to a strong influence of convection on it in the weld pool. However, some attempts have been made to measure them experimentally [95–98] with significant success. Computational modelling may also be a useful technique in this regard. Growth or solidification rate may be defined as the rate at which the solid-liquid interface advances in weld pool. The growth rate varies considerably at different locations of weld pool. During steady state welding, it is correlated to speed and direction of moving heat source in weld pool [83, 84, 99, 100] of different shape.

In presence of pre-existing solid-liquid interface in fusion zone, growth of solid occurs by addition of atoms from liquid to solid. The kinetics of growth may be controlled to a large extent by the interface structure on atomic scale. The growth may occur two-dimensionally on a screw dislocation, or by uniform continuous growth [101, 102]. The stability of interface critically determines the microstructural characteristics of weld metal. Shape of the interface is determined by thermal and constitutional conditions in its immediate vicinity. These factors determine mode of growth as planer, cellular or dendritic. Many theories have been developed to explain the interface stability under equilibrium conditions for normal solidification or non-equilibrium conditions of rapid solidification [102, 103]. They clearly describe the fundamentals of dendritic solidification [104] of weld pool. In case of pure metal having no solute effect, the thermal gradients only control stability of the interface. Whereas in alloys, the stability of solidification front becomes relatively more complex due to partitioning of solute elements that arrives during solidification in between the solid and liquid. Thus, here the effect of composition gradient ahead of the interface must be considered. Effect of formation of solute at the interface and its redistribution on morphological stability of solidification front has been examined by considering the concept of constitutional super cooling [102, 103]. As the growth conditions leave the planar stability, the interface morphology relevantly changes from the planar to cellular to dendritic depending upon distribution characteristics of chemistry and thermal state in solidifying matrix. Under favourable conditions, dendrites may exhibit secondary and tertiary arms in crystallographically predetermined preferred growth directions.

The concept of constitutional super cooling is useful for broad understanding of development of microstructure and influence of process parameters on weld microstructure but, it is not sufficient to describe precisely the morphological stability at solidification front. The under-cooling, defined as how far a liquid alloy of given composition is cooled below its equilibrium liquidus temperature, is a more critical parameter. It controls the microstructure and segregation effect of weld pool through solute redistribution during its solidification. It happens in association with nucleation and/or growth during solidification. Unlike in castings, weld solidification proceeds from pre-existing solid substrate, and thus, there is little or no nucleation barrier. Therefore, for growing solid in weld pool the effects of constitutional under-cooling and under-cooling due to curvature at dendrite tip are important to consider. This is because it significantly influences the solidification morphology and extent of micro segregation of elements and produces much finer microstructure in weld as compared to that of casting. The effect of increased under-cooling at dendrite tips reduces the micro segregation.

Redistribution of solute that promotes segregation of elements during solidification, can significantly affect microstructure, cracking susceptibility and properties of weld. This is an important phenomenon often observed in welding of alloy. Segregation in relatively finer scale with dendrite arm spacing in the range of 10–100 μm is called micro-segregation. But, in case of macro-segregation the scale of arm spacing increases up to several hundreds of micrometers. The mode of metal transfer in GMAW may significantly affect the solute redistribution due to its influence on thermal characteristics of solidifying weld pool. Using different models to describe solute redistribution that occurs during solidification of castings, some studies have been carried out in the area of welding with very little theoretical and experimental understanding on it in the matrix during solidification of weld pool. Compositional effect of solidification of alloy can be divided into three categories [101, 102]. First one is the equilibrium solidification of melt with complete diffusion of solute in liquid, which generally does not apply in case of welding. The second one assumes little or no diffusion of solid, while a thorough mixing in liquid by convection and diffusion process is practically apparent. This condition leads to micro segregation effects in weld microstructures. The third category assumes no diffusion of solid and limited diffusion of liquid with no convection. This case allows build up of a solute boundary layer ahead of the interface which may be used to describe morphological stability for microstructure development in welds. The region of plane front solidification along the fusion line of weld with limited diffusion in liquid is most appropriate to analyze the solute redistribution. In welds the solute redistribution under dendritic growth conditions at the dendrite tip and in inter dendritic regions both should be considered. Solute redistribution at dendrite front characterising its under-cooling can be significantly [105] realised by knowing the change in core composition of the primary solid.

Extensive solid-state phase transformations occur during post solidification cooling of fused zone to ambient temperature, whereas in HAZ it occurs during heating and cooling segments of relevant part of weld thermal cycle[106]. Depending on the heating and cooling rates as well as the peak temperature of any location attained during thermal cycle of welding, the extent and type of phase transformations varies across the weld joint. Accordingly the HAZ is often divided into several sub zones in reference to their distance from the fusion line. It is not an exaggeration to say that most of the mechanisms of phase transformations known in metallurgical practices can be found either in fusion zone or in HAZ of a typical weldment. These mechanisms include solidification, chemical segregation, solid-state phase transformation, recrystallization, recovery, grain growth, precipitation, coarsening of precipitates and stress induced transformations. But, most of these processes in welding occur at non-equilibrium and non-isothermal conditions under much faster thermal transients especially during heating and cooling as compared to those occur in case of controlled isothermal annealing or continuous cooling conditions of conventional heat treatments. Therefore, a thorough evaluation of phase transformation and microstructure development during welding requires consideration of both the heating and cooling behaviours of the process.

The phase transformation may involve a change in crystal structure of parent phase to form a product phase, which can be achieved by distortion of the parent phase as it occurs in case of martensite transformation from austenite. The externally applied stresses and/or residual stresses in weldment also influence the solid-state phase transformation by providing a positive or negative influence on thermodynamics and kinetics of phase transformation and its growth by influencing random reconstructive atomic motions. Thus it significantly affects the microstructure of weld and HAZ [107]. Various models have been developed [108] to predict fusion zone microstructure based on physical metallurgy principles, for computing the HAZ microstructures by considering transient heat transfer [109] and to predict the grain growth in HAZ using [110] Monte Carlo technique.

Estimation of phase transformation in weld conveniently starts by computational modelling for simulation of thermal behaviour of molten weld pool and base plate followed by coupling it with appropriate model of phase transformation. Such a coupled quantitative model may be useful to understand the characteristics of phase transformation under effective variables. Significant efforts have been made to estimate heat and mass transfer in weld as stated earlier, but a relatively little effort has been made [106–109, 111, 112] to develop appropriate models to characterize phase transformation and to predict microstructure of weld joint. However, such modelling is quite complicated due to involvement of number of interacting factors, such as welding process, procedure and parameters. They determine stable and multiple thermal cycles governing the shape and size of weld pool, mode of metal deposition, weld chemistry, solidification morphology, segregation, inclusion characteristics and grain size.

Development of residual stress and stress gradient due to thermal effects and phase transformations take place during welding make the analysis of welded structure more complicated. It happens because of their uniqueness in combination with the compositional and microstructural variations in the vicinity of fusion weld. The usual treatment of engineering mechanics of material behaviour assumes homogeneous properties of matrix. This may not be appropriate to use in case of weldment due to existence of such gradients in behaviour of matrix of weld joint. Thus, it is necessary to develop coordination amongst the fundamental understanding of welding processes and knowledge of evolution of microstructure and properties for intelligent process control to produce defect free structurally sound weld.

1.4.3 Weld Defects

In GMAW process an inappropriate variation of welding parameters with respect to welding procedure and position adversely affect the weld quality due to their strong influence on thermal and transfer behaviours of metal. Weld defects commonly encountered in this process are the lack of fusion and penetration and porosity content. For commonly used down hand flat position of welding the effect of various welding parameters on weld quality is fairly well known and with this knowledge often the control of this process in positional welding is analysed.

In the light of descriptions of various aspects of GMAW process learned above it is well understood that the use of a high welding current up to an optimum level of spray mode of metal transfer gives comparatively better weld quality. Increase of welding current reduces the duration of presence of superheated molten droplet in arc cavern. Thus, it reduces the absorption of gases in depositing liquid metal and consequently minimizes porosity content of weld deposit [113–115]. However, in this process arc voltage primarily governs the arc length and affects size and shape of weld fusion zone. A long arc length gives flat and shallow deposit. It also allows the arc to wander which increases spattering and may also cause porosity in weld deposit due to air aspiration in shielding gas jacket [116]. However, the use of long arc length increases arc plasma perimeter resulting in a wider weld pool, reduces heat concentration directly below the welding electrode and gives a sluggish response of weld pool to weave [10]. A shorter arc length improves wetting characteristics and control of weld pool but, a too short arc length may allow the electrode to touch and short circuit the weld pool. It may result low base metal melting, narrow weld deposit and variation in heat input causing possibility of developing weld defects. Hence, it is important to select proper arc voltage/arc length, especially at low arc voltage for a good arc initiation and stability, to obtain appropriate weld pool control for desired weld characteristics. Thus, especially when selecting a power source for short circuit GMAW, it is recommended to use one with good voltage control at low end for good arc start and arc stability.

An increase of electrode extension from contact tube to work piece reduces arc stability and often increases the risk of porosity in weld metal [14]. Longer extension results in enhanced resistive heating and consequently reduces the balance of energy available to melt the filler wire and base metal, which gives rise to lack of fusion [14]. During GMA welding the solidification behaviour dictated by welding parameters controls solute redistribution, microstructure, shape and size of grains, distribution of inclusions as well as the defects such as porosity and hot cracking.

References

1. Gourd, L.M. : Principles of Welding Technology, p. 32. English Language Book Society, Edward Arnold Ltd., (1986)
2. Mathison, Jerry : Understanding transfer modes for GMAW. Pract. Weld. Today **14** (2008)
3. Kim, Y.S., Eagar, T.W.: Analysis of metal transfer on gas metal arc welding. Weld. J. **72**(7), 269s–278s (1993)
4. Heald, P.R., Madigan, R.B., Siewert, T.A., Lin, S.: Mapping the droplet transfer modes for an ER-1000 S-1 GMAW electrode. Weld. J. Res. Suppl. **73**, 38s–43s (1994)
5. Johnson, J.A., Carlson, N.M., Smart, H.B., Clark, D.E.: Process control of GMAW: sensing of metal transfer mode. Weld. J. Res. Suppl. **70**, 91s–99s (1991)

6. Liu, S., Siewert, T.A.: Metal transfer in gas metal arc welding: droplet rate. Weld. J. **68**, 52s–58s (1989)
7. Lyttle, K.A.: GMAW—a versatile process on the move. Weld. J. **62**(3), 15–23 (1983)
8. Wilson, Eric.: Welding technology data sheet, 60 metal inert gas arc welding—Part 4. Weld. Met. Fab. **14**, 5, 165–172 (1982)
9. Craig, E.: A unique mode of GMAW transfer. Weld. J. **66**(9), 51–55 (1987)
10. Matthews, J.R., Lassaline, E.E., Porter, J.F., Leewis, K.G.: Evaluating frequency modulated GMA welding for HY 80 steel. Weld. J. **71**(9), 49–53 (1992)
11. Norrish, J., Richardson, I.F.: Metal transfer mechanisms. Weld. Met. Fab. **20**(1), 17–22 (1988)
12. Waszink, J.H., Graat, L.H.J.: Experimental investigation of the forces acting on a drop of a weld metal. Weld. J. **62**(4), 108s–116s (1983)
13. Essers, W.G., Van Gompal, M.R.M.: Arc control with pulsed GMA welding. Weld. J. **63**(6), 26–32 (1984)
14. Stenbacka, N., Persson, K.A.: Shielding gases for gas metal arc welding. Weld. J. **68**(11), 41–47 (1989)
15. Loh, P.C., Loper, C.R., Gregory, J.T.: Gas metal arc welding of HK 40 steel. Weld. J. **70** (11), 31–36 (1991)
16. Rhee, S., Kannatey–Asibu, E.: Observation of metal transfer during gas metal arc welding. Weld. J. **71**(10), 381s–386s (1992)
17. Quinn, T.P., Madigan, R.B., Siewert, T.A.: An electrode extension model for gas metal arc welding. Weld. J. **73**(10), 241s–247s (1994)
18. Amin, M., Ahmed, N.: Synergic control in MIG welding. Met. Constr. **19**(1), 22–28 (1987)
19. Lancaster, J.F.: The Physics of Welding, IIW, 1st edn, pp. 134–230. Pergamon Press, New York (1984)
20. Smart, H.B., Einerson, C.J.: A model for heat and mass input control in GMAW. Weld. J. **72** (5), 217s–229s (1993)
21. Allum, C.J., Quintino, L.: Control of fusion characteristics in pulsed current MIG welding. Part-II, Simple model of fusion characteristics. Met. Constr. **17**(5), 314R–317R (1985)
22. Colombieer, L., Hochmann, L. : Stainless Heat Resisting Steels, pp. 68–69. Edward Arnold Publishers Ltd., (1967)
23. Kent, Von Horn R.: Aluminium Properties and Physical Metallurgy, pp. 17–181. American Society for Metals, Metals Park, Ohio (1967)
24. Maruo, H., Hirata, Y.B.: Study of Pulsed MIG Welding, IIW Doc. SG 212-585-84. Welding Department, Osaka University, Japan, July (1984)
25. Metals Handbook, vol. 3, 9th Ed.. ASM International, Metals Park, Ohio (1979)
26. Metals Handbook : Properties and Selection: Nonferrous Alloys and Pure Metals. ASM International, Metal Park, Ohio, 10th Ed., **2**, 1099 (1979)
27. Tekriwal, P., Mazumder, J.: Finite element analysis of 3-Dimensional transient heat transfer in GMA welding. Weld. J. **67**, 150s–156s (1988)
28. Waszink, J.H., Piena, M.J.: Experimental investigation of drop detachment and drop velocity in GMAW. Weld. J. **65**(11), 289s–298s (1986)
29. Waszink, J.H., Van Den Heuvel, G.J.P.M.: Heat generation and heat flow in the filler wire in GMAW welding. Weld. J. **61**(8), 269s–282s (1982)
30. Kim, Y.S., Mceligot, D.M., Eagar, T.W.: Analysis of electrode heat transfer in gas metal arc welding. Weld. J. **70**(1), 20s–31s (1991)
31. Jilong, Ma., Apps, R.L.: Analysing metal transfer during MIG welding. Weld. Metal Fab. **15** (4), 119–128 (1983)
32. Rosenthal, D.: The theory of moving sources of heat and its application to metal treatments. ASME Trans. **68**, 849–866 (1946)
33. Goldak, J., Chakravarti, A., Bibby, M.: A new finite element model for welding heat sources. Met. Mater. Trans. **15B**, 299–305 (1984)
34. Jeong, S.K., Cho, H.S.: An analytical solution to predict the transient temperature distribution in fillet arc welds. Weld. J. **76**, 223s–232s (1997)

35. Komanduri, R., Hou, Z.B. : Thermal Analysis Of The Arc Welding Process: Part I. General solutions. Met. Mater. Trans. **31B**, 1353–1370 (2000)
36. Nguyen, N.T., Ohta, A., Matsuoka, K., Suzuki, N., Maeda, Y.: Analytical solutions for transient temperature of semi infinite body subjected to 3–D moving heat sources. Weld. J. **78**, 265s–274s (1999)
37. Wells, A.A.: Heat flow in welding. Weld. J. **31**, 263s–267s (1952)
38. Apps, R.L., Milner, D.R.: Heat flow in argon arc welding. Br. Weld. J. **34**, 475–485 (1955)
39. Paley, Z., Lunch, J.N., Adams Jr., C.M.: Heat flow in welding heavy steel plate. Weld. J. **43**, 71s–79s (1964)
40. Christensen, N., de Davies, L., Gjermundsen, K.: Distribution of temperatures in arc welding. Br. Weld. J. **12**(2), 54–75 (1965)
41. Pavelic, V., Tanbakuchi, R., Uyehara, O.A., Myers, P.S.: Experimental and computed temperature histories in gas tungsten arc welding of thin plates. Weld. J. **48**, 295s–305s (1969)
42. Kou, S.: Simulation of heat flow during welding of thin plates. Met. Mater. Trans. **12A**, 2025–2030 (1981)
43. Kou, S., Kanavsky, T., Fyfitch, S.: Welding thin plates of aluminium alloys—a quantitative heat flow analysis. Weld. J. **61**, 175s–181s (1982)
44. Wilson, E.L., Nickell, R.E.: Application of the finite element method to heat conduction analysis. Nucl. Eng. Design **4**, 277–286 (1966)
45. Paley, Z., Hibbert, P.D.: Computation of temperature in actual weld design. Weld. J. **54**, 385s–392s (1975)
46. Friedman, E. : Thermodynamic analysis of the welding process using the finite element method. Trans. ASME J. Press. Vessel Technol. 206–213 (1975)
47. Friedman, E., Glickstein, S.S.: An investigation of the thermal response of stationary gas tungsten arc welds. Weld. J. **55**, 408s–420s (1976)
48. Tsai, C.L., Lee, S.G., Shim, Y.L., Jaeger, J.J., Chasten, C.P.: ASME Heat Transfer in Materials Processing, (JHTD), **224**, 9–17 (1992)
49. Tsai, C.L. : American Society of Metals Conference on Trends in Welding Research in the United States, New Orleans, LA, pp. 91–108. ASM International, Metal Park, OH (1982)
50. Nunes Jr., A.C.: An extended Rosenthal weld model. Weld. J. **62**(6), 165s–170s (1983)
51. Eagar, T.W., Tsai, N.S.: Temperature fields produced by travelling distributed heat sources. Weld. J. **62**(12), 346s–355s (1983)
52. Na, S.J., Lee, S.Y. : A study of the three dimensional analysis of the transient temperature distribution in gas tungsten arc welding. In: Proceeding of Institute of Mechanical Engineering (London), Part B 201, pp. 149–156 (1987)
53. Prasad, N., Silva, Sankara, Narayanan, T.K.: Finite element analysis of temperature distribution during arc welding using adaptive grid technique. Weld. J. **75**, 123s–128s (1996)
54. Tekriwal, P., Mazumder, J.: ASM International Conference on Proceedings of Trends in welding research, Gatlinburg, TN. ASM International, Metal Park, OH, (1986)
55. Boo, K.S., Cho, H.S. : Transient temperature distribution in arc welding of finite thickness plates. In: Proceeding of Institute of Mechanical Engineering Part B 204, pp. 175–183 (1990)
56. Ule, R.L., Joshi, Y., Sedy, E.B.: A new technique for three dimensional transient heat transfer computations of autogenous arc welding. Met. Mat. Trans. **21B**, 1033–1047 (1990)
57. Kasuya, T., Yurika, N.: Prediction of welding thermal history by a comprehensive solution Weld. J. **72**, 107s–115s (1993)
58. Kasuya, T., Yurika, N.: Determination of necessary preheat temperature to avoid cold cracking under varying ambient temperature. ISIJ Int. **35**(10), 1183–1189 (1995)
59. Zacharia, T., Eraslan, A.H., Aidun, D.K.: Modeling of autogenous welding. Weld. J. **67**, 53s–62s (1988)
60. Zacharia, T., Eraslan, A.H., Aidun, D.K.: Modeling of non autogenous welding. Weld. J. **67**, 18s–27s (1988)

61. Saha, S.C., Ghosh, A.K., Malhotra, S.L.: Heat transfer in welding—a numerical approach. Indian Weld. J. **26**(4), 8–12 (1993)
62. Nguyen, N.T., Mai, Y.W., Simpson, S., Ohta, A.: Analytical approximate solution for double ellipsoidal heat source in finite thick plate. Weld. J. **83**, 82s–93s (2004)
63. Pathak, A.K., Datta, G.L.: 3-D finite element analysis on heat flow in welding under varying arc lengths. Indian Weld. J. **33**, 4, 24–29 (2000)
64. Pathak, A.K., Datta, G.L.: Three dimensional finite element analysis of heat flow in arc welding. Indian W. J. **32**(3), 32–38 (1999)
65. Christensen, K.H., Sorensen, T., Al-Erhayem, O.: A neural network approach for GMA butt joint welding. In: Proceeding of 11th International Conference on Joining of Materials, JOM Institute for the Joining of Materials, p. 8. Gilleleje, Denmark (2003)
66. Oreper, G.M., Eagar, T.W., Szekely, J.: Convection in arc weld pools. Weld. J. **62**, 307s–312s (1983)
67. Oreper, G.M., Szekely, J.: Heat and fluid flow phenomena in weld pools. J. Fluid Mech. **147**, 53–79 (1984)
68. Kou, S., Wang, Y.H.: Weld pool convection and its effect. Weld. J. **65**, 63s–70s (1986)
69. Hong, K., Weckman, D.C., Strong, A.B., Zheng, W.: Modelling turbulent thermofluid flow in stationary gas tungsten arc weld pools. Sci. Technol. Weld. Joining **7**(3), 125–136 (2002)
70. Sahoo, S.K., Roy, G.G.: Mathematical modelling of fluid flow and heat transfer during fusion welding: Some case studies, Procd. Seminar on Advances in Welding Technology (Weld Tech–2003), pp. 26–41. IIT Kharagpur, 14–15 March 2003
71. Zhang, W., Roy, G.G., Elmer, J.W., DebRoy, T.: Modeling of heat transfer and fluid flow during GTA spot welding of 1005 steel. J. App. Phys. **93**, 5, 1, 3022–3033 (2003)
72. Tsao, K.C., Wu, C.S.: Fluid flow and heat transfer in GMA weld pools. Weld. J. **67**, 70s–75s (1988)
73. Ushio, M., Wu, C.S.: Mathematical modelling of three dimensional heat and fluid flow in a moving Gas Metal Arc weld pool. Met. Mater. Trans. **28B**, 509–516 (1997)
74. Kim, I.S., Basu, A.: A mathematical model of heat transfer and fluid flow in gas metal arc welding process. J. of Mater. Process. Technol. **77**, 17–24 (1998)
75. Sun, J.S., Wu, C.S., Gao, J.W.: Modelling the weld pool behaviours in GMA welding. Int. J. Joining Mater. **11**(4), 112–117 (1999)
76. Kumar, S., Bhaduri, S.C.: Theoretical investigation of penetration characteristics in gas metal arc welding using finite element method. Met. Mater. Trans. **26B**, 611–624 (1995)
77. Kumar, S., Bhaduri, S.C.: Three dimensional finite element modeling of gas metal arc welding. Met. Mater. Trans. B **25B**, 435–441 (1994)
78. Pardo, E., Weckman, D.C. : The interaction between process variables and bead shape in GMA welding: a finite element analysis, recent trends in welding science and technology. In: 2nd International Conference of Welding Research, pp. 391–395, 14–18 May 1989
79. Essers, W.G., Walter, R.: Heat transfer and penetration mechanisms with GMA and plasma-GMA welding. Weld. J. **60**(2), 37–42s (1981)
80. Dimbylow, C.S.: Bead geometry and properties of inconel 625 overlays on steel. Metal Constr. **17**(1), 35–39 (1985)
81. McGlone, J.C.: Weld bead geometry prediction—a review. Metal Constr. **14**(7), 378–384 (1982)
82. David, S.A., Liu, C.T.: High-power laser and arc welding of thorium-doped iridium alloys. Weld. J. **61**, 157–163s (1982)
83. Davies, G.J., Garland, J.G.: Solidification structures and properties of fusion welds. Int. Met. Rev. **20**, 83–105 (1975)
84. Savage, W.F.: Solidification, segregation and *weld* imperfections. Weld. World **18**, 89–133 (1980)
85. Kerr, H.W., Villafuerte, J.C.: Metal Science of Joining (Cieslak, M.J., Perepezko, J.H., Kang, S., Glicksman, M.E. eds.), p. 11. The Minerals, Metals and Materials Society, Warrendale, PA (1992)

86. David, S.A., Vitek, J.M., Rappaz, M., Boatner, L.A.: Microstructure of stainless steel single crystal electron beam welds. Metall. Trans. **21A**, 1753–1766 (1990)
87. Rappaz, M., David, S.A., Vitek, J.M., Boatner, L.A.: Development of microstructures in Fe–15 Ni–15 Cr single-crystal electron beam welds. Metall. Trans. A **20A**, 1125–1138 (1989)
88. Rappaz, M., David, S.A., Vitek, J.M., Boatner, L.A.: Analysis of solidification microstructures in Fe–Ni–Cr single crystal welds. Metall. Trans. A **21A**, 1767–1782 (1990)
89. Easterling, K.: Introduction to Physical Metallurgy of Welding. Butterworth Publications (1985)
90. Harrison, P., Ferrar, Roy: Microstructural development and toughness of C–Mn and C–Mn–Ni weld metal; Part—I. Metal Constr. **19**(7), 392–399R (1987)
91. Lancaster, J.F., Metallurgy of Welding, pp. 144–229. Allen and Unwin (Publishers) Ltd (1987)
92. Reed-Hill, Robert E.: Physical Metallurgy Principles. Affiliated East West Press Pvt. Ltd., New Delhi (1973)
93. David, S.A., Vitek, J.M.: Int. Mater. Rev. **34**, 213–245 (1989)
94. DebRoy, T., David, S.A.: Physical processes in fusion welding. Rev. Mod. Phys. **67**(1), 85–112 (1995)
95. Garland, G.J.: The control of weld pool solidification. Ph.D. thesis, University of Cambridge, UK, (1972)
96. Matsuda, F., Hashimato, T., Senda, T.: Fundamental investigations of solidification structure in weld metal. Trans. Natl. Res. Inst. Met. (Jpn) **11**(1), 43–58 (1969)
97. Nippes, E.F., Savage, W.F., Grotke, G.: Further studies of the hot-ductility of high-temperature alloys. Weld. Res. Council. Bull. **33**, 1 (1957)
98. Nippes, E.F., Wawrousek, H., Fleishman, W.L.: The heat-affected zone in arc-welded Type 347 stainless steel. Weld. J. **34**(4), 169–182s (1955)
99. Rappaz, M., Carrupt, B., Zimmerman, M., Kurz, W.: Numerical-simulation of eutectic solidification in the laser treatment of materials. Helv. Phys. Acta **60**, 924–936 (1987)
100. Rappaz, M., Gremaud, M., Dekumbis, R. and Kurz, W.: Laser Treatment of Materials (Mordike, B.J., ed.) DGM Informationsgesellschaftverlag, Bad Nauheim, FRG, (1987)
101. Biloni, H.: Physical Metallurgy, Part I (Chan, R.W., Haasen, P., eds.), pp. 478–579. Elsevier, New York (1983)
102. Flemings, M.C.: Solidification Processing. McGraw Hill, New York (1974)
103. Kurz, W., Fisher, D.J.: Fundamentals of solidification. Trans Tech Publications, Aedermannsdorf, Switzerland (1986)
104. Glickman, M.E.: Principles of Solidification and Materials Processing, Vol. I, p. 11 (Trivedi, R., Sekhar, J.A., Mazumder, J., eds.). Oxford & IBH, Publishing Co. Pvt. Ltd., New Delhi (1989)
105. Brooks, J.A., Baskes, M.I.: Advances in Welding Science and Technology (David, S.A., ed.), p. 93. ASM International, Metals Park, OH (1986)
106. Easterling, K.E.: Mathematical Modelling of Weld Phenomena (Cerjak, H., Easterling, K.E., eds.), p. 163. The Materials Society, London (1992)
107. Bhadeshia, H.K.D.H.: Bainite in Steels. Institute of Materials, London (1992)
108. Watt, D., Coon, L., Bibby, M., Goldak, J., Henwood, C.: An algorithm for modeling of micro structural development in weld heat affected zones (Part A) reaction kinetics. Acta Metall. **36**, 3029–3035 (1988)
109. Henwood, C., Bibby, M., Goldak, J., Watt, D.: Coupled transient heat transfer—microstructure weld computation (Part B). Acta Metall. **36**, 3037–3046 (1988)
110. Shen, Y., Radhakrishnan, B., Thompson, R.G.: International Trends in Welding Science and Technology (David, S.A., Vitek, J.M. eds.), p. 259. ASM International, Materials Park, OH (1993)
111. Bhadeshia, H.K.D.H., Svensson, L.E.: Mathematical Modeling of Weld Phenomena (Cerjak, H., Easterling, K.E. eds.), p. 109. The Materials Society, London (1992)

112. Bhadeshia, H.K.D.H.: Recent Trends in Welding Science and Technology (David, S.A., Vitek, J.M. eds.), p. 189. ASM International, Materials Park, OH (1990)
113. Ghosh, P.K., Gupta, P.C.: Influence of pulsed current MIG welding on the characteristics of Al–Zn–Mg alloy weldments. Trans. Indian Inst. Met. **44**(4), 317–326 (1991)
114. Ghosh, P.K., Sharma, Vijay: Chemical composition and microstructure in pulsed MIG welded Al–Zn–Mg alloy weldments. Mater. Trans. JIM **32**(2), 145–150 (1991)
115. Ghosh, P.K., Gupta, P.C., Somani, R.: Influence of pulse parameters on the porosity formation in pulsed MIG weld deposit of aluminium alloy. Int. J. Joining Mater. **3**(2), 49–54 (1991)
116. Johnson, P.G., Szekely, J., Madigan, R.B., Quinn, T.P.: Power characteristics in GMAW: experimental and numerical investigations. Weld. J. **74**(3), 93–102 (1995)

Chapter 2
Concept of Pulse Current Gas Metal Arc Welding Process

Abstract The unique features of pulsing the current flow in GMAW process that led to development of pulse current gas metal arc welding (P-GMAW) have been discussed. The concept of the P-GMAW process has been explained in terms of its basic process variables. Fundamentals of the control of pulse parameters that can precisely influence the welding process primarily in terms of variation in phenomena of metal transfer has been thoroughly described. The mechanism of drop detachment as a function of pulse parameters that classifies the use of P-GMAW process in two different segments of one drop transfer per pulse and multiple drop transfer pulse has been critically explained with respect to its utility. In the light of all such variations the thermal characteristics of the P-GMAW process have been stated in reference to the wire melting and thermal behaviour of depositing droplet as a function of pulse parameters.

Keywords P-GMAW · Process parameters · Fundamentals of control · Metal transfer · Drop detachment · Thermal characteristics · Wire melting · Thermal nature of droplet

Using conventional GMAW is not favoured in various applications due to its inherent limitation in control of the characteristics of weld deposition. The stable spray mode of metal transfer that offers maximum ease of operation with high weld quality is possible to employ only at relatively higher welding current. Thus, metal transfer in conventional GMAW cannot be regulated as independent of heat input. In pulse current gas metal arc welding (P-GMAW) process, the current is periodically modulated between a relatively low base current and high peak current, as schematically shown by a typical square wave pulse (Fig. 2.1). The mean current level generally falls within the range normally associated with the globular transfer and the peak current level far exceeds the transition value giving axial spray transfer of smaller droplets. By repeated application of current pulses a synthetic spray mode of metal transfer is produced at a comparatively low level of heat input primarily due to intermittent pulses of high current [1, 2] for short duration. Due to this synthetic spray mode of metal transfer of intermittent nature there is a distinct

P.K. Ghosh, *Pulse Current Gas Metal Arc Welding*, Materials Forming, Machining and Tribology, DOI 10.1007/978-981-10-3557-9_2

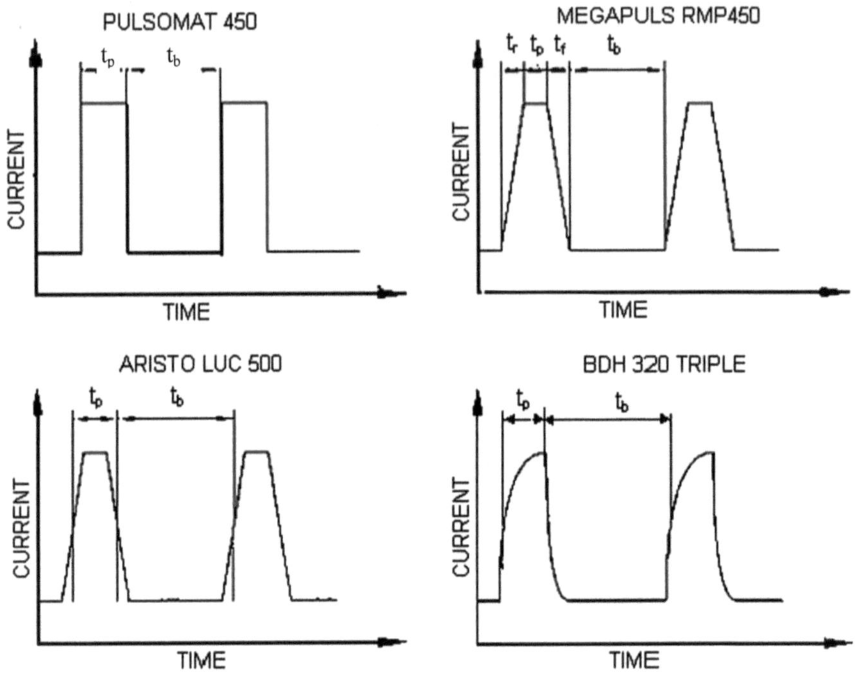

Fig. 2.1 Typical modes of pulsation of various power sources

possibility of availing more precisely controlled weld thermal cycle and metal transfer in this process than that available in conventional GMAW process. The process offers greater ease of operation for welding in any position, for root passes without backing and in case of welding of heat sensitive materials and thin sheets [3]. The use of pulsing enables control of drop size and frequency of its transfer to eliminate spatter and modify penetration without changing average or mean current [4]. The mean current of P-GMAW as of welding current of GMAW is primarily controlled by regulation of feeding rate of filler wire which is correlated almost linearly within the working range of parameters for various ferrous and non ferrous materials (Fig. 2.2a–e). The process facilitates the use of comparatively lower diameter of filler wire and reduces the wire feed problem at relatively high current of spray mode of metal transfer. Due to such operational ease the P-GMAW process is widely accepted in weld fabrication of various ferrous and non-ferrous metals requiring high weld quality.

The P-GMAW process is suitable for depositing high quality welds at relatively low heat input [5] but, the favourable characteristics of this process are primarily governed by the pulse parameters. It introduces additional welding parameters such as, mean current (I_m), peak current (I_p), base current (I_b), pulse frequency (f), peak current duration (t_p) and base current duration (t_b) in the process over those involved in conventional GMAW. It has also been realized that a change in any

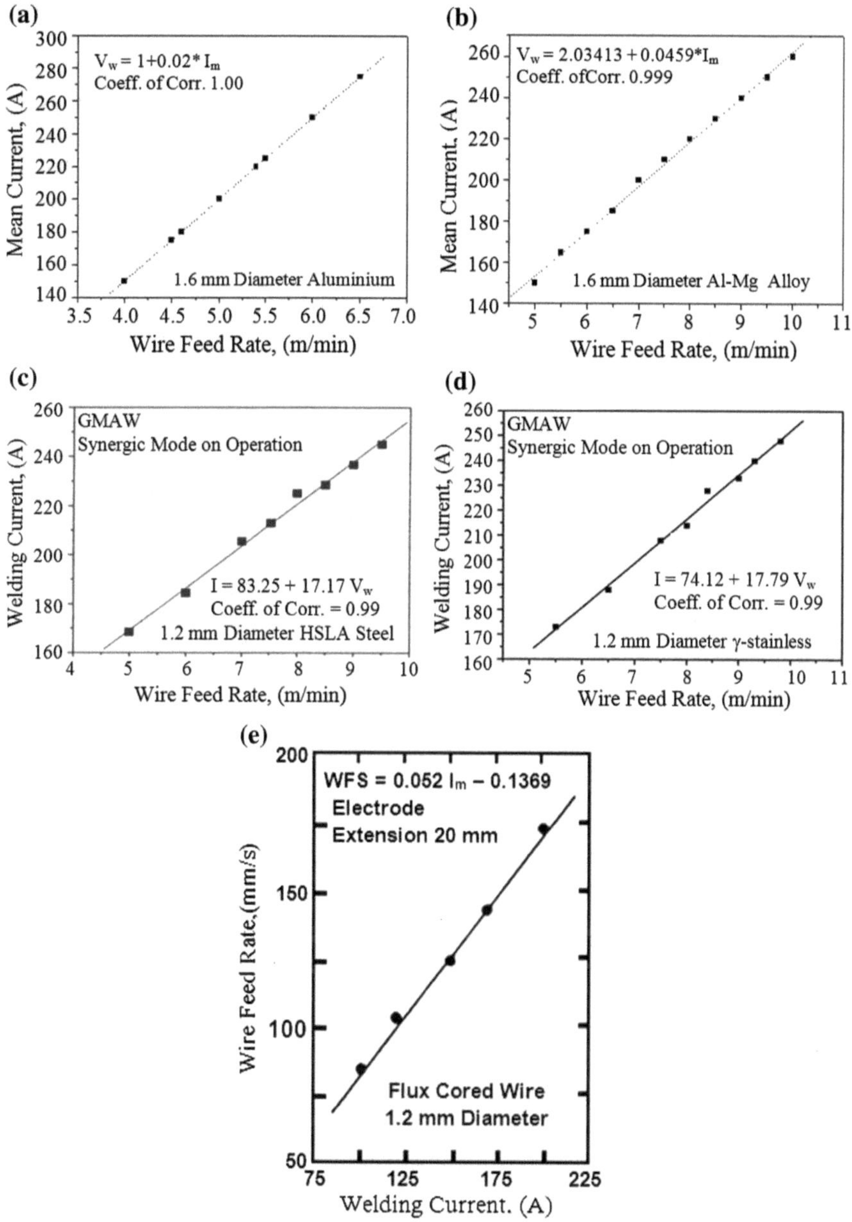

Fig. 2.2 Effect of wire feed rate on mean (P-GMAW) and welding (GMAW) current in case of **a** 1.6 mm diameter aluminium, **b** 1.6 mm diameter Al–Mg alloy and **c** 1.2 mm HSLA steel, **d** 1.2 mm γ-stainless steel and **e** 1.2 mm diameter FCW steel filler wires

pulse parameter simultaneously affects the other under the concept of energy balance [6] and hence influences the weld characteristics to a great extent [7–9]. The simultaneous influence of pulse parameters on each other in operation makes it difficult to select optimum combination of pulse parameters for any welding condition. Detailed relationships amongst the pulse parameters and their effect on metal transfer and weld bead characteristics are not yet well established [10–13]. Criticality noted in selection of pulse parameters possesses difficulties to operate this process in wide and versatile application of weld fabrication with ease and satisfaction. The limiting criteria for successful operation of P-GMAW primarily lies on the process control to avoid [3] arc extinction at too low base current, globular metal transfer at an insufficient peak current and arc burning back at too high peak current.

The limitations of short circuit and spray mode of metal transfer of GMAW process are successfully overcome by using P-GMA welding with its excellent directional capabilities [14, 15]. The process has better and more precise control on metal transfer and is capable of welding at a comparatively low heat input and hence more useful as compared to GMAW [16]. The optimum region for pulse amplitude and duration are well defined [17]. However, a success of the process to produce sound weld is largely governed by the pulse parameters, which simultaneously interact among themselves during welding and dictate weld quality. The interaction of pulse parameters is largely independent of the type of power source commonly in use [18]. Thus, the selection of right pulse parameters is quite critical and needs knowledge of correlation amongst the pulse parameters, behaviour and thermal nature of metal transfer and weld characteristics [3, 10–13].

2.1 Basic Process Variables and Control

In P-GMAW process the mean current, I_m, for a square wave pulse is expressed [10, 19] as

$$I_m = \left(I_p\, t_p + I_b\, t_b\right) \cdot t_{pul}^{-1} \tag{2.1}$$

where, t_b is base current duration and t_{pul} is pulse cycle time expressed as

$$t_{pul} = t_p + t_b \tag{2.2}$$

So, the I_p can be resolved as

$$I_p = \left(\frac{t_{pul}}{t_p}\right) \cdot I_m - \left(\frac{t_{pul}}{t_p} - 1\right) \cdot I_b \tag{2.3}$$

At a given wire feed rate (V_w) and droplet volume (V_d), a stable parametric zone containing all feasible combinations of pulse parameters such as I_p, I_b, t_p and f can be obtained [10] by evaluating the Eq. 2.3 under the following constraints,

$$I_m = m V_w + C \tag{2.4}$$

where, m is the slope, and C is the intercept. The restriction of required droplet volume, V_d, estimated by considering transfer of one drop per pulse (Number of droplets transfer per second equal to the frequency of pulse current) can be expressed as

$$V_d = \frac{A_w \cdot V_w}{f} \tag{2.5}$$

In terms of pulse frequency, f, the t_{pul} as a function of the characteristics of filler wire and volume of drop transfer is expressed as

$$f = \frac{1}{t_{pul}} \tag{2.6}$$

$$t_{pul} = \frac{V_d}{A_w \cdot V_w} \tag{2.7}$$

The optimum condition of drop detachment parameter is necessary to maintain smooth, reproducible spray transfer at all operating conditions. Based on the experimental observations, optimum condition for a given filler wire diameter and material ensuring one or multi drop transfer per pulse can be achieved over a range of combinations of I_p and t_p defined as

$$I_p^2 \, t_p = D_n \tag{2.8}$$

where, D_n is detachment parameter denoting number of drops detached in each duration of pulse. However, in every case the condition for minimum base current for stable arc, below which arc extinguishes is observed as

$$I_b \geq \beta \tag{2.9}$$

where, β is minimum base current limit for stable arc. The range of possible combinations of pulse parameters is quite restricted at low wire feed rate or I_m but, it extends with the increase in it [3]. At a specific arc length an approximately linear relationship exists between the peak and base currents for a given combination of wire feed rate, V_w, pulse frequency, f, and peak current duration, t_p, [3].

$$I_p \propto I_b \tag{2.10}$$

At a given peak current duration and a peak current in excess, I_e, over the base current, both the base current and pulse frequency vary in direct proportion to the wire feed rate.

$$I_e = I_p - I_b \tag{2.11}$$

$$I_b \text{ and } f \propto V_w \tag{2.12}$$

At a given pulse frequency and I_e, both the base current and peak current duration also vary directly with wire feed rate.

$$I_b \text{ and } t_p \propto V_w \tag{2.13}$$

From the Eqs. 2.12 and 2.13 it may be understood that at a given I_e, both the base current and either pulse frequency or peak current duration vary directly with wire feed rate. This relationship remains valid for the entire operating range of wire feed rate from 0.5 to about 10 m/min under shielding of Ar + 5% CO_2 with high degree of arc stability and stable mode of metal transfer, as confirmed in case of using 1.2 mm diameter mild steel filler wire. Thus, the Eq. [2.6] can also be expressed as,

$$I_m = I_b + I_e\, t_p\, f \tag{2.14}$$

A solution of the Eqs. 2.4 and 2.14, as given below implies that for a synergic operation both the base current and the product of peak current in excess over the base current, peak current duration and pulse frequency must vary directly with wire feed rate [20].

$$m \cdot V_w + C = I_b + I_e\, t_p\, f \tag{2.15}$$

2.2 Phenomena of Metal Transfer

It is reported [17] that under identical conditions of welding, transition current for P-GMAW is comparatively higher than that of GMAW. It may often happen due to presence of stationary pendant droplet at the tip of the electrode during base current, which requires additional current to overcome inertia of droplet. The transition current for 1.2 mm diameter steel electrode under the shielding of Ar + 20% CO_2 has been found 380 A in P-GMAW process as compared to 275 A observed in GMAW process [17, 21]. However, in spite of this it is possible to achieve spray mode of metal transfer in P-GMAW at a mean current that is much lower than the required welding current of conventional GMAW process for this purpose. It offers greater ease of operation for welding in all positions, for root passes without backing, as well as for joining thinner plates with consistent quality.

2.2.1 Mechanism of Drop Detachment

The duration of entire mechanism of detachment of a droplet from electrode tip in P-GMAW process primarily combines four stages as time required for heating and melting, formation and growth of drop, necking at the root of growing drop and detachment of droplet, while the current amplitude varies during each stage [22] of action. Under the influence of peak current, a continuous spray of drops may transfer if the pulse duration is prolonged. Time required to form and detach a droplet is inversely proportional to the amplitude of peak current but independent of its duration. Once necking starts, the process requires certain period to detach the droplet, which primarily depends upon peak current and wire diameter, irrespective of the current level at the time of its detachment. Necking as a plastic deformation of heated electrode happens due to the Lorentz force and melting of filler wire occurs under the neck. The detachment of the drop is induced by vaporisation of molten metal at the neck due to resistance heating. The speed of droplet detachment is determined by the rate at which the fused metal is compressed into droplet [4]. It is well known that in P-GMAW, the metal transfer characteristics and thermal behaviour of weld deposit are governed by the pulse parameters, which interact amongst themselves during welding and dictate the characteristics of weld deposit [7, 8, 23, 24].

2.2.2 Nature of Metal Transfer

A steady state of metal transfer at a low level of mean current of P-GMAW creates a high Lorentz force for short duration by superimposition of the pulses of high current resulting into drop detachment and its accelerated movement in axial direction [25]. Each pulse detaches one or more droplets depending on its amplitude and duration. The behaviour of metal transfer in pulsed current gas metal arc welding is better understood by dynamic analysis [26]. The metal transfer in gas metal arc welding under modified pulsed current condition is also studied by numerical analysis [27].

At a given duration of a peak current the mean current as well as base current along with its duration has insignificant influence on droplet transfer. The droplet detachment takes place within a range of combination of I_p and t_p [10, 12, 14, 28, 29] according to $I_p^2\, t_p$ = Const., where the constant depends up on properties of filler wire. Although metal transfer primarily depends on peak current and its duration, the volume of metal detached per pulse is generally governed by the ratio of wire feed rate and pulse frequency. Thus, the nature of metal transfer such as the diameter, velocity and number of droplets transferred per pulse is largely governed by the combined influence of pulse parameters. Type of metal transfer in P-GMAW is broadly classified into two categories as one drop transfer per pulse and multi drop transfer per pulse. The one drop transfer per pulse is popularly used in thin sheet welding with relatively low heat input. But the multi drop transfer per pulse is gaining wide application in welding of various kinds of engineering components due to use of relatively high current giving rise to higher deposition rate at controlled heat build-up in weld.

2.2.2.1 One Drop Transfer Per Pulse

The detachment parameter, D_n, (Eq. 2.8) for one drop transfer per pulse is primarily determined by selection of proper combination of peak current and its duration. The same value of detachment parameter may be obtained using different combinations of peak current and its duration such as higher I_p with shorter duration or lower I_p with relatively longer duration. For one drop per pulse the droplet volume, V_d, can be obtained [10, 14] by Eq. 2.5 when the wire feed rate (burn off rate), V_w, is expressed as

$$V_w = m\, I_m \tag{2.16}$$

where, the slope m depends up on type of filler wire. Using the Eqs. 2.1 and 2.16 the Eq. 2.5 can be resolved as

$$V_d = A_w\, m\, (I_p\, t_p + I_b\, t_b) \tag{2.17}$$

The control of droplet transfer can also be made through estimation of pulse frequency, f′, [30] by using the following expression.

$$f' = \frac{A_w\, V'_w}{V'_d} \tag{2.18}$$

where, V'_w is estimated wire melting rate obtained from the sum of molten mass at I_p and I_b, and V'_d is estimated volume of a droplet obtained from the predicted droplet diameter using static balance theory at the peak current. When the ratio of applied pulse frequency to the theoretical pulse frequency is equal to unity, each pulse tends to produce one drop [31].

2.2.2.2 Multi Drop Transfer Per Pulse

The metal transfer undergoes through a instability condition that can be analysed on the basis of the pinch instability theory [32–34] to estimate a critical wavelength

(λ_c) of a cylindrical current conductor liquid column, which tends to grow and cause the column to break up into droplets. This theory postulates that the pinch force being most active at I_P, the liquid column breaks up due to self-induced electromagnetic force. Electrode tip may develop a tapering depending on its chemical composition and welding current, which reduces the effective fluid cylinder diameter by a factor δ. The diameter of droplet, D, may be estimated by considering the electrode tapering coefficient, δ, and using the concept of energy balanceEnergy balance concept in detachment dynamics of liquid metal droplets as follows.

$$D = [4r/(1 + 3\theta_t)/16] \tag{2.19}$$

where,

$$r = R_w\,\delta \tag{2.20}$$

$$\theta_t = \mu_o\, I_p^2/(\gamma\pi^2\, r) \tag{2.21}$$

μ_o and γ are the permeability of free space and coefficient of surface tension of liquid filler metal respectively and R_w is radius of filler wire. Number of droplets transferred per pulse, N_D, may be estimated by mass transfer per pulse divided by mass of each droplet as follows.

$$N_D = \left[\frac{(A_W\, V_W\, \rho_W)}{f}\right] \Big/ \left[\frac{\pi\, D^3\, \rho_d}{6}\right] \tag{2.22}$$

where, ρ_w is the density of filler wire and ρ_d is the density of molten filler metal [35]. The Eq. 2.22 implies that at a given I_b and I_m, number of droplets transfer per pulse decreases with increase of pulse frequency [36]. At the ratio of applied pulse frequency to the theoretical optimum pulse frequency less than unity, it tends to promote a situation of multi drop transfer per pulse [31]. In case of multi drop transfer the detachment parameter, D_n, is significantly higher than that of one drop transfer per pulse [14].

2.2.3 *Role of Pulse Parameters*

Influence of pulse parameters on nature of metal transfer is in principle similar in case of both the one or multiple drop transfer per pulse with respect to energy input and phenomena of wire melting. But it differs significantly in reference to the mechanism of drop detachment and its characteristics of transfer to weld pool. The role of pulse parameters in controlling one drop transfer per pulse is well known from the inception of the P-GMAW process in practice. But, effect of pulse parameters on multi drop transfer per pulse is yet to be fully understood primarily with respect to its thermal nature and transfer behaviour affecting the weld characteristics. Various aspects of this process have been discussed in latter chapters more in detail. However, some basic issues of the influence of pulse parameters on nature of metal transfer in reference to one drop transfer per pulse may be realized now.

At a given base and peak currents along with their duration the pulse frequency directly varies the mean current [1]. Thus, there is an optimum frequency for any welding condition and around that critical pulse frequency the metal transfer takes place as one drop per pulse. At a given mean current increase of peak current shifts critical frequency range towards higher side for one drop transfer per pulse. But, at a lower value of pulse frequency the spray mode of metal transfer occurs during the period of peak current [36]. Consideration of the rate of ascent and descent of pulse or peak current is important because a sharp rise and fall of current helps in restricting the width of arc plasma and produces a relatively stiffer arc [11]. A high but narrow pulse tends to produce a more constricted arc than a lower one. Thus, even a wide pulse can be used with high pulse for tight joints. The dynamic effects developed during peak current duration are responsible for metal transfer. However, the time for formation and detachment of a droplet is inversely proportional to peak current, but it is independent of its duration [22]. At a given peak current some critical level of its duration is always necessary to transfer a droplet within the pulse peak [14]. At a given mean and peak currents, a too short duration of peak current does not allow the drop detachment to take place within t_p rather it may be pushed to occur at the half way along the down slope of the pulse. At this situation the drop detachment occurs under a comparatively smaller Lorentz force with lower drop velocity, which reduces the depth of penetration. However, at a relatively higher peak current duration the main drop may be detached just at the end of the pulse followed by a detachment of second drop during base current period without proper direction causing deposition out of the weld pool [25].

In P-GMAW the role of base current is to retain the welding arc between the pulses [37–39]. An appropriate base current significantly facilitates the control over weld pool and consequently on weld bead shape. To keep the overall heat input at low level the base current is often set to a minimum, which may result a high crowned weld bead with poor side wall fusion. The drop detachment is hardly influenced by wire melting phenomenon during base current duration. The volume of droplet detached from filler wire can be significantly influenced by the base current and its duration, although an appropriate value of peak current and its duration set a condition for certain form of drop detachment [14]. During variation in mean current with the change of pulse frequency at a given peak current and its duration, the droplet volume may be fixed by maintaining constant electrode heating at base current duration. It can be made constant with enhancement of base current as its duration decreases with increase of pulse frequency [21]. At a given peak current and its duration, increase of base current duration increases the detachment time of droplet, whereas magnitude of base current is having an insignificant effect on detachment time. Effect of base current duration on drop detachment time is comparatively more pronounced at low peak current, where by employing a higher value of peak current the influence of base current on droplet detachment may be reduced [29]. At an appropriate drop detachment parameter, D_n, the detachment of one drop during base current can also be achieved with smooth weld metal transfer giving good bead appearance [29].

2.3 Thermal Characteristics of the Process

The understanding of thermal characteristics of P-GMAW process basically starts from the wire melting or burn off rate of filler wire as a function of pulse parameters. During transfer of molten filler metal in a size of relatively big drops or being fragmented into a stream of tiny droplets based on the pulse parameters, their thermal behaviour at the time of detachment followed by deposition in weld pool largely governs the thermal characteristics of the process. From the time of its inception in practice the fundamentals of thermal characteristics of P-GMAW process is largely understood with respect to one drop transfer per pulse as described below. However, considerable variation in metal transfer by multiple drop modes in entire pulse system affecting the thermal characteristics of the process has also been studied in advanced applications of P-GMAW process, which is discussed latter.

2.3.1 Wire Melting

The burn off rate or melting rate of filler wire in P-GMAW process is expressed [4, 12] as

$$V_{w(pc)} = \int_0^{t_{pul}} V_w \, t_{pul} \, dt_{pul} \tag{2.23}$$

For a square pulsed current waveform it is stated as

$$V_{w(pc)} = (V_{wp} \, t_p + V_{wb} \, t_b) \, t_{pul}^{-1} \tag{2.24}$$

where, $V_{w(pc)}$ is the overall wire burn off rate and V_{wp} and V_{wb} are the wire burn off rates during t_p and t_b respectively. In consideration of the expressions for the V_{wp} and V_{wb} the $V_{w(pc)}$ may be expressed as follows.

$$V_{wp} = A \, I_p + B \, E_w \, I_p^2 \tag{2.25}$$

$$V_{wb} = A\,I_b + B\,E_w\,I_b^2 \tag{2.26}$$

$$V_{w(pc)} = [A(I_p\,t_p + I_b\,t_b) + B\,E_w(I_p^2\,t_p + I_b^2\,t_b)]t_{pul}^{-1} \tag{2.27}$$

Considering the Eqs. 2.6 and 2.8 and by neglecting the ohmic heating during base current period in the context of $I_p^2 t_p \gg I_b^2 t_b$ the overall wire burn off rate $V_{w(pc)}$ as a function of men current and pulse frequency can be expressed as

$$V_{w(pc)} = A\,I_m + B\,E_w\,D_n\,f \tag{2.28}$$

In the light of the Eq. 2.5 considering the wire feed rate V_w in place of equivalent linear function of wire burn off rate $V_{w(pc)}$ the Eq. 2.28 can be expressed [5] as a function of the droplet volume V_d as follows,

$$V_d = A\,A_w\left(\frac{I_m}{f}\right) + A_w\,E_w\,B\,D \tag{2.29}$$

The Eq. 2.29 shows that for any wire feed rate, by fixing the ratio of mean current (I_m) to pulse frequency (f) the droplet size, especially for one drop transfer per pulse, can be held constant for a given wire diameter, electrode extension and detachment parameter. A value of 2 for the ratio of the mean current to pulse frequency gives satisfactory droplet transfer in case of using 1.2 mm diameter steel filler wire [40] when the volume of droplet remains insensitive to I_m. By following the same logic as stated earlier and by considering ohmic heating during base current duration the Eq. 2.27 reduces to

$$V_{w(pc)} = A\,I_m + E_w\,B\,I_{eff}^2 \tag{2.30}$$

where,

$$I_{eff} = \left\{k_p\,I_p^2 + (1 - k_p)I_b^2\right\}^{1/2} \tag{2.31}$$

and k_p is pulse duty cycle defined as the ratio of t_p to t_{pul}. The effective current I_{eff} may also be expressed as

$$I_{eff}^2 = I_m^2 + k_p.(1 - k_p).I_e^2 \tag{2.32}$$

Solving the Eqs. 2.30 and 2.32 and considering it further as a function of wire feed rate $V_{w(cc)}$ for conventional GMA welding (Eq. 1.4) the matter may be resolved as follows.

$$V_{w(pc)} = V_{w(cc)equiv.} + E_w\,B\,k_p(1 - k_p)I_e^2 \tag{2.33}$$

The expression given in Eq. 2.33 reveals [14] that the burn off rate of pulse current gas metal arc welding is comparatively higher than that of the continuous current welding under the welding current equivalent to mean current of the pulse process. The pulsed structure influences the burn off rate for a given mean or effective current and maximum burn off rate can be achieved when $k_p = 1/2$. Thus, it happens at equal peak and base current time and at largest peak current in excess over the base current.

2.3.2 *Thermal Behaviour of Droplet*

The P-GMAW has been employed in number of potential applications exploring its unique features of low heating of weld pool along with highly directional ability of metal transfer [41]. A relatively low temperature [42] of metal transfer in droplets favours better control of weld pool. In this welding process characteristics of metal transfer and thermal behaviour of weld deposit are largely governed by pulse parameters [9, 43] such as, I_m, I_p, I_b, t_p and f, which simultaneously interact among themselves during welding and governs the weld characteristics. At a given mean current and pulse duration, heat content of droplet at the time of detachment enhances significantly with the increase in pulse frequency [43]. The heat content and temperature of the droplets at the time of deposition in weld pool reduces with increase of I_m significantly [43]. The thermal or process efficiency of GMA welding becomes slightly lower with the use of rapid arc and pulsed current [44, 45].

References

1. Collard, J.E.: Adaptive pulsed GMAW control: The digipulse system. Weld. J. **67**(11), 35–38 (1988)
2. Craig, E.: A unique mode of GMAW transfer. Weld. J. **66**(9), 51–55 (1987)
3. Amin, M.: Synergic pulsed MIG welding. Met. Constr. **13**(6), 349–353 (1981)
4. Cornu, Jean: Advanced welding system. IFS Publication Ltd. UK **2**, 127–165 (1988)
5. Lambentt, J.A.: Assessment of the pulsed GMA technique for tube attachment welding. Weld. J. **68**(2), 35–43 (1989)
6. Ghosh, P.K.: Decide pulse parameters for desired properties of pulsed current GMAW weld. In: International Welding Conference (IWC 99) on Welding and Allied Technology—Challenges in 21st century, New Delhi, pp. 18–28. 15–17 Feb 1999
7. Ghosh, P.K., Rai, B.K.: Characteristics of pulsed current bead on plate deposit in flux cored GMAW process. ISIJ Int. **36**(8), 1036–1045 (1996)
8. Ghosh, P.K.: An analysis of weld characteristics as a function of pulse current MIG welding parameters. Int. J. Joining Mater. **8**(4), 157–161 (1996)
9. Randhawa, H.S., Ghosh, P.K., Gupta, S.R.: Geometric characteristics of pulsed current positional GMA weld. ISIJ Int. **38**(4), 276–284 (1998)
10. Amin, M.: Pulse current parameters for arc stability and controlled metal transfer in arc welding. Met. Constr. **15**(5), 272–287 (1983)

11. Pan, J.L., Zhang, R.H., Ou, Z.M., Wu, Z.Q., Chen, Q.: Adaptive control GMA welding—a new technique for quality control. Weld. J. **68**(3), 73–76 (1989)
12. Smati, Z.: Automatic pulsed MIG welding. Weld. J. **65**(1), 38s–44s (1986)
13. Amin, M., Ahmed, N.: Synergic control in MIG welding. Met. Constr. **19**(1), 22–28 (1987)
14. Alum, C.J.: MIG welding—time for a reassessment. Met. Constr. **15**, 6, 347–353 (1983)
15. Craig, E.F.: Pulse spray welding—a mode of weld metal transfer that should revolutionize the GMAW process. Weld. J. **62**, 11, 79 (1983)
16. Weber, Jeff: Pulsed GMAW- a plus for mid western metalworker. Weld. J. **61**(11), 51–52 (1982)
17. Ueguri, S., Hara, K., Komura, H.: Study of metal transfer in pulsed GMA welding. Weld. J. **64**(8), 242–250s (1985)
18. Ghosh, P.K., Dorn, L., Goecke, S.F.: Universality of correlationships among pulse parameters for different MIG welding power sources. Int. J. Joining Mater. **13**(2), 40–47 (2001)
19. Choi, S.K., Ko, S.H., Yoo, C.D., Kim, Y.S.: Dynamic simulation of metal transfer. Part II: Short circuit transfer modes. Weld. J. **77**(1), 45s–51s (1998)
20. Amin, M.: Microcomputer control of synergic pulsed MIG welding. Met. Constr. **18**(4), 216–221 (1986)
21. Norrish, J., Richardson, I.F.: Metal transfer mechanisms. Weld. Met. Fab. **20**(1), 17–22 (1988)
22. Jilong, Ma., Apps, R.L.: MIG transfer discovery of importance to industry. Weld. Met. Fab. **14**(9), 307–316 (1982)
23. Ghosh, P.K., Gupta, P.C.: Use of pulse current MIG welding improves the weld characteristics of Al-Zn-Mg alloy. Indian Weld. J. **29**(2), 24–32 (1996)
24. Ghosh, P.K., Gupta, P.C., Somani, R.: Influence of pulse parameters on bead geometry and HAZ during bead on plate deposition by MIG welding process. Z. Metallkde **82**(10), 756–762 (1991)
25. Essers, W.G., Van Gompal, M.R.M.: Arc control with pulsed GMA welding. Weld. J. **63**(6), 26–32 (1984)
26. Choi, S.K., Yoo, C.D., Kim, Y.S.: The dynamic analysis of metal transfer in pulsed current gas metal arc welding. J. Phys. D Appl. Phys. **31**, 207–215 (1998)
27. Wang, G., Huang, P.G., Zhang, Y.M.: Numerical analysis of metal transfer in gas metal arc welding under modified pulsed current conditions. Met. Mater. Trans. B **35B**(10), 857–866 (2004)
28. Jacobsen, N.: Monopulse investigation of drop detachment in pulsed gas metal arc welding. Phys. D Appl. Phys. **25**, 783–797 (1992)
29. Trindade, F.M., Allum, C.J.: Characteristics in steady and pulsed current GMAW. Weld. Met. Fab. **16**(9), 264–271 (1984)
30. Easterling, K.E.: Mathematical modelling of weld phenomena (Cerjak, H., Easterling, K.E., eds.), p. 163.The Materials Society, London, (1992)
31. Kim, Y.S., Eagar, T.W.: Metal transfer in pulsed current gas metal arc welding. Weld. J. **72** (7), 279s–287s (1993)
32. Choi, S.K., Yoo, C.D., Kim, Y.S.: Dynamic simulation of metal transfer. Part I: Globular and spray transfer modes. Weld. J. **77**(1), 38s–44s (1998)
33. Kim, Y.S., Eagar, T.W.: Analysis of metal transfer on gas metal arc welding. Weld. J. **72**(7), 269s–278s (1993)
34. Lancaster, J.F.: The Physics of Welding, 1st edn, pp. 134–230. IIW, Pergamon Press, New York (1984)
35. Jeong, S.K., Cho, H.S.: An analytical solution to predict the transient temperature distribution in fillet arc welds. Weld. J. **76**, 223s–232s (1997)
36. Metals Handbook: Properties and selection: Nonferrous alloys and pure metals, vol. 2, 10th edn, p. 1099. ASM International, Metal Park, Ohio (1979)
37. Allum, C.J., Quintino, L.: Control of fusion characteristics in pulsed current MIG welding—part I. Dependence of fusion characteristics of process parameters. Met. Constr. **17**(4), 242R–245R (1985)

38. Needham, J.C., Carter, A.W.: Material transfer characteristics with pulsed current. Br. Weld. J. **12**(5), 229–240 (1965)
39. Needham, J.C.: Pulse controlled consumable electrode welding arcs. Br. Weld. J. **12**(4), 191–197 (1965)
40. Quintino, L., Allum, C.J.: Pulsed GMAW: Interactions between process parameters. Part I Weld. Met. Fab. **16**(4), 126–129 (1984)
41. Ghosh, P.K., Dorn, L.: Thermal behaviour of pulsed MIG Al-Zn-Mg weld—an analytical model analysis. Int. J. Joining Mater. **5**(4), 143–150 (1993)
42. Waszink, J.H., Piena, M.J.: Experimental investigation of drop detachment and drop velocity in GMAW. Weld. J. **65**(11), 289s–298s (1986)
43. Ghosh, P.K., Goyal, V.K., Dhiman, H.K., Kumar, M.: Thermal and metal transfer behaviours in pulsed current GMA weld deposition of Al-Mg alloy. Sci. Technol. Weld. Joining **11**(2), 232–242 (2006)
44. Pépe, Nuno, Egerland, Stephan, Colegrove, Paul A., Yapp, David, Leonhartsberger, Andreas, Scotti, Americo: Measuring the process efficiency of controlled gas metal arc welding processes. Sci. Technol. Weld. Joining **16**(5), 412–417 (2011)
45. Bosworth, M.R.: Effective heat input in pulsed current gas metal arc welding with solid wire electrodes. AWS, Weld. J. **70**(5), 111s–117s (1991)

Chapter 3
Basic Nature of Pulse Current GMA Welding

Abstract The governance of pulse characteristics to exploit the merits of P-GMAW in welding of various ferrous and nonferrous metals to produce superior quality welds to those prepared by the conventional GMAW process has been discussed. Control of the simultaneously interacting pulse parameters by considering their summarised influence addressed through a hypothetically developed dimensionless factor ϕ has been critically described. Considering this factor as a primary feature that governs the basic nature of process characteristics of P-GMAW, the behaviour of arcing with respect to its profile, stiffness, efficiency and stability during welding of different metals has been thoroughly discussed. The nature of shielding environment as a function of ϕ during welding of different metals has been addressed. The behaviour of metal transfer has been theoretically analysed and correlated to the factor ϕ in order to realise its desired control by regulating the pulse parameters. The transfer behaviour has been finally discussed taking example of using aluminium, stainless steel and mild steel.

Keywords Governance of P-GMAW · Pulse characteristics · Summarised influence of pulse parameters · Dimensionless factor · Process control · Arc characteristics · Shielding environment · Metal transfer · Different filler metals

3.1 Governance of Pulse Characteristics

The benefits of P-GMAW over GMAW in improving the weld characteristics and properties of various ferrous and non-ferrous metals primarily lie on its ability for more precise regulation of thermal and transfer behaviour of depositing metal by control of pulse parameters as I_m, I_p, I_b, t_b, and f. This is because at a given heat or energy input of certain I_m, the thermal distribution in P-GMAW process significantly varies with a change in I_p, I_b, t_b, and f affecting the thermal behaviour and nature of transferring droplet. So the successful use of P-GMAW for improved weld quality largely depends upon proper dealing of pulse parameters. In order to address

P.K. Ghosh, *Pulse Current Gas Metal Arc Welding*, Materials Forming, Machining and Tribology, DOI 10.1007/978-981-10-3557-9_3

this issue synergic GMA welding set was introduced. It gives unit current pulses to transfer molten droplets of pre-estimated volume from the electrode wire, maintaining the other parametric relationships necessary for stable arcing. It is the basic variation of pulse current GMAW where the pulses are unique function of diameter and material of filler wire. A modern GMA welding power source offers built-in programme with options for choosing the pulse parameters for welding of different materials. The programme is primarily developed on data base of statistical nature of empirical understanding. They are largely useful to the regular welding of commonly known jobs. But, universal application of pulsed current GMA welding for desired weld characteristics in fabrication of any section of different materials in flat and out of position welding with the help of such built-in programme is practically impossible. This is because the control of pulse parameters to regulate the thermal and metal transfer behaviours becomes complicated due to simultaneous influence of the large number of parameters as I_p, I_b, t_b, and f on each other at a given mean current (I_m) of P-GMAW process. This phenomenon of interactive nature of pulse parameters makes the selection and control of pulse parameters highly critical. It deeply handicaps the wide spread use of pulsed current GMA welding in spite of its unique potentialities. However, the difficulties of controlling the pulse parameters with respect to thermal behaviour and quality of weld can be well addressed by considering a balance of energy input and energy consumed in the process as a function of pulse parameters. Thus, at a given time t the energy supplied to the process (E_P) by pulse current is expressed as

$$E_P = \left[(I_b t_b) + I_p[(1/f) - t_b]\right] ft \tag{3.1}$$

Assuming that the energy is primarily consumed in melting of filler wire at a given time t, it may be equated to E_F denoted as energy required for melting of filler wire at time t, which is estimated as

$$E_F = \left[(V_w t)(\pi d^2/4)\rho\right] C_P T_m \tag{3.2}$$

An understanding of the assumption expressed as $E_P = E_F$ may be resolved as

$$f t_b \left(I_b - I_p\right) + I_p = V_w k_c \tag{3.3}$$

$$\left[\left(I_b/I_p\right) f t_b\right] - f t_b + 1 = \left(V_w/I_p\right) k_c \tag{3.4}$$

where, k_c is constant estimated as $[(\pi d^2/4)\rho]C_P T_m]$ for a consumable filler wire. Considering $[(I_b/I_p)ft_b] = \phi$ and V_w is substituted by I_m due to its direct proportionality to mean current, the Eq. 3.4 can be resolved as

$$\phi = \left[\left[\left(I_m/I_p\right)k_c\right] + f t_b - 1\right] \tag{3.5}$$

and

$$\phi = [(k_c - 1)/(1 - k_c)] \tag{3.6}$$

where,

$$I_m = \left[\left(I_p t_p + I_b t_b\right)/\left(t_p + t_b\right)\right] \tag{3.7}$$

The Eq. 3.6 describes that the dimensionless factor ϕ is constant for a given condition of P-GMAW process, which may also be conceived as a summarised influence of pulse parameters. Thus, it is hypothetically inferred that the summarised influence of pulse parameters [1–3] derived from energy balance concept of the process can systematically characterise the P-GMAW. The summarised influence of pulse parameters hypothetically derived as the factor ϕ is expressed as follows.

$$\phi = \left[\left(I_b/I_p\right) f.t_b\right] \tag{3.8}$$

where, $t_b = [(1/f) - t_p]$. The factor ϕ helps in working out parameter selection programme based on clear understanding over correlation of pulse parameters with arc characteristics, nature of gas shielding as well as thermal and transfer behaviours of metal deposition. Selection of pulse parameters for a desired condition is made in conformation of the function $k_c \approx 1$ when, by solving the Eqs. 3.3 and 3.7 as substitution of V_w, the k_c is expressed as

$$k_c = \left[\left[I_p + f t_b\left(I_b - I_p\right)\right]\left(t_p + t_b\right)\right]/\left(I_p t_p + I_b t_b\right) \tag{3.9}$$

The Eq. 3.8 shows that the summarised influence of pulse parameters is a function of two components referring to energy introduced (I_b/I_p) and energy distributed ($f.t_b$) in P-GMAW process. Both the components are function of machine parameters and their control may depend upon the characteristics of power source interacting with different pulse parameters. But, as the relationship amongst the pulse parameters is primarily governed by energy balance within the pulse system, so it may remain valid for any P-GMAW power source in spite of their different type of response to control the pulse parameters. The empirical expressions of various aspects of pulse characteristics as (I_b/I_p), ($f.t_b$) and t_b with rest of the pulse parameters, as functionally described below, are worked out with satisfactory coefficient of correlations within large amount of data collected [4–7, 8] from variety of P-GMAW power sources (Fig. 3.1) and verified [9] in operation on different power sources.

$$I_b/I_p = F(I_m, f, t_b) \tag{3.10}$$

$$t_b = F\left(I_b, I_p, I_m, f\right) \tag{3.11}$$

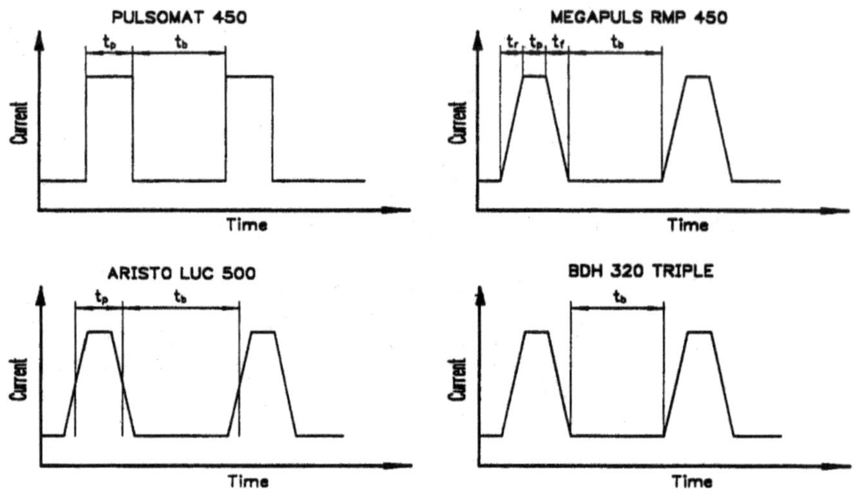

Fig. 3.1 Typical pulse current characteristics in different P-GMAW power sources

$$t_b = F(I_b, I_p, I_m, f) \tag{3.12}$$

The empirical expressions are functional within wide range of I_m, f and t_b varying in the range of 135–265 A, 25–100 Hz and 1–37.5 ms respectively for weld deposition of 1.6 mm diameter Al–Mg filler wire using commercial argon (98–99.9%) gas shielding at a flow rate of 16–18 $l.min^{-1}$ with electrode extension of 15–18 mm and arc voltage of 9–36 V. At different pulse parameters the characteristics of pulse current vary with different behaviour of metal transfer as typically shown in Fig. 3.2a, b. About 92, 89 and 72% of the experimental data regarding (I_b/I_p), ($f.t_b$) and t_b respectively, obtained from five different commercial

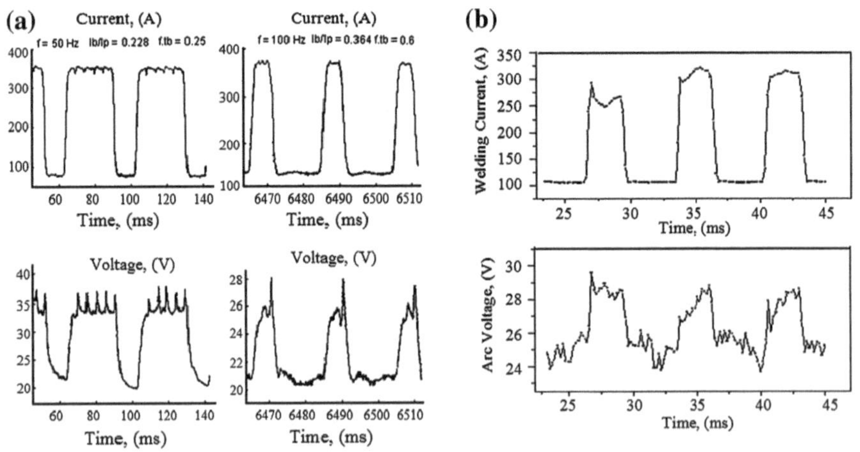

Fig. 3.2 Typical variation in pulse characteristics and transfer behaviour of metal **a** Al–Mg alloy and **b** stainless steel at different pulse parameters

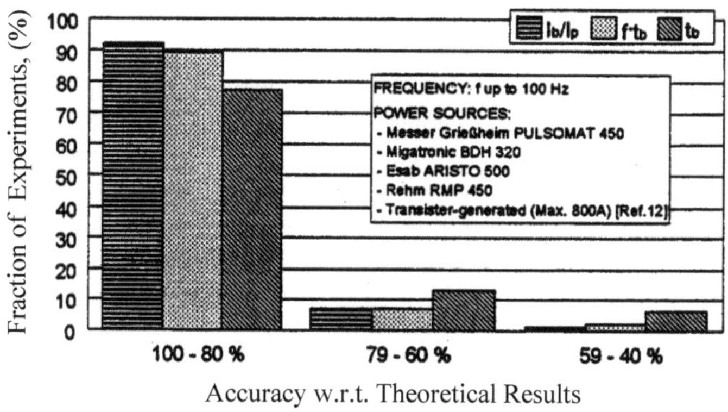

Fig. 3.3 Accuracy of theoretical estimation of (Ib/Ip), (f.tb) and tb in reference to experimental data compared for different P-GMAW power sources

power sources, has been found to come under 80–90% accuracy (Fig. 3.3) with respect to the theoretical results.

The universality of these empirical equations to various power sources, operating at different modes of control over pulsation, may be realized by the energy balance concept. The required energy input for certain rate of metal deposition is almost constant for a given welding condition, which is primarily related to electrode size, material and shielding environment. Variation in mode of pulsation with a change in pulse parameters and characteristics of power sources differ the pattern of distribution of energy input per unit time. Thus, the empirical equations of (I_b/I_p), ($f.t_b$) and t_b as a function of other pulse parameters (Eqs. 3.10–3.12), can be used efficiently for any set of pulse parameters and pulse current GMA welding power sources. The fairly good agreement of the experimental data of wide range of arc voltage with the theoretical results of pulse parameters further reveals that the arc voltage is having insignificant influence on the energy ratio (I_b/I_p), pulse-off-time fraction ($f.t_b$) and pulse off time (t_b) for a stable arc welding. In view of these facts it is considered that the factor ϕ as a function of (I_b/I_p) and ($f.t_b$) can be applied to any condition of pulsed current GMA welding irrespective of the type of power source used. Many investigations [10, 1, 2, 11, 9, 12–21] amply justify the applicability of ϕ in different conditions of pulsation while studying its effect on different weld characteristics using ferrous and non-ferrous materials.

3.2 Arc Characteristics in Welding of Different Materials

Arc characteristics and behaviour of metal transfer strongly influence thermal nature of weld metal affecting the chemistry, microstructure, weld geometry and stresses in weld joint. Porosity formation in P-GMA weld is often understood with respect to air aspiration in arc environment. Puncture in shielding gas jacket due to a negative pressure created by turbulent flow developed within it caused by fluctuation of arc pressure under pulse current is primarily responsible for air aspiration from the outer environment. Occurrence of such detrimental arc characteristics and behaviour of metal transfer under certain pulse parameters distract the benefits of employing P-GMAW process for improvement of weld quality. Number of studies have developed [22, 23, 24] significant knowledge in this area. But, for proper operation of P-GMAW process to get improved weld quality, knowledge of correlation of arc characteristics with the summarised influence of pulse parameters is very much necessary. It may help in right selection of pulse parameters through proper choice of ϕ to achieve desired weld quality.

The arc characteristics of P-GMAW process are primarily considered in two alternative phases of relatively strong and weak arc of pulse on and pulse off periods respectively. The characteristics may be primarily defined by root diameter (D_R), projected diameter (D_P), length (L) and deflection (A_D^0) of arc as shown in Fig. 3.4a, b which largely indicates its degree of constriction and stiffness influencing the efficient shielding. The pulse parameters and arc voltage primarily influence the D_R, D_P and L of arc profile and the stability of shielding of arc environment and nature of droplets transferred during welding that affect weld quality. During pulsed current GMA welding the influence of different parameters on arc profile of the pulse on and pulse off periods and behaviour of metal transfer have been studied by mathematical modelling. The understandings are also confirmed on high speed (10^4 frames/sec) video-graphs of welding of both the non-ferrous [17] and ferrous [18–21] metals. The arc profile observed in video-graphs may not be considered as true profile of the arc due to coverage by glare of plasmatic part of shielding gas around it. But, in spite of this which may cause an over estimation in measurement of various features of arc profile, the basic understanding acquired on qualitative nature of response of arc characteristics to welding parameters can at least be a matter of great interest to realise its influence on weld quality. The arc characteristics as function of summarized influence of

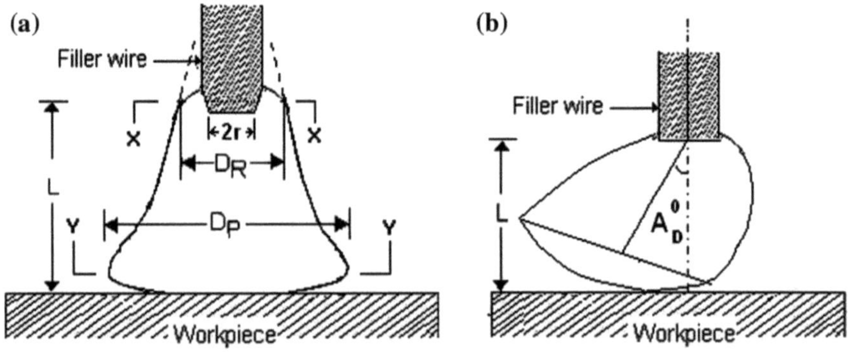

Fig. 3.4 Schematic diagram of **a** arc geometry and radius of electrode tip and **b** arc deflection from electrode axis

pulse parameters are discussed for weld deposition of aluminium using 1.6 mm diameter filler wire as well as stainless and mild steels using 1.2 mm diameter filler wire under different gas shielding.

3.2.1 Arc Profile of Pulse on Period

Arc profile of pulse on period most significantly affects the weld characteristics of P-GMAW because the welding process remains primarily active at peak current. The pulse parameters along with arc voltage significantly affect the energy concentration in weld pool as well as its shielding from atmospheric environment. However, influence of various pulse parameters on arc profile of pulse on period becomes different for different materials.

3.2.1.1 Aluminium Alloy

The characteristics of stable arc vary [17] as a function of P-GMAW parameters as shown in Table 3.1 for weld deposition using Al–Mg filler wire with DCEP at electrode extension of 15 mm in argon shielding. At a given close range of I_m or ϕ of pulse current the increase of any one of them significantly influences (Fig. 3.5) the D_R, D_P and L, when the arc voltage remains constant at about 20–21 V. However, the increase of arc voltage significantly enhances the D_R especially

Table 3.1 Pulse parameters giving stable arc at different ϕ with 1.6 mm diameter Al–Mg filler wire

ϕ	I_m (A)	I_b (A)	I_p (A)	$[I_b/I_p]$	t_p (ms)	t_b (ms)	f (Hz)	Arc voltage (V)
0.053	159	33	240	0.137	9.1	5.9	66	21
0.1	159	53	296	0.179	7.0	9.1	60	20
0.198	160	92	328	0.280	4.2	10.0	70	21
0.248	161	107	327	0.327	3.3	10.4	73	22
0.430	168	145	310	0.468	0.98	12.6	73	21
0.062	220	54	315	0.171	21.1	12.2	30	20
0.112	215	80	368	0.217	14.9	16.0	32	22
0.204	213	122	410	0.297	7.6	17.1	40	22
0.254	212	144	417	0.345	4.6	14.1	53	23
0.429	212	191	411	0.465	1.2	14.8	62	21
0.059	258	72	332	0.217	9.9	3.8	72	20
0.108	263	98	387	0.253	8.8	6.6	64	25
0.190	254	141	444	0.317	5.7	7.6	75	23
0.238	265	161	456	0.353	4.1	8.2	79	26
0.385	262	220	473	0.465	1.8	9.6	88	23

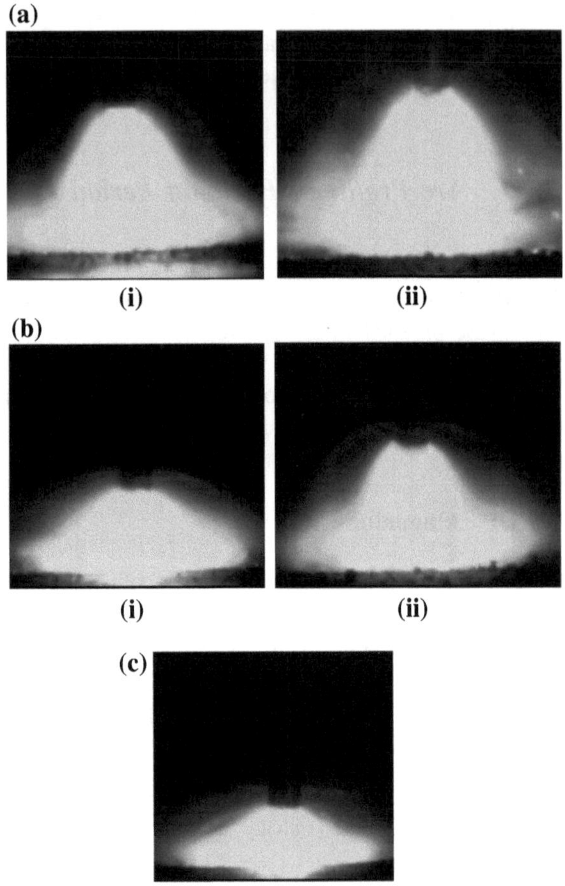

Fig. 3.5 At a given arc voltage of 20–21 V typical arc characteristics of Al–Mg filler wire at different I_m and ϕ respectively of; **a** (i) 160 A, 0.05 and (ii) 168 A, 0.4, **b** (i) 220 A, 0.06 and (ii) 212 A, 0.43, and **c** 258 A and 0.06

beyond about 22 V, where the influence of ϕ and I_m practically remains insignificant (Fig. 3.6) within a range of their functioning as 0.05–0.4 and 160–260 A respectively. The typical arc characteristic at high arc voltage and I_m of 25 V, 260 A respectively but at low ϕ of 0.1 is shown in Fig. 3.7. Nevertheless, at lower arc voltage of about 20–21 V the increase of ϕ up to about 0.2 relatively enhances the D_R followed by an insignificant variation in it with a further increase of ϕ up to 0.4 (Fig. 3.8) whereas, at a given ϕ the increase of I_m almost linearly reduces the D_R (Fig. 3.9).

During deposition of Al–Mg alloy the D_P also enhances appreciably (Fig. 3.10) with increase of arc voltage beyond about 22 V but with relatively less sensitivity than D_R, which may cause some irregularities in arc characteristics. It is reported [17] that at low arc voltage of 20–21 V the ϕ does not influence the D_P appreciably but at a given ϕ increase of I_m significantly enhances the D_P almost linearly (Fig. 3.11). It is understood that the I_m has comparatively more influence on D_P

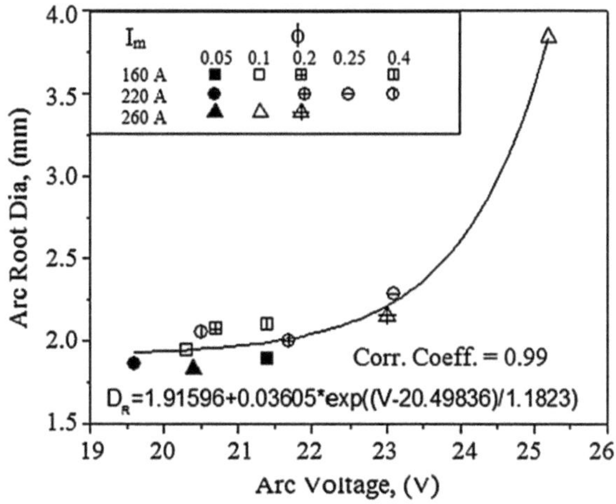

Fig. 3.6 In Al–Mg weld deposition effect of arc voltage on arc root diameter at different ϕ and mean currents

Fig. 3.7 Arc characteristics of Al–Mg filler wire at I_m = 260A, ϕ = 0.1 and arc voltage of 25 V

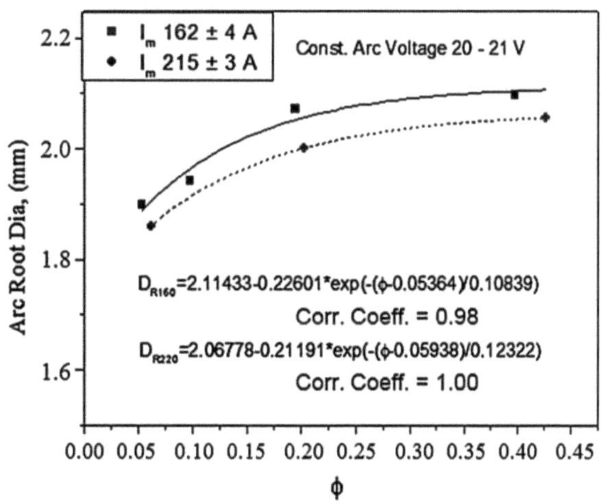

Fig. 3.8 In Al–Mg weld deposition at a given arc voltage of about 20 V effect of ϕ on arc root diameter at different mean currents

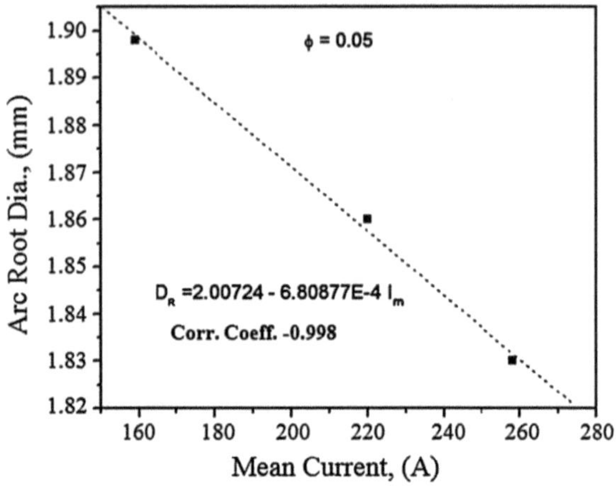

Fig. 3.9 In Al–Mg weld deposition at a given $\phi = 0.05$ and arc voltage of about 20 V effect of mean current on arc root diameter

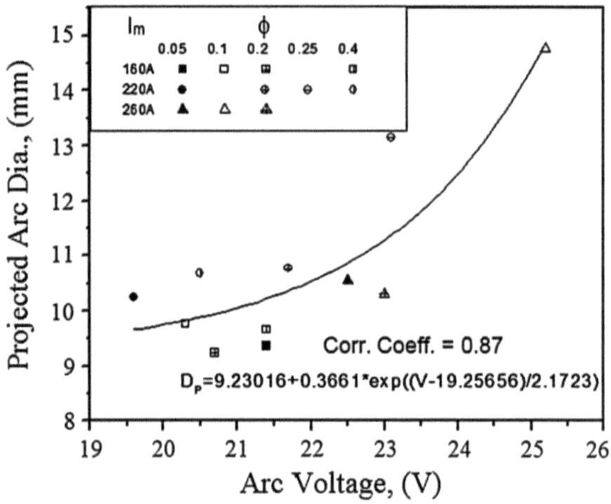

Fig. 3.10 In Al–Mg weld deposition effect of arc voltage on projected arc diameter at different ϕ and mean currents

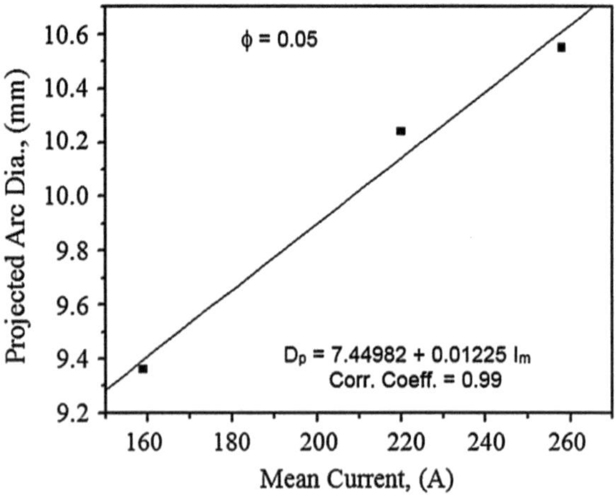

Fig. 3.11 In Al–Mg weld deposition at a given $\phi = 0.05$ and arc voltage of about 20 V effect of mean current on projected arc diameter

than D_R whereas, the other pulse parameters (considered under ϕ) are having relatively more influence on D_R than D_P.

At a given arc voltage (20–21 V) the arc length marginally enhances with the increase of ϕ but, significantly reduces almost linearly with the increase of I_m (Fig. 3.12) at a given ϕ. The empirical expression of L as a function of I_m resolves that under a given condition of welding the arc extinguishes (L = 0) at about 426 A of I_m primarily due to high wire feed rate when it burns off or touches the job without allowing enough time to transfer droplet from the electrode. However, under the influences of ϕ and I_m varying in the range of (0.05–0.4) and (160–260 A) respectively, the arc voltage (V) predominantly enhances the arc length (L) (Fig. 3.13). According to the empirical expression of L as a function of V, the arc extinguishes (L = 0) at about 14 V under a given condition of welding

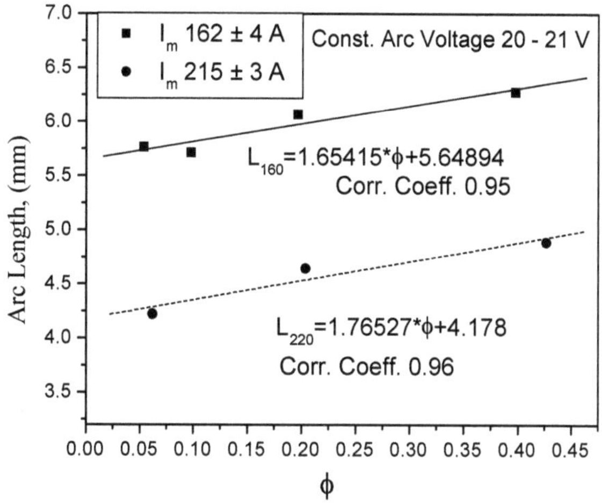

Fig. 3.12 In Al–Mg weld deposition at given arc voltage of about 20 V effect of ϕ on arc length at different mean currents

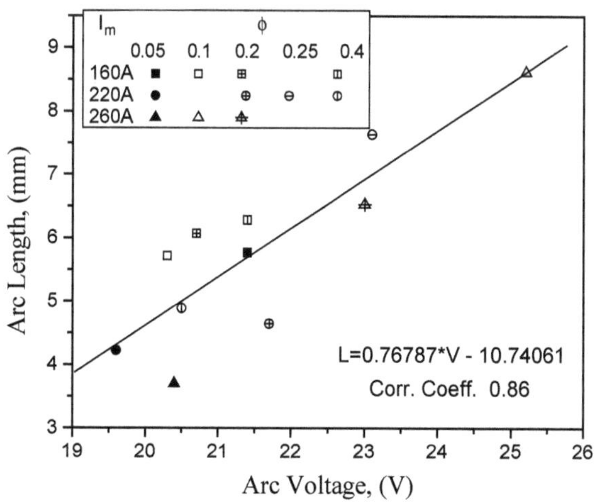

Fig. 3.13 In Al–Mg weld deposition effect of arc voltage on arc length at different ϕ and mean currents

using 1.6 mm diameter Al–Mg filler wire with argon shielding. Even though the minimum arc voltage required for arc stability largely depends upon type of power source and the welding current, the correlation of arc length with arc voltage broadly satisfies the general requirement of GMAW of aluminium [25]. Such understandings about the P-GMAW are very much useful for precise control of the process for desired weld characteristics in critical applications of welding.

3.2.1.2 Stainless Steel

The characteristics of stable arc also vary (Fig. 3.14) with a change of pulse parameters (Table 3.2) of P-GMAW process as it is observed during weld deposition using 1.2 mm diameter austenitic stainless steel (SS) filler wire with DCEP at electrode extension of 12 mm in argon shielding [18]. The pulse parameters I_m, I_p, I_b, t_p and f of varying ϕ lying in the range of 0.053–0.27 at different arc voltages are measured on current and voltage plots of the pulse captured by transient recorder as typically shown in Fig. 3.2b. At a similar order of I_p (355–362 A) the variation of ϕ, arc voltage and mean current significantly influences the arc characteristics of pulse on period. It possibly occurs due to change of energy distribution in entire pulse system when a lower arc voltage and I_b of 18 V and 40 A respectively at longer t_p of 6.36 ms (Table 3.2) gives a relatively weak arc (Fig. 3.15a) than that observed at comparatively higher arc voltage and I_b of 25 V and 157 A respectively at shorter t_p of 3.4 ms (Fig. 3.15b). At relatively higher (Fig. 3.16a–c) and lower (Fig. 3.17a–c) ranges of arc voltage of the order of 24 $\pm$ 1 and 20 $\pm$ 1 V respectively the L, D_R and D_p enhances significantly with the increase of ϕ irrespective of variation in I_m within a range of about 250–167 A. Whereas at a given ϕ and arc voltage increase of I_m reduces the D_R and D_p but, enhances the L appreciably. However, at a given close range of I_m, as for example 204–206 A, the rate of increase of L, D_R and D_p with respect to ϕ varies significantly with the change of arc voltage from 24 to 20 V. The arc length becomes comparatively more sensitive to ϕ and I_m at the lower arc voltage of 20 V, while an opposite behaviour occurs for the D_R and Dp. Thus, the arc characteristics can be appreciably regulated by varying ϕ, I_m and arc voltage following their empirical correlations [18] as stated in the Figs. 3.16 and 3.17 in order to control the stiffness and spreading of arc on the weld.

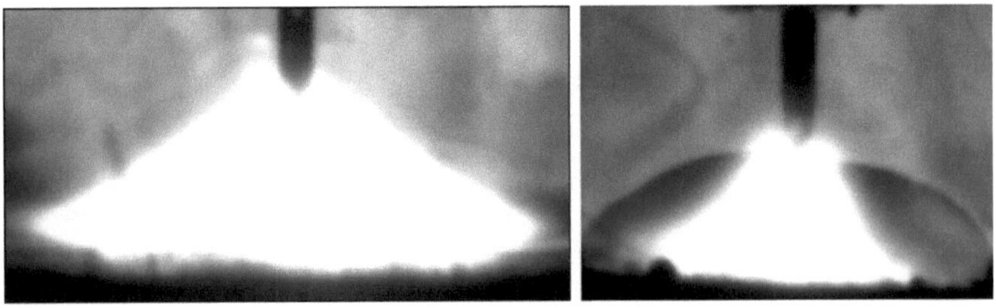

Fig. 3.14 Characteristics of stable arc at different pulse parameters

Table 3.2 Pulse parameters giving stable arc at different ϕ with 1.2 mm diameter stainless steel filler wire

Wire feed rate (mm/sec)	ϕ	I_m (A)	I_p (A)	I_b (A)	$[I_b/I_p]$	t_p (ms)	t_b (ms)	f (Hz)	Arc voltage (V)
141.7	0.18	243	374	108	0.289	2.7	4.44	140	24
158.3	0.20	248	320	106	0.331	2.7	3.97	150	24
158.3	0.24	255	300	102	0.34	2.29	5.71	125	25
92	0.27	248	355	157	0.442	3.4	5.45	113	25
92	0.08	206	430	58	0.135	3.65	4.82	118	23
141.7	0.09	212	410	59	0.144	2.8	4.34	140	23
125	0.10	205	376	57	0.151	2.8	4.89	130	24
125	0.15	204	331	80	0.241	2.8	4.95	129	24
92	0.18	207	380	93	0.245	2.7	6.82	105	24
125	0.21	220	320	109	0.340	2.9	4.62	133	25
92	0.25	200	310	109	0.351	2.6	6.57	109	25
141.7	0.06	211	319	32	0.100	2.6	4.65	138	19
125	0.12	208	313	58	0.185	2.8	4.89	130	20
125	0.17	214	274	84	0.306	3.86	5.04	110	21
92	0.05	181	362	40	0.110	6.36	5.54	84	18
125	0.06	199	333	32	0.096	2.7	4.99	130	19
92	0.11	182	354	57	0.161	2.9	6.19	110	19
108.3	0.16	189	305	76	0.249	2.9	5.16	124	20
92	0.20	179	303	85	0.280	2.7	6.82	105	21
92	0.26	188	310	110	0.355	2.6	6.75	107	21

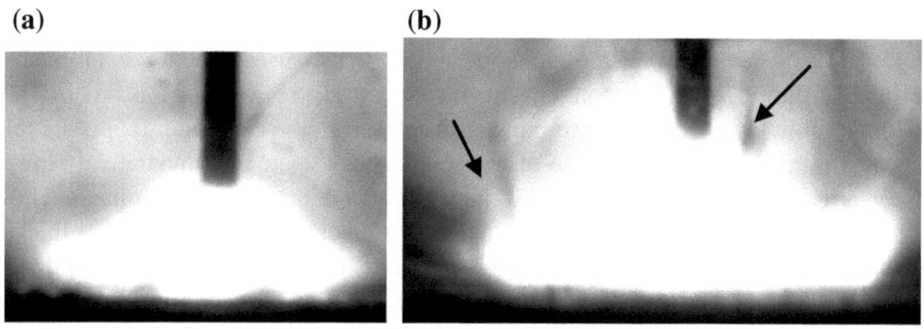

Fig. 3.15 Typical change in arc characteristics of austenitic SS at different welding conditions of **a** ϕ = 0.05, I_m = 181 A, V = 18 V and **b** ϕ = 0.27, I_m = 248 A, V = 25 V

From the empirical correlations of arc length with ϕ and I_m it appears that at relatively lower and higher ϕ (Table 3.2) of about 0.05 and 0.25 respectively the arc extinguishes (L = 0) at I_m of 728 and 515 A respectively when the arc voltage is kept around 24 V. Similar behaviour is also observed at relatively lower I_m of

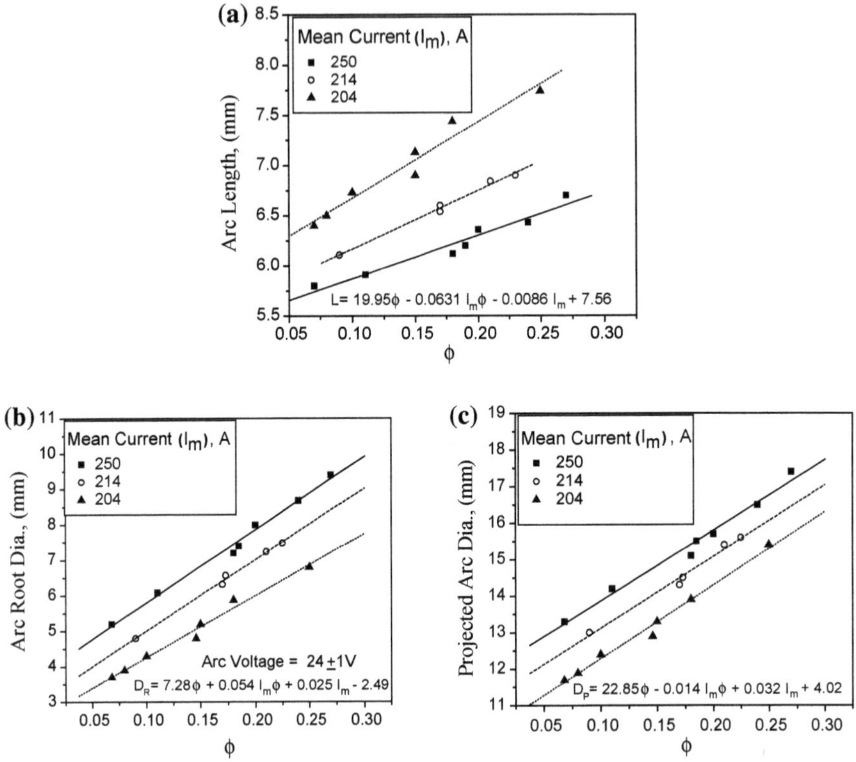

Fig. 3.16 At different I_m effect of ϕ on **a** arc length **b** arc root diameter and **c** projected arc diameter during pulse on time at a given arc voltage of the order of 24 V

691–504 A when a lower arc voltage of about 20 V is used. The arc extinguishes primarily due to high burn off rate of filler electrode or when it touches the job without allowing enough time to transfer droplet from the electrode. It appears that at higher arc voltage the increase of ϕ extinguishes the arc at comparatively lower I_m whereas, the situation becomes opposite in case of lower arc voltage. Thus, keeping a lower ϕ and higher arc voltage may be beneficial for welding of thicker section with stable arc at higher wire feed rate (higher I_m) but, maintaining a relatively higher ϕ at lower arc voltage is more useful. It possibly happens because at higher arc voltage the change of arc length becomes more sensitive to ϕ at lower mean current (Fig. 3.16a) whereas, at lower arc voltage the characteristic change of arc length becomes more sensitive to ϕ at higher mean current (Fig. 3.17a). The physical implication of these behaviours to a great extent justifies their use for control of arc characteristics in specific application of P-GMAW in case of relatively thin and thick sections.

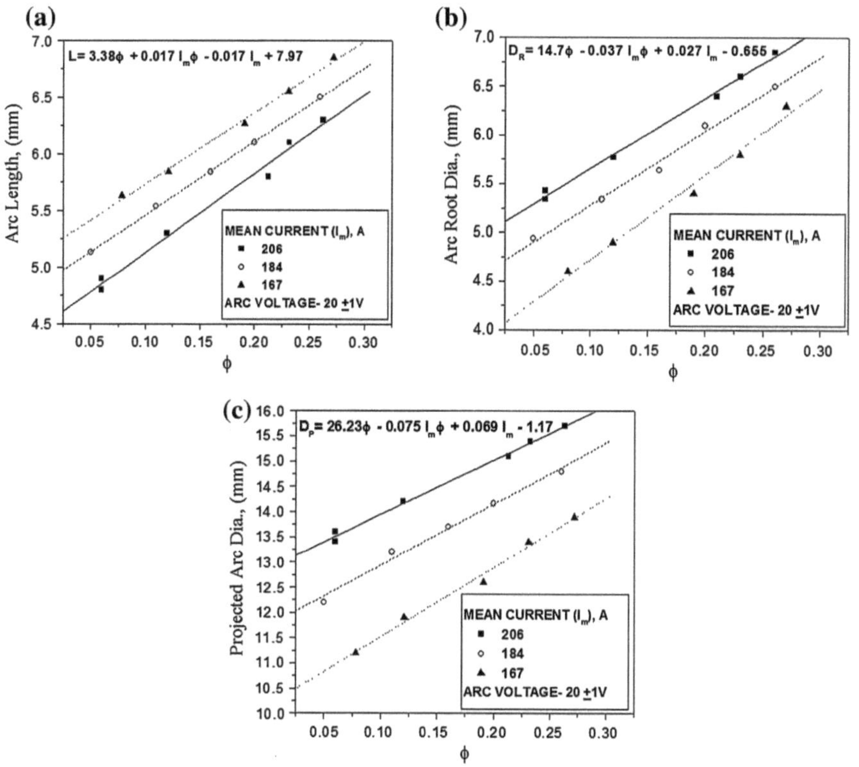

Fig. 3.17 At different Im effect of ϕ on **a** arc length **b** arc root diameter and **c** projected arc diameter during pulse on time at a given arc voltage of the order of 20 V

3.2.1.3 Mild Steel

The characteristics of stable arc also significantly vary in pulse on period (t_p) as a function of (I_m) and ϕ (Table 3.3) during weld deposition of mild steel [19] by P-GMA welding at a given arc voltage (Fig. 3.18a–i). In any case of relatively low and high I_m in the range of about 160–230 A an increase of ϕ at a given arc voltage significantly enhances the D_R, D_P and L while 1.2 mm diameter filler wire is used with DCEP at electrode extension of 15 mm. Whereas, at a given ϕ and arc voltage the increase of I_m reduces the D_R and L but, enhances the D_P appreciably (Fig. 3.19a–c). This is similar to the nature of arc observed in this regard in case of aluminium alloy and stainless steel. Thus the arc characteristics with respect to its shape that spreads on the weld under Ar + 2% CO_2 gas shielding can be significantly controlled up to certain extent by varying ϕ and I_m following the empirical correlation given in Fig. 3.19. In consideration of the correlation of L with ϕ and I_m it appears that during welding under a relatively low and high value of ϕ (Table 3.3) as 0.07 and 0.23 respectively the arc extinguishes (L = 0) at I_m of 655 and 820 A respectively when the arc voltage is kept as 24 ± 1 V. As discussed

Table 3.3 Pulse parameters of different ϕ under various shielding gases during welding with 1.6 mm diameter mild steel filler wire

Shielding gas	Wire feed speed (m/min)	ϕ	I_m (A)	I_p	I_b	I_b/I_p	f (Hz)	t_b (sec)	t_p (sec)	Arc voltage (V)
Ar + 2% CO_2	5	0.07	159	280	40	0.143	50	0.01	0.01	24
	5	0.11	160	312	60	0.192	50	0.012	0.008	23
	5	0.20	162	352	96	0.273	50	0.015	0.005	24
	5	0.23	161	372	108	0.290	50	0.016	0.004	25
	6	0.11	176	344	64	0.186	100	0.006	0.004	24
	6	0.20	177	392	104	0.265	50	0.015	0.005	24
	6	0.23	176	416	116	0.279	50	0.016	0.004	25
	7	0.11	204	356	76	0.213	150	0.0037	0.003	24
	7	0.15	205	372	88	0.237	100	0.006	0.004	24
	7	0.23	205	420	128	0.305	50	0.015	0.005	25
	8	0.07	233	344	60	0.174	100	0.004	0.006	24
	8	0.11	234	380	80	0.211	100	0.005	0.005	25
	8	0.23	231	440	140	0.318	100	0.007	0.003	25
Ar + 18% CO_2	6	0.19	179	408	105	0.257	50	0.015	0.005	25
	6	0.19	180	408	105	0.257	50	0.015	0.005	31
	6	0.19	180	408	104	0.255	50	0.015	0.005	33
	8	0.06	227	394	50.6	0.128	100	0.005	0.005	25
	8	0.08	230	395	64	0.162	100	0.005	0.005	24
	8	0.10	232	388	80	0.206	100	0.005	0.005	25
	8	0.19	230	390	123	0.316	100	0.006	0.004	24
	8	0.35	227	410	184	0.449	200	0.004	0.001	24
	8	0.35	231	407	184	0.451	200	0.004	0.001	30
	8	0.35	227	413	181	0.438	200	0.004	0.001	33
Argon	5	0.07	158	280	40	0.143	50	0.01	0.01	24
	5	0.15	160	324	72	0.222	50	0.013	0.007	23
	5	0.23	160	372	108	0.290	50	0.016	0.004	23
	7	0.07	200	316	52	0.165	150	0.0037	0.003	24
	7	0.15	202	372	88	0.236	100	0.006	0.004	25
	7	0.23	201	420	128	0.304	50	0.015	0.005	23
	8	0.07	234	344	60	0.174	100	0.004	0.006	25
	8	0.10	232	388	80	0.206	100	0.005	0.005	25
	8	0.15	229	420	104	0.248	50	0.012	0.008	25
	8	0.23	230	440	140	0.318	100	0.007	0.003	25

earlier it also primarily happens due to high wire feed rate when it either burns off or touches the job without allowing enough time to transfer droplet from the electrode. Such a logical agreement of the equation with its physical implications to

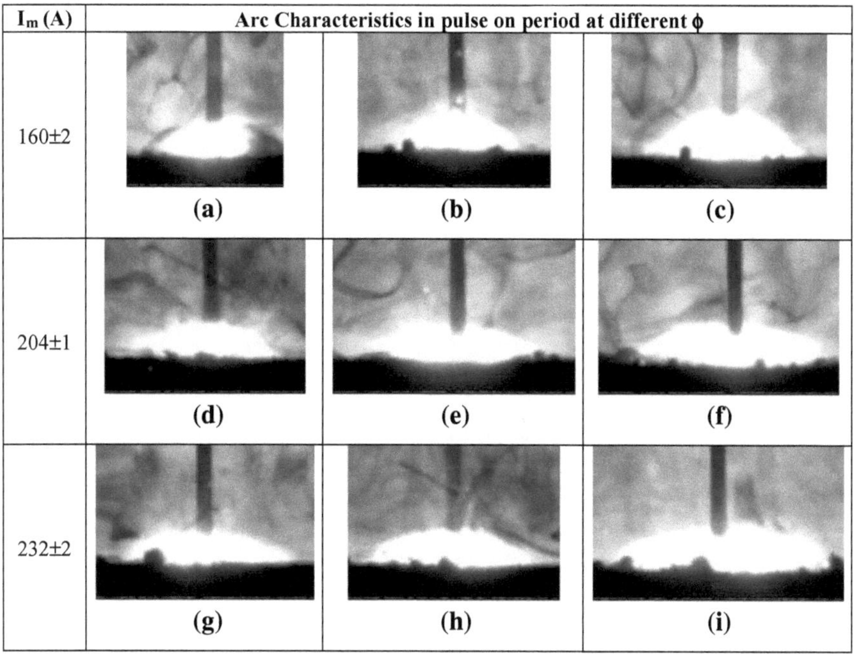

Fig. 3.18 At a given arc voltage of 24 ± 1 V and Ar + 2% CO_2 gas shielding typical change in arc characteristics in pulse on period at different I_m with varying ϕ of **a** 0.07 **b** 0.20 **c** 0.23 **d** 0.11 **e** 0.15 **f** 0.23 **g** 0.07 **h** 0.11 and **i** 0.23

a great extent justifies its use for control of arc characteristics in specific application of P-GMAW to mild steel. However, appropriate selection of I_m within its range of stable welding process shall depend upon other criterion primarily related to the thermal behaviour of the weld [26, 27, 3].

Under a different Shielding environment of Ar + 18% CO_2 typical change in arc characteristics with varying ϕ from 0.06 to 0.35 at given arc voltage and I_m of 24 ± 1 V and 229 ± 2 A respectively has been shown in Fig. 3.20a–c. Similarly typical change in nature of arc with an increase of arc voltage from about 24 to 33 V at a given ϕ and I_m of 0.35 and 229 ± 2 A respectively and at a relatively lower ϕ and I_m of 0.19 and 179 ± 0.5 A respectively are shown in Figs. 3.21a–c and 3.22a–c respectively. The video photographs reveal that the intensity of the influence of parameters on arc characteristics is (Fig. 3.20) relatively different from that noted in case of Ar + 2% CO_2 gas shielding (Fig. 3.18). This may have primarily occurred due to difference in thermal conductivity of shielding gases, which produces a variation in arc shape [28, 29]. Nevertheless, in agreement to the behaviour of arc under the Ar + 2% CO_2 gas shielding (Fig. 3.19) here also at given mean current and arc voltage of 229 ± 2 A and 24 ± 1 V respectively the

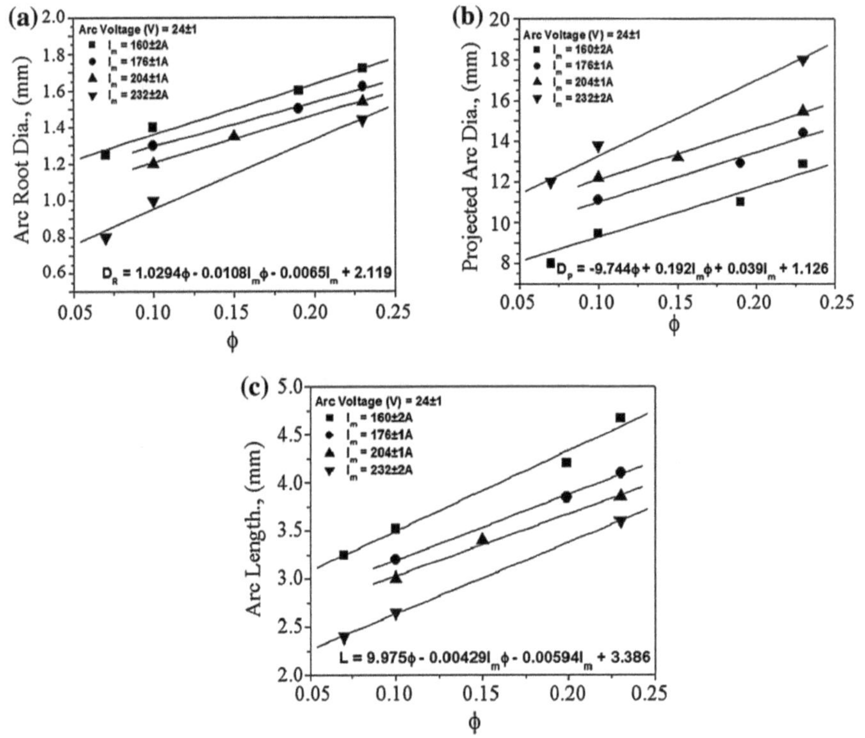

Fig. 3.19 Under Ar + 2% CO_2 gas shielding at different Im effect of ϕ on **a** arc root diameter **b** projected arc diameter and **c** arc length during pulse on period at a given arc voltage of 24 ± 1 V

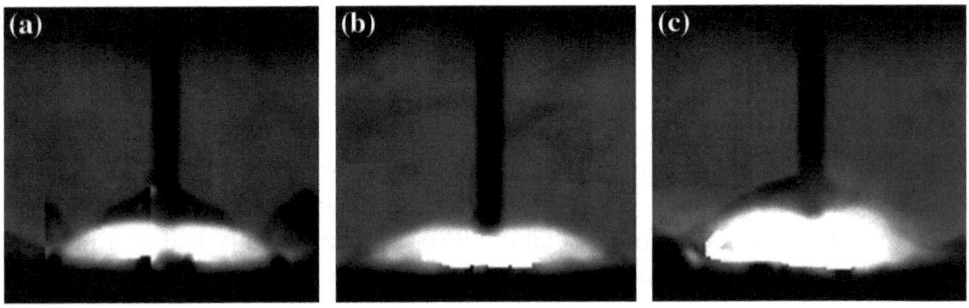

Fig. 3.20 Under Ar + 18% CO_2 gas shielding at a given arc voltage and I_m of 24 ± 1 V and 229 ± 2 A typical change in arc characteristics in pulse on period with varying ϕ of **a** 0.06 **b** 0.19 and **c** 0.35

D_R, D_P and L enhances significantly with the increase of ϕ (Fig. 3.23a–c). It appears that D_R, D_P and L enhances almost linearly as a function of ϕ but, at a relatively lower and higher rates in lower and higher range of ϕ respectively with a criticality at about 0.1. It seems to be also true in other cases as well. The criticality

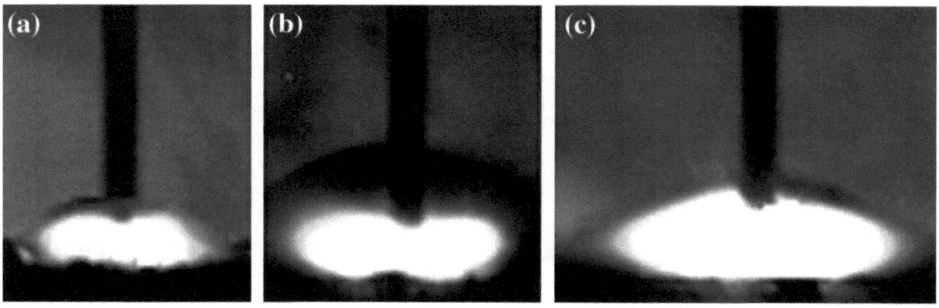

Fig. 3.21 Under Ar + 18% CO_2 gas shielding at a given I_m and ϕ of 229 ± 2 A and 0.35 typical change in arc characteristics in pulse on period with varying arc voltage of **a** 24 ± 1 V, **b** 30 V and **c** 33 V

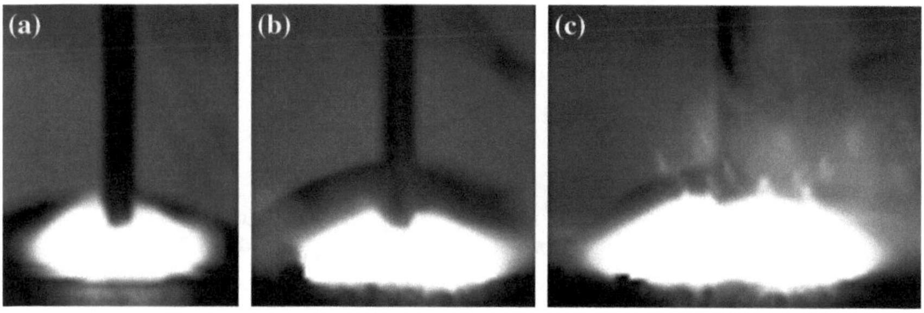

Fig. 3.22 Under Ar + 18% CO_2 gas shielding at a given Im and ϕ of 179 ± 0.5 A and 0.19 typical change in arc characteristics in pulse on period with varying arc voltage of **a** 24 ± 1 V, **b** 30 V and **c** 33 V

of ϕ playing a crucial role in governing the arc characteristics may influence the thermal behaviour, geometry and properties of weld significantly with change in pulse parameters as it is noted in case of the P-GMA welding of aluminium alloy [27, 30, 31]. The appearance of such critical point should be understood more deeply in order to explore further beneficial use of P-GMAW process. Under a given condition of welding D_R decreases but, D_P and L enhances with the increase of arc voltage. At given arc voltage increase of I_m and ϕ relatively reduces the D_R, D_P and L (Fig. 3.24a–c). It happens primarily because at given mean current and ϕ increase of arc voltage enhances the arc pressure and tends to constrict the arc root that consequently strengthens magnetic field at it [32, 28]. The variation of arc length basically happens because it is directly proportional to arc voltage, where a longer arc at higher arc voltage consumes more energy under a given mean current [32, 33–37]. Under a given condition of welding at relatively lower and higher combinations of I_m and ϕ respectively of 179 ± 0.5 A and 0.19 and 229 ± 2 A and 0.35 the arc extinguishes (L = 0) at about 13 and 14 V respectively. The minimum arc voltage required for arc stability largely depends also upon type of power source and welding current. Nevertheless in consideration of minimum arc

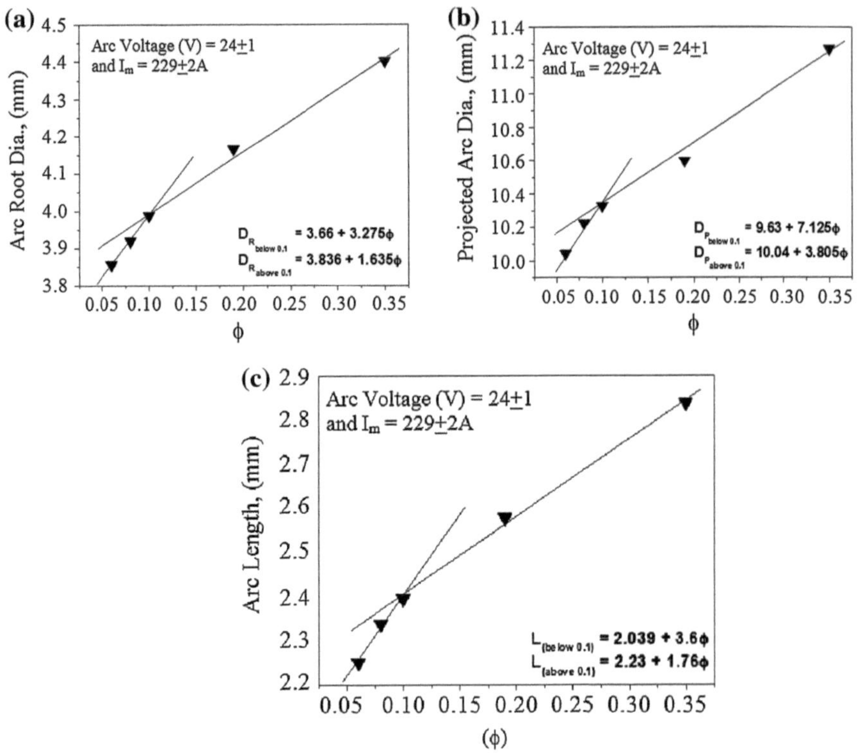

Fig. 3.23 Under Ar + 18% CO_2 gas shielding at a given arc voltage and I_m of 24 ± 1 V and 229 ± 2 A effect of ϕ on **a** arc root diameter **b** projected arc diameter and **c** arc length

voltage required as 21–22 V for arc stability under Ar + 18% CO_2 gas shielding using 1.2 mm diameter mild steel filler wire [36], the solution of the empirical equations (Fig. 3.24c) for L as a function of arc voltage broadly satisfies the prescribed general requirement of P-GMAW of mild steel.

3.2.2 Arc Profile of Pulse off Period

Similar to that observed in case of pulse on period the arc profile of pulse off period also varies with pulse parameters. But unlike to that occurs during pulse on period here arc characteristics vary from the point of practically non existence to significant presence primarily depending upon the magnitude of I_b as well as some other pulse parameters. In case of the existence of weak but recognizable arc geometry in pulse off period its profile plays a significant role on process stability and thus also affects the weld quality. However, depending upon level of I_b in combination with other pulse parameters and arc voltage it may also influence the thermal behaviour

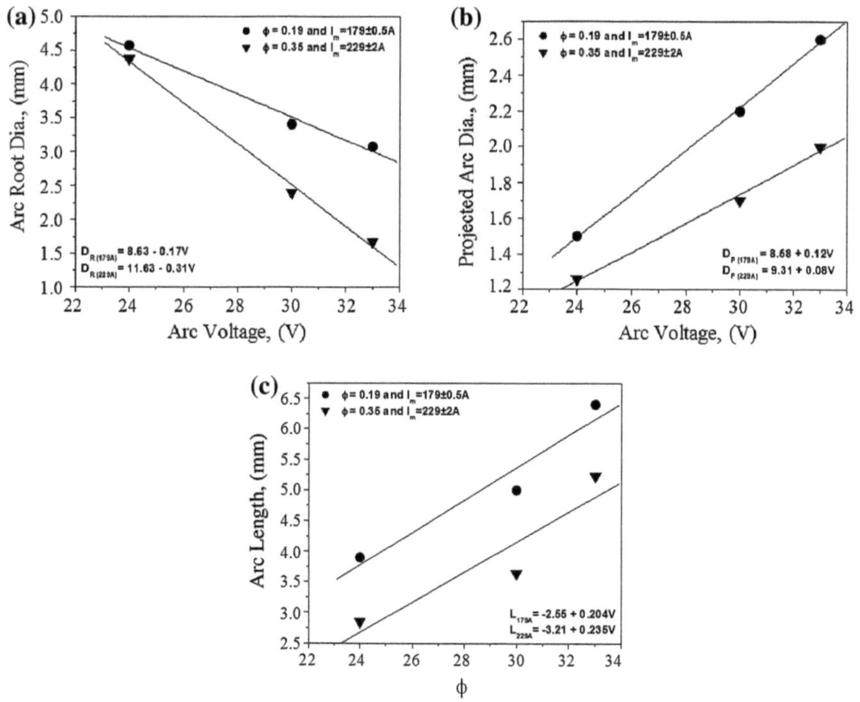

Fig. 3.24 Under Ar + 18% CO_2 gas shielding at a given combination of relatively higher and lower ϕ and I_m of 0.35, 229 ± 2 A and 0.19, 179 ± 0.5 A respectively, effect of arc voltage on **a** arc root diameter **b** projected arc diameter and **c** arc length

of weld of different materials. Thus, during pulse off time (t_b) the knowledge of arc profile as a function of welding parameters of P-GMAW process is a matter of considerable interest.

3.2.2.1 Aluminium

During P-GMA welding of aluminium at a comparatively low I_b in the range of 33–92 A an increase of ϕ up to 0.2 extinguishes the arc in spite of increase of I_b up to 92 A at a large pulse off time of about 10 ms. But, at higher I_b of about 145 A the arc remains appreciably strong even at a higher ϕ of 0.4, where the mean current and arc voltage are kept in a close range of 159–168 A and 20–21 V respectively (Fig. 3.25a–d) in case of using 1.6 mm diameter filler wire under argon shielding. This behaviour of predomination of I_b in maintaining a strong arc during pulse off time also corroborate the characteristics of a comparatively higher I_m of about 213 A (Fig. 3.26). Here at a given ϕ of about 0.2 an existence of appreciably strong arc has been marked at I_b of 122 A in comparison to that observed (Fig. 3.25c) at a relatively low I_b of 92 A, when the arc voltage was also kept in the range of

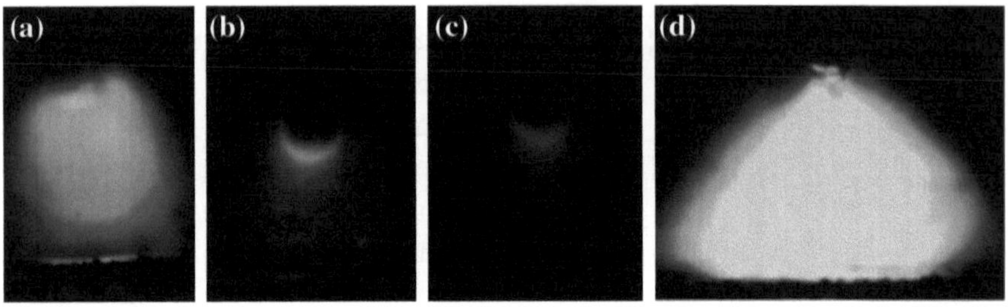

Fig. 3.25 At given close range of mean current (159–168 A) and arc voltage (20–21 V) typical arc characteristics of pulse off time in Al–Mg electrode at different I_b and ϕ respectively of **a** 33 A, 0.053, **b** 53 A, 0.1, **c** 92 A, 0.2 and **d** 145 A, 0.4

Fig. 3.26 Typical arc characteristic of pulse off time in Al–Mg electrode at arc voltage of 22 V, I_m = 213 A, I_p = 410 A, I_b = 122 A and ϕ = 0.2

21–22 V. At a practically constant ϕ and arc voltage of about 0.05 and 21 V respectively a significant increase of I_m even up to about 258 A may not hold an appreciably strong arc (Fig. 3.27) at pulse off time if the I_b becomes too low as 72 A. Thus, it primarily infers that the pulse off time can have a strong arc not by keeping a high I_m but by maintaining a minimum level of base current beyond about 92 A. This phenomenon is necessary to understand for arc stability, where ϕ

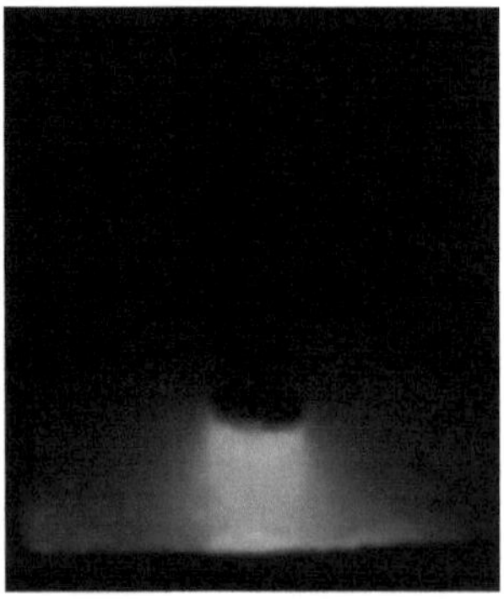

Fig. 3.27 Arc characteristic of pulse off time in Al–Mg electrode at arc voltage of 20 V, I_m = 258 A, I_p = 332 A, I_b = 72A and ϕ = 0.059

may have relatively insignificant influence on it [17]. But, at comparatively lower I_b, the ϕ may significantly influence the arc stability, which may be suitably explored in manoeuvring the P-GMAW process to control the weld quality.

3.2.2.2 Stainless Steel

At the I_b, t_b and arc voltage of the order of 109 A, 4.62 ms and 25 V respectively (Table 3.2) a short arc of insignificant extension can exist during pulse off time in between the 1.2 mm diameter SS electrode tip and austenitic SS base material (Fig. 3.28) under argon shielding. However, by selection of appropriate I_b, t_b and arc voltage an arc of recognisable geometry can be maintained in pulse off period [18]. Thus, knowledge of arc profile as a function of pulse parameters is necessary in order to understand the thermal nature of P-GMAW process governing the weld quality.

At a given I_m of 215 A an arc of appreciable geometry exists in pulse off period having I_b and ϕ lying in the range of 59–109 A and 0.09–0.25 respectively as shown in Fig. 3.29. The stability of arc profile understood by its undeformed bright part becomes comparatively better at relatively lower ϕ of less than about 0.1 irrespective of the relatively low or high I_b [18]. Unlike that observed in case of P-GMA welding of aluminium alloy [17], the arc at base current in P-GMA welding of stainless steel deflects from its vertical axis. It appears as arc blow which is a function of ϕ and I_b. Arc deflection enhances almost linearly with the increase of ϕ but, at a given higher value of ϕ an increase of I_b reduces the arc deflection significantly [18]. At a moderate I_m of the order of 215 A the increase of (I_b/I_p) ratio in the range of 0.351–0.443 enhances the arc blow significantly at both the high and low arc voltages. However, under similar order of I_m at any voltage of arcing no such arc blow takes place at low (I_b/I_p) in the range of 0.1–0.115 [18]. An increase of ϕ enhances the (I_b/I_p) ratio and thus promotes arc blow as shown in Fig. 3.29b. Such biasness in directionality of arcing may also appear in relatively stronger arc of pulse on period at high (I_b/I_p), especially at high arc voltage of about 24 V. Such

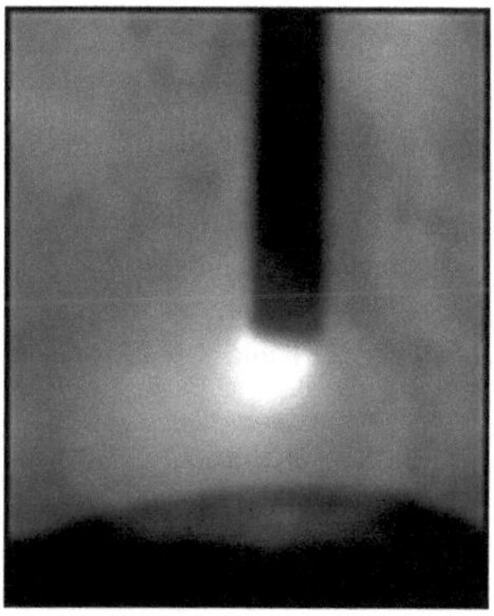

Fig. 3.28 Typical appearance of insignificant arc during austenitic SS welding at pulse off time of $\phi = 0.21$, $I_m = 220$ A and V = 25 V

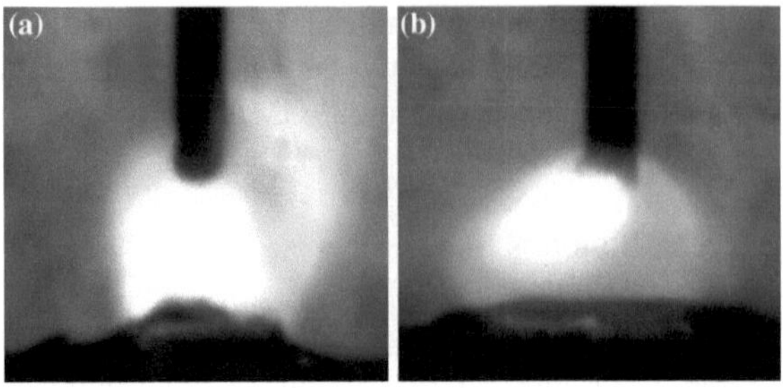

Fig. 3.29 During austenitic SS welding at a given I_m = 215 A and arc voltage of 24 V the arc characteristics of pulse off time under different pulse parameters of **a** ϕ = 0.09, I_b = 59 A and **b** ϕ = 0.25, I_b = 109 A

behaviour of arc, primarily caused by a fluctuation of arc pressure at different pulse parameters, may create irregularities in heat distribution and thermal characteristics of weld joint.

3.2.2.3 Mild Steel

An arc of noticeable geometry may exist in pulse off period depending upon magnitude of I_b, t_b and arc voltage. Thus, knowledge of arc profile as a function of pulse parameter is necessary to understand the thermal behaviour of P-GMAW process [19]. At a given arc voltage and I_m a non deflected arc of appreciable geometry at 1.2 mm diameter electrode under Ar + 2% CO_2 exists in pulse off period having I_b and ϕ lying in the range of 60–140 A and 0.07–0.23 respectively (Fig. 3.30a–c) with a comparatively better stability at relatively higher ϕ. This is similar to that happens in case of aluminium alloy [17] but, in contrast to this a comparatively better arc stability with insignificant deflection appears at relatively lower ϕ of the order of 0.06 in case of stainless steel [18]. The phenomenon of arc deflection may be to a great extent understood by arc stiffness in terms of arc pressure as a function of arc profile and nature of metal transfer. Under Ar + 18% CO_2 shielding a relatively higher combination of ϕ, I_m and of 0.35, 229 ± 2 A and 33 V respectively also improves the arc stability Fig. 3.31a–f.

I_m (A)	Arc Characteristics in pulse off period at different ϕ		
232±2	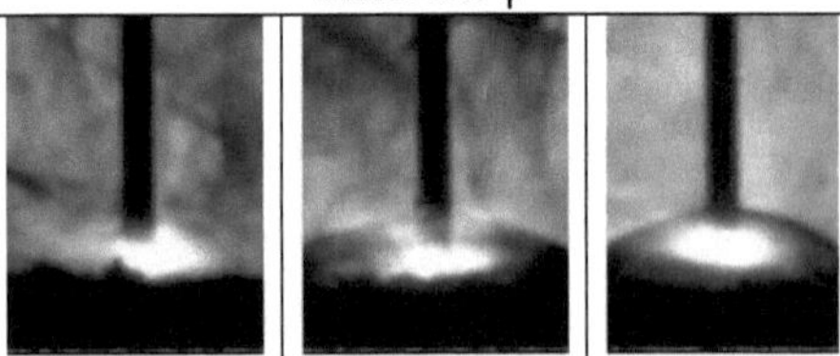		
	(a)	(b)	(c)

Fig. 3.30 Under Ar + 2% CO_2 gas shielding at a given arc voltage and I_m of 24 ± 1 V and 232 ± 2 A typical change in arc characteristics in pulse off period with varying ϕ of **a** 0.07 **b** 0.11 and **c** 0.23

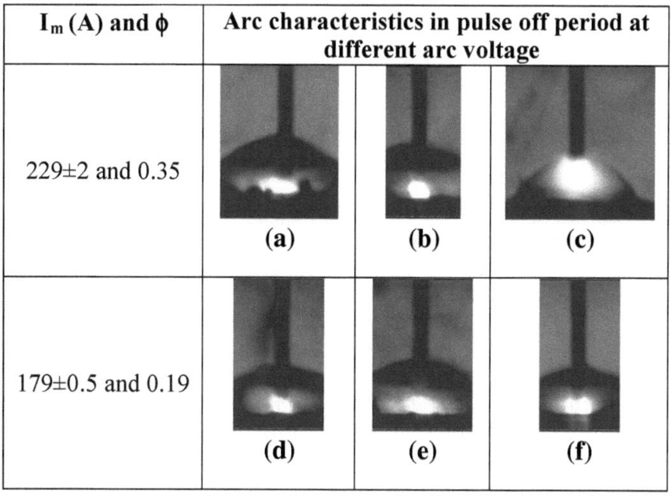

Fig. 3.31 Under Ar + 18% CO_2 gas shielding at a given combination of relatively higher and lower ϕ and Im of 0.35, 229 ± 2 A and 0.19, 179 ± 0.5 A respectively, typical change in arc characteristics in pulse off period with varying arc voltage of **a** 24 ± 1 V **b** 30 V **c** 33 V **d** 24 ± 1 V **e** 30 V and **f** 33 V

3.2.3 Arc Stiffness

The deflection of arc from its central axis which adversely affects the energy concentration during welding is largely governed by the arc stiffness as a function of welding parameters. The arc stiffness is generally considered as direct function of arc pressure. The arc pressure (P_a) is often estimated [38] by an equation derived from total pressure distribution at perturbed boundary of solid-liquid interface. It assumes arc as a hollow conducting fluid cylinder of inner and outer radius equal to equilibrium radius of the molten metal (R) and arc root radius (R_a) respectively. The P_a is expressed as follows.

$$P_a = \frac{\mu_0 J_a^2}{4}\left[R_a^2 - R^2 - 2\varepsilon_o R\cos(\omega t)\cos(k_w z)\right] \tag{3.13}$$

where, μ_o is the permeability of free space, $J_a = \frac{I_p^2}{\pi R_a^2}$ is the arc current density during pulse on period, ω is angular frequency, k_w is a wave number and ε_o is amplitude of perturbation parameter. The Eq. 3.13 can be resolved [38] as follows with the help of expression of pressure (P_1) due to surface tension attributed to a cylindrical radius (R_1) at the perturbed boundary as proposed earlier.

$$P_1 = \frac{\gamma}{R_1} = \frac{\gamma}{R}\left(1 - \frac{\varepsilon_0}{R}\cos(\omega t)\cos(k_w z)\right) \tag{3.14}$$

$$\varepsilon_0 \cos(\omega t)\cos(k_w z) = R - \frac{R^2}{R_1} \tag{3.15}$$

Considering the Eqs. 3.13 and 3.14 the expression for estimation of arc pressure P_a in GMAW is finally derived as

$$P_a = \frac{\mu_0 I_P^2}{4\pi^2 R_a^4}\left[R_a^2 - 3R^2 + \frac{2R^3}{R_1}\right] \tag{3.16}$$

where, R and R_1 are considered as size of droplet radius (D/2) and effective radius (r) of tapering of electrode respectively. Performance of P-GMAW process is primarily characterised by its pulse current where the back ground current maintains continuity of the process. The arc pressure P_a as a measure of arc stiffness is primarily considered as function of I_P and the geometry of effective part of the arc along its vertical axis corroborating the metal transfer as shown in the Eq. (3.16). However, in case of an arc of noticeable geometry at low current (I_b) of pulse off time the P_a can also be estimated by the Eq. (3.16) for comparatively higher R_a, R and R_1 at negligible electrode tapering. In case of welding with or without metal transfer in pulse off time the R may be assumed as equal to R_1 and accordingly the P_a at I_b becomes appreciably lower than that of I_p. The intensity of fluctuation in P_a under the pulsed current depends upon the ratio of (I_b/I_p) at different I_m and ϕ. However, effect of such fluctuation of arc pressure on arc environment is comparatively more significant in case of relatively softer long arc at high arc voltage than in stiffer arc of low arc voltage.

3.2.3.1 Aluminium

During welding under argon shielding at a flow rate of 18 l/min using 1.6 mm diameter Al–Mg filler wire at mean currents lower (160 A) and higher (212 A) than the transition current of spray transfer, the change of ϕ at a given arc voltage significantly affects the arc pressure (P_a) (Fig. 3.32). In both the levels of I_m the increase of ϕ up to about 0.2 appreciably enhances the P_a followed by a relative decrease in it with a further increase of ϕ to 0.4. This behaviour is found relatively more pronounced in case of higher I_m. However, at a given ϕ increase of I_m considerably increases the P_a.

3.2.3.2 Stainless Steel

The arc pressure (P_a) or stiffness at I_p decreases significantly with increase of ϕ at any mean current within 167–250 A as shown in Fig. 3.33a, b for comparatively

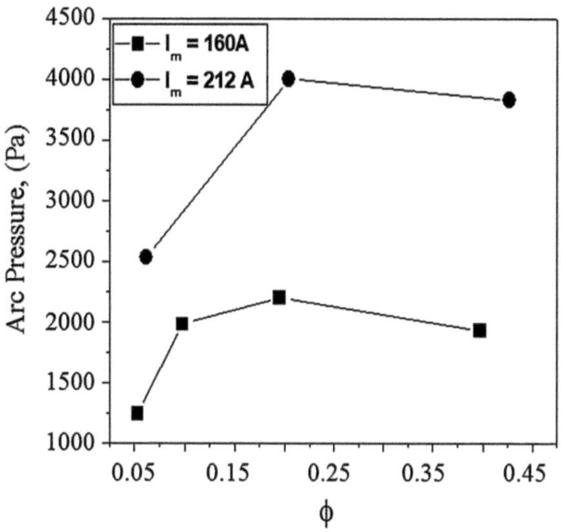

Fig. 3.32 At a given mean current effect of ϕ on arc pressure during welding using Al–Mg filler wire

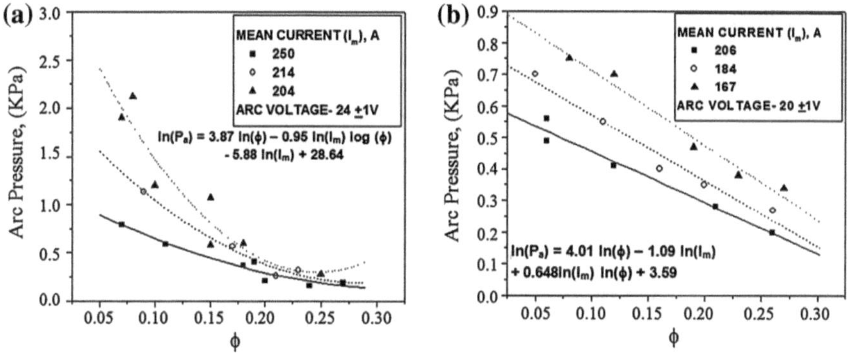

Fig. 3.33 At different Im the effect of ϕ on arc pressure (Pa) during pulse on time at a given arc voltage of the order of **a** 24 V and **b** 20 V

high and low arc voltages of about 24 and 20 V respectively. It is also observed that at a given ϕ a significant enhancement in arc stiffness occurs with increase of I_m. But, at a given I_m keeping higher arc voltage gives rise to larger arc stiffness than that observed in case of lower arc voltage when the ϕ is kept constant. Thus, at a given arc voltage the arc stiffness in terms of arc pressure at pulse on time is satisfactorily correlated to ϕ and I_m [18, 39].

3.2.3.3 Mild Steel

At any mean current variation of pulse parameter at a given arc voltage appreciably affects arc pressure (P_a). It is well understood in welding at a mean current beyond or below the transition current of 1.2 mm diameter filler wire under Ar + 2% CO_2 gas shielding while the arc voltage is kept constant. Here it is found that a change of pulse parameters increasing the ϕ up to about 0.15 enhances the P_a followed by a relative decrease in it with a further increase of ϕ to 0.23 [20] (Fig. 3.34). But, at a given ϕ the increase of I_m relatively enhances the $P_{a.}$ Thus, at a given I_m the increase of ϕ and at a given ϕ the increase of I_m during both the pulse on period (t_p) and pulse off period (t_b) reduces the deflection (A_D^0) as shown in Fig. 3.35a, b respectively. Arc deflection becomes comparatively stronger in pulse off period, which is in agreement to that observed in case of stainless steel [39]. In contrast to this behaviour, the P_a always reduces with the increase of ϕ under Ar + 18% CO_2 gas shielding following a relatively sharper rate during its escalation up to about 0.1 (Fig. 3.36). However, in similar range of I_m the magnitude of P_a under Ar + 2% CO_2 gas shielding is considerably higher than that observed in case of the Ar + 18% CO_2 shielding. A considerable increase in magnitude of D_R under Ar + 18% CO_2 gas shielding may adversely affect the arc pressure possibly by enhancing radial diffusion of ion and anode drop which reduces efficiency of ionization [40, 41]. Consequently the arc deflection (A_D^0) reduces almost linearly with reduction of ϕ but, at different rates at its upper and lower ranges in reference to the critical point of about 0.1 (Fig. 3.37). Although it is clearly identified in various occasions, still the basic reasons behind the critical influence of ϕ around its value of 0.1 on various characteristics of arc and weld in P-GMA welding is not well understood. To establish this phenomenon more precisely in the interest of better utilisation of P-GMAW, it should be studied further based on arc physics.

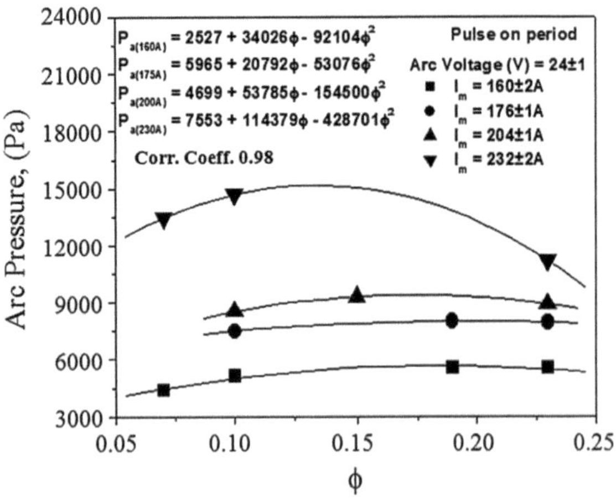

Fig. 3.34 Under Ar + 2% CO_2 gas shielding at different I_m effect of ϕ on arc pressure (Pa) during pulse on period at a given arc voltage of 24 ± 1 V

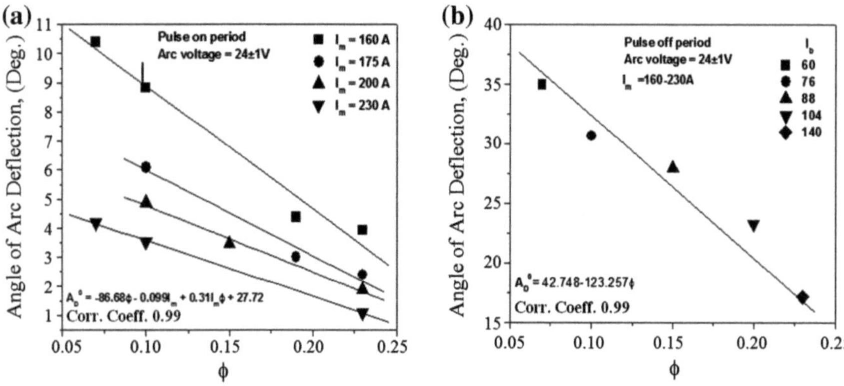

Fig. 3.35 Under Ar + 2% CO_2 gas shielding at a given arc voltage of 24 ± 1 V effect of ϕ on arc deflection during **a** pulse on period at different I_m and **b** pulse off period at different I_b

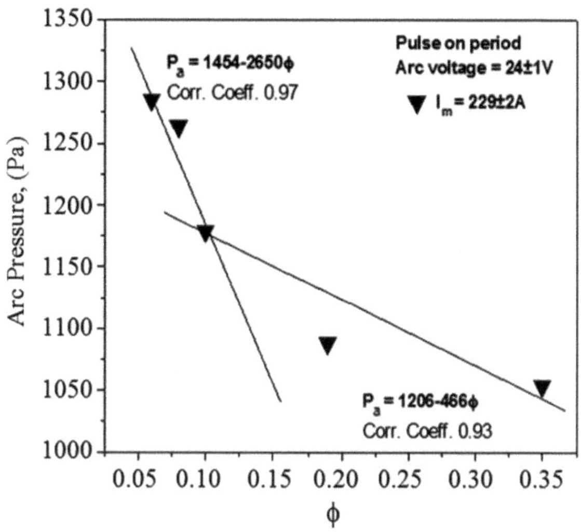

Fig. 3.36 Under Ar + 18% CO_2 gas shielding the effect of ϕ on arc pressure (Pa) during pulse on period at a given arc voltage and I_m of 24 ± 1 V and 229 ± 2 A respectively

Increase of arc voltage enhances the arc pressure being relatively more significant in case of higher values of ϕ and I_m (Fig. 3.38). At lower arc voltage it is relatively higher at lower ϕ and I_m whereas, at higher arc voltage beyond about 30 V the difference in arc pressure gradually becomes insignificant with the change of I_m. In corroboration to the nature of variation of arc pressure affecting the arc deflection, it reduces significantly in pulse on period with increase of arc voltage at a given ϕ and I_m (Fig. 3.39).

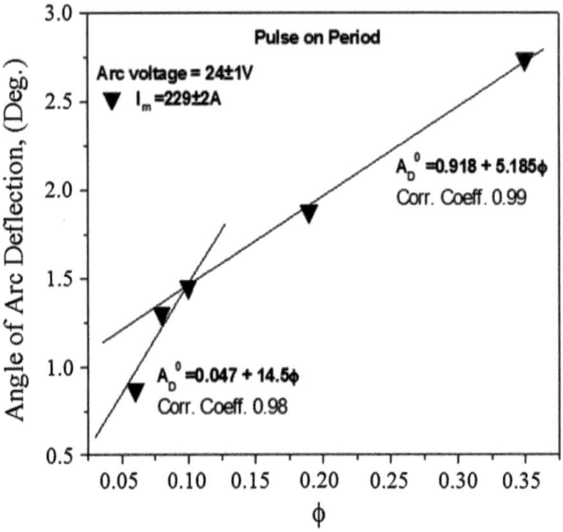

Fig. 3.37 Under Ar + 18% CO_2 gas shielding effect of ϕ on arc deflection during pulse on period at a given arc voltage and I_m of 24 ± 1 V and 229 ± 2 A

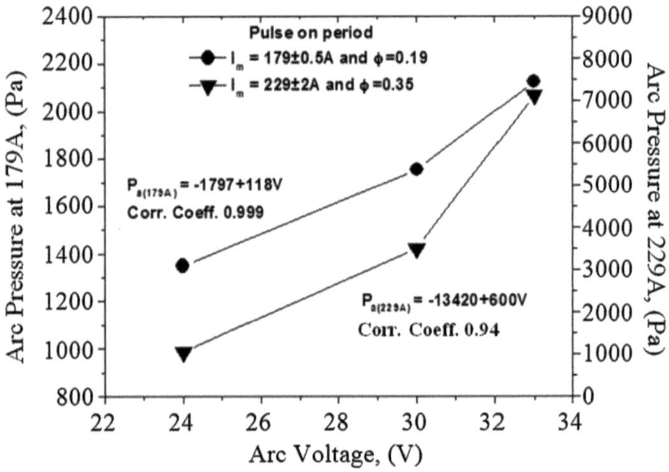

Fig. 3.38 Under Ar + 18% CO_2 gas shielding effect of on arc pressure during pulse on period at a given combination of relatively *lower* and *higher* ϕ and I_m of (0.19, 179 ± 0.5 A) and (0.35, 229 ± 2 A) respectively

3.2.4 Arc Efficiency

Successful application of GMAW in preparation of different type and size of joints at varying position and groove design requires an appropriate control of welding

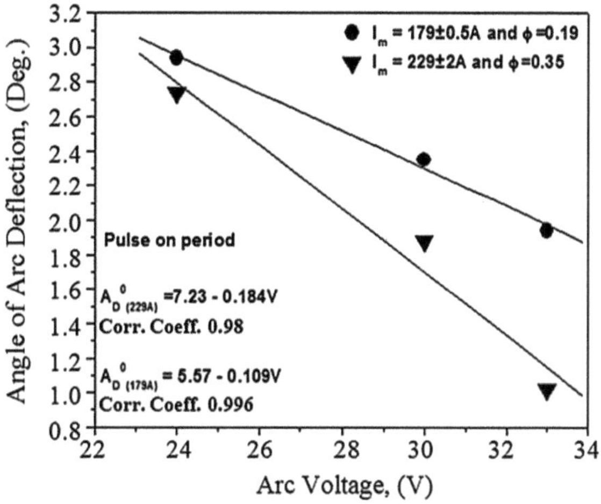

Fig. 3.39 Under Ar + 18% CO_2 gas shielding effect of arc voltage on arc deflection during pulse on period at a given combination of relatively lower and higher ϕ and I_m of (0.19, 179 ± 0.5 A) and (0.35, 229 ± 2 A) respectively

process based on thorough understanding of nature of variation in power and efficiency of arc. The arc efficiency (η_a) refers to the fraction of arc energy that is delivered or transferred to work piece [32, 37]. Thus, at any given time (t) the energy (E) input to work piece as a function of arc voltage, welding current and arc efficiency is expressed as [32].

$$\frac{dE}{dt} = \eta_a VI \tag{3.17}$$

The change in arc profile with variation in pulse parameters [17–19] consequently affects the arc power and efficiency as well as energy distribution in P-GMAW process. The influence of pulse parameters on these aspects becomes quite complex due to their simultaneously interactive nature but, it may be largely well understood as a function of summarized influence of pulse parameters defined by the dimensionless hypothetical factor ϕ. In consideration of the primary characteristics of an active arc (Fig. 3.4) of GMAW process, the arc efficiency (η_{ae}) as function of heat loss from the arc column due to convection (q_{cov}), conduction (q_{cod}) and radiation (q_{rad}) can be expressed as

$$\eta_{ae} = 1 - [q_{\mathrm{cov}} + q_{cod} + q_{rad}]/VI_{eff} \tag{3.18}$$

where, V is arc voltage and I_{eff} is effective current (root mean square value of the pulsed current wave form) estimated as [25, 42, 43].

$$\mathrm{I_{eff}} = \sqrt{\left[\mathrm{k_p.I_p^2} + (1 - \mathrm{k_p}).\mathrm{I_b^2}\right]} \tag{3.19}$$

where, pulse duty cycle,

$$k_p = t_p/t_{pul} \tag{3.20}$$

The q_{cov}, q_{cod} and q_{rad} can be estimated [32, 44] as follows.

$$q_{cov} = h_c A_{SA}[T_{AR} - T_{AP}] \tag{3.21}$$

$$q_{cod} = K_{Ar} A_{SA}[T_{AR} - T_{AP}]/L \tag{3.22}$$

$$q_{rad} = \varepsilon \sigma_{sbz} A_{SA} T_{AT}^4 \tag{3.23}$$

where, T_{AR} is average temperature of arc root, T_{AP} is average temperature of arc projection, T_{AT} is temperature of outer arc surface, ε is emissivity of argon plasma, σ_{sbz} is the Stefan Boltzmann constant, K_{Ar} is thermal conductivity of argon, L is arc length, A_{SA} is surface area of arc column and h_c is average convection heat transfer coefficient [32, 44, 45]. The shape of effective arc column is assumed as a frustum of a cone (Fig. 3.40) cut off parallely by a plane XX, passing through the points defined by tangential interception of projected lateral surface of the arc to the curvature of arc root and the base plane YY as shown in Fig. 3.4. The A_{SA} has been geometrically estimated as a function of arc root diameter (D_R), projected arc diameter (D_P) and arc length (L) as follows.

$$A_{SA} = \frac{\pi[D_P^2 + D_R^2 + 2S(D_P + D_R)]}{4} \tag{3.24}$$

where, the length of lateral surface of the arc column denoted by S is expressed as

$$S = \sqrt{L^2 + \left(\frac{D_P}{2} - \frac{D_R}{2}\right)^2} \tag{3.25}$$

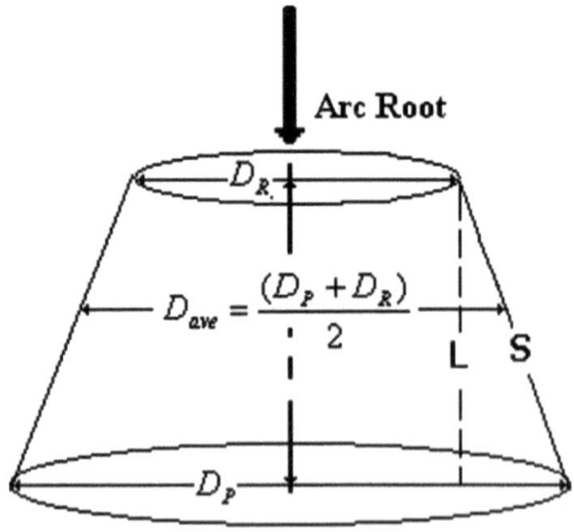

Fig. 3.40 Geometrical concept of arc shape analysis

Average convection heat transfer coefficient [32, 44],

$$h_c = N_u K_{Ar}/D_{ave} \tag{3.26}$$

Nusselt number [3, 44],

$$\mathrm{N_u} = 2.0 + 0.6\mathrm{R_e^{1/2}}\ \mathrm{P_r^{1/3}} \tag{3.27}$$

Average diameter of arc column (Fig. 3.40),

$$D_{ave} = \frac{(D_P + D_R)}{2} \tag{3.28}$$

Reynolds number [32, 44],

$$R_e = V_{eff} D_{ave}/v \tag{3.29}$$

Prandtl number [32, 44],

$$P_r = C_p v/K_{Ar} \tag{3.30}$$

where, C_p is specific heat of argon plasma and v is kinematic viscosity of ionized shielding gas. The effective velocity, V_{eff} of plasma has been estimated [32, 3, 27] as follows.

$$V_{eff} = 0.5\left[\frac{\mu_0 I_p J_g}{2\pi\rho_g}\right]^{1/2} \tag{3.31}$$

where, the μ_0 is permeability of free space and J_g and ρ_g are current density and mass density of plasma in arc column respectively [32].

Efficiency of arc can be largely realized by its capacity to create different size of weld pool in the work-piece. It primarily results as a function of heat generation in arc column and heat loss from the arc environment and transferring metal droplet. Thus, the arc efficiency (η_{ae}) estimated by the model Eq. 3.18, can be validated by comparing it with the arc efficiency (η_{aw}) evaluated by considering arc heat (Q_{AW}) transferred to weld pool and heat loss in arc column during P-GMA welding [27].

$$\eta_{aw} = \frac{Q_{AW}}{[VI_{eff} - \psi I_{eff}]} \tag{3.32}$$

where, ψ is effective melting potential at anode [32, 27]. The Q_{AW} is estimated by assuming average temperature (T) of weld pool [16, 27] measured at depth of about 2.5 mm from its surface and estimated heat transfer to weld pool by super heated (Q_f) droplets of filler metal. It is considered that the depth of measurement of

T representing average temperature of liquid weld pool remains valid in close approximation for all the ferrous and non ferrous metals in question.

$$Q_{AW} = \frac{\left[\left\{T - \frac{Q_f}{2\pi k} e^{-\lambda v \xi}\left\langle \frac{e^{-\lambda v R}}{R} + \sum_{n=1}^{n=\infty}\left(\frac{e^{-\lambda v R_n}}{R_n} + \frac{e^{-\lambda v R'_n}}{R'_n}\right)\right\rangle\right\} - T_0\right] \times \rho c}{\left[\int_0^t \left\{\frac{dt'}{\sqrt{\left(12a(t-t') + a_h^2\right)}\cdot\sqrt{\left(12a(t-t') + b_h^2\right)}}\right\} \times \left\{\frac{A'}{\sqrt{12a(t-t') + c_{hf}^2}} + \frac{B'}{\sqrt{12a(t-t') + c_{hb}^2}}\right\}\right]} \quad (3.33)$$

$$Q_f = Q_{de}\, m_t\, f \quad (3.34)$$

The R, R_n, R'_n, A′ and B′ are estimated as stated earlier in Sect. 1.3.1 where, the r_f and r_b are the proportion coefficients in front and behind the heat source. Where, a_h, b_h, c_{hf} and c_{hb} are the ellipsoidal heat source parameters defined by a location having at least a power density of 5% to that exists at the centre on surface of the ellipsoid [27]. The values of a_h, b_h, c_{hf} and c_{hb} obtained by measurement of weld pool geometry are considered as 0.008, 0.002 and 0.008 [27, 46] respectively.

The qualitative nature of correlations of arc efficiency (η_{ae}) with various welding parameters studied on the basis of measured geometrical characteristics of welding arc [17–19] gives significant understanding of control of the P-GMAW process. In this regard observations on arc characteristics of 1.6 mm diameter Al–Mg (5086) alloy electrode and 1.2 mm diameter stainless steel (308L) electrode under argon gas shielding and the arc characteristics of 1.2 mm diameter mild steel electrode under argon, Ar + 2% CO_2 and Ar + 18% CO_2 gas shielding provides significant information. This is in order to broadly realise the justification for variation in estimated arc efficiency (η_{ae}) with change in parameters of P-GMAW process which is latter on compared to the efficiency (η_{aw}) evaluated on the basis of measured weld pool temperature. The heat transfer efficiency of P-GMAW process, estimated on the basis of arc power as a product of the average current and voltage, is significantly high of the order of 82% and it varies from 68 to 72% over a wide range of pulse parameters [47]. The estimated nature of variation of arc efficiency with change of pulse parameters has also been verified with the corresponding change in relevant characteristics of weld bead on plate, which are predominantly governed by thermal behaviour of welding largely as a function of arc efficiency.

3.2.4.1 Al- Mg Alloy

An increase of surface area of arc column with the increase of ϕ significantly reduces the arc efficiency (Fig. 3.41a, b) however, at a given ϕ the increase of I_m from globular to spray mode of metal transfer considerably enhances the surface area of arc column but marginally reduces the arc efficiency especially at higher ϕ. This may have primarily attributed to the behaviour of metal transfer where a high I_m significantly reduces heat loss in spray mode of metal transfer [3, 27, 48]

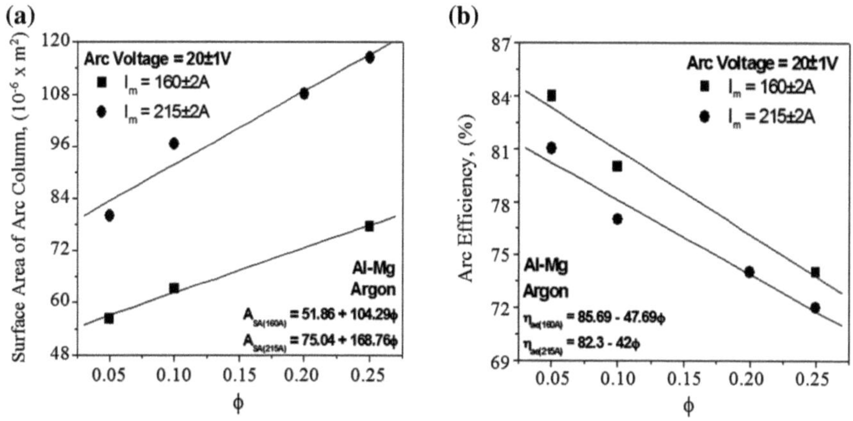

Fig. 3.41 41 At a given arc voltage effect of ϕ on **a** surface area of arc column and **b** arc efficiency during bead on plate weld deposition of Al–Mg alloy at different Im

minimizing its adverse influence on arc efficiency. Similarly a decrease of arc voltage also reduces the arc efficiency due to increase of surface area of arc column (Fig. 3.42). During welding within a given range of arc voltage, I_m and ϕ, the estimated arc efficiency (η_{ae}) appears to be well in agreement to the evaluated arc efficiency (η_{aw}). But it gives ±3–5% under estimation of η_{ae} than the η_{aw} (Fig. 3.43) primarily due to over estimation of A_{SA} under the glare of arc column giving rise to higher estimation of heat loss by q_{cov}, q_{cod} and q_{rad}.

At a given I_m the estimated decrease of arc efficiency with the increase of ϕ justifies the observations [49] to make the welding thermally less efficient at a given heat input reducing the area of fusion in bead on plate weld deposition. However, in contrast to this observation the area of fusion of base metal significantly enhances

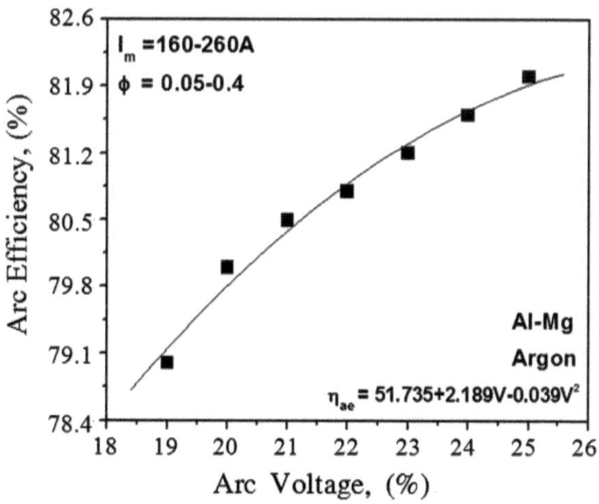

Fig. 3.42 At a given range of ϕ and I_m effect of arc voltage on arc efficiency during bead on plate weld deposition of Al–Mg alloy

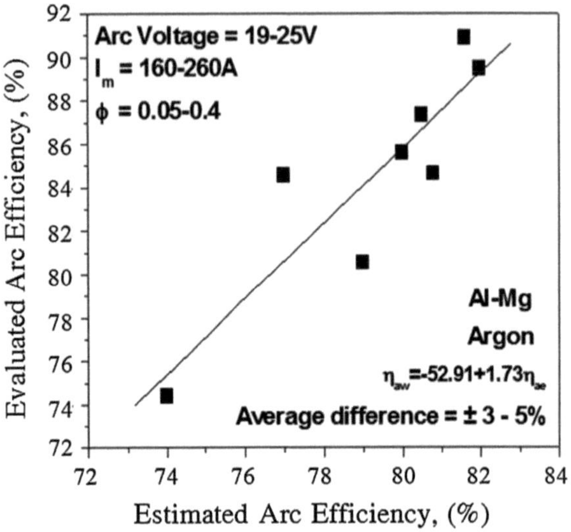

Fig. 3.43 Comparison of estimated and evaluated arc efficiency in case of weld deposition of Al–Mg alloy

with the increase of I_m at a given ϕ (Fig. 3.41b) in spite of lowering of arc efficiency. It primarily happens possibly due to predominant role of increase of cathode surface temperature (T_{cath}) and cathode heating rate (Q_{cath}) with the increase of I_m vis-à-vis the increase of I_{eff} at a given ϕ in pulse current welding, where the T_{cath} and Q_{cath} are estimated as [32, 49].

$$T_{cath} = \sqrt{\frac{J_{cath}}{Ue^{\left(\frac{-\xi_W}{K_B T_{cath}}\right)}}} \tag{3.35}$$

$$Q_{cath} = \left[V_{cath} - \left(\xi_w + \frac{3K_B T_{cath}}{2e}\right)\right] I_{eff} \tag{3.36}$$

where, U and K_B are the Richardson's and Boltzmann constant [32], ξ_w is work function of the cathode surface [50], J_{cath} is current density of cathode, V_{cath} is cathode fall voltage [32, 50] and e is the charge of electron.

3.2.4.2 Mild Steel

In case of mild steel also at a given I_m the increase of ϕ enhances the surface area of arc column under different compositions of argon base shielding gas as shown in Fig. 3.44a–c for argon, Ar + 2% CO_2 and Ar + 18% CO_2 respectively. It shows that irrespective of the composition of shielding gas the A_{SA} increases with the increase of I_m at a given ϕ. The estimated arc efficiency (η_{ae}) as usual reduces with increase of ϕ at a given arc voltage and I_m in case of using different shielding gases as argon, Ar + 2% CO_2 and Ar + 18% CO_2 (Fig. 3.45a–c respectively). It is noted

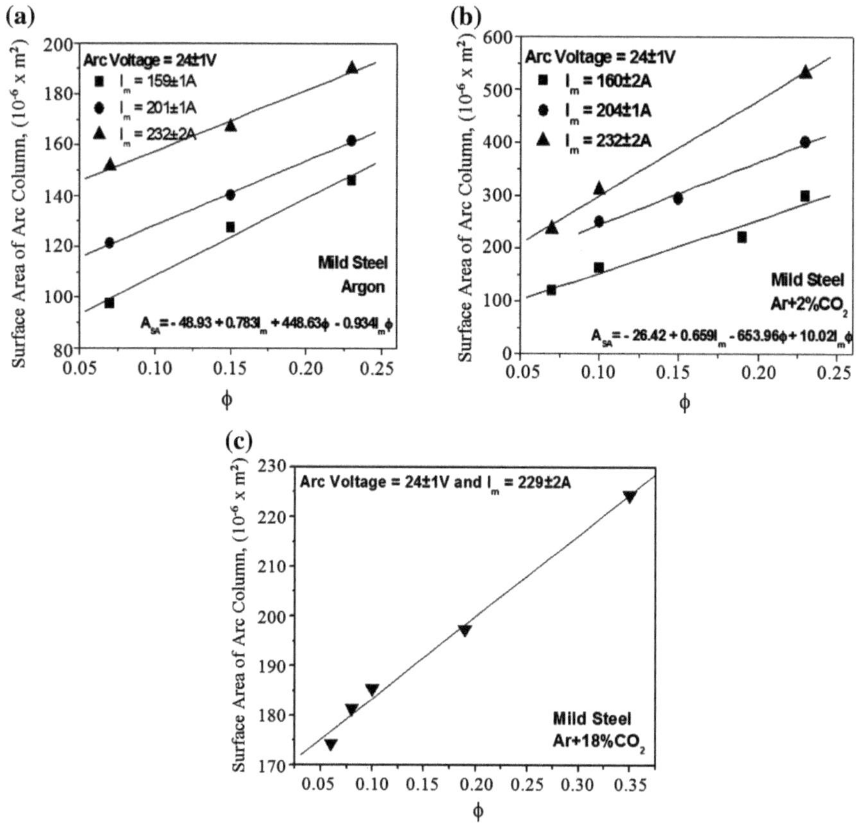

Fig. 3.44 At a given arc voltage effect of ϕ and I_m on surface area of arc column under different gas shielding of **a** argon, **b** Ar + 2% CO_2 and **c** Ar + 18% CO_2 during bead on plate weld deposition of mild steel

that at a similar order of I_m and arc voltage the A_{SA} at a given ϕ relatively enhances with the change in shielding gas composition from argon to Ar + 2% CO_2 but, it comparatively reduces with more CO_2 content (Ar + 18% CO_2) of the shielding gas. This behaviour of variation in A_{SA} with the change in shielding gas composition as discussed in earlier section may have influenced the extent of heat loss and accordingly the η_{ae}.

The estimated arc efficiency (η_{ae}) also gives good agreement to the arc efficiency (η_{aw}) evaluated on the basis of arc heat (Q_{AW}) transferred to weld pool and heat loss in arc column (Fig. 3.46). At given arc voltage and heat input (Ω) variation of I_m and ϕ influences area of fusion of base plate (Fig. 3.47) while at a given I_m an increase of ϕ reduces it. But at a given ϕ the increase of I_m increases the same. This may primarily happens because fusion of base plate depends upon heat delivered

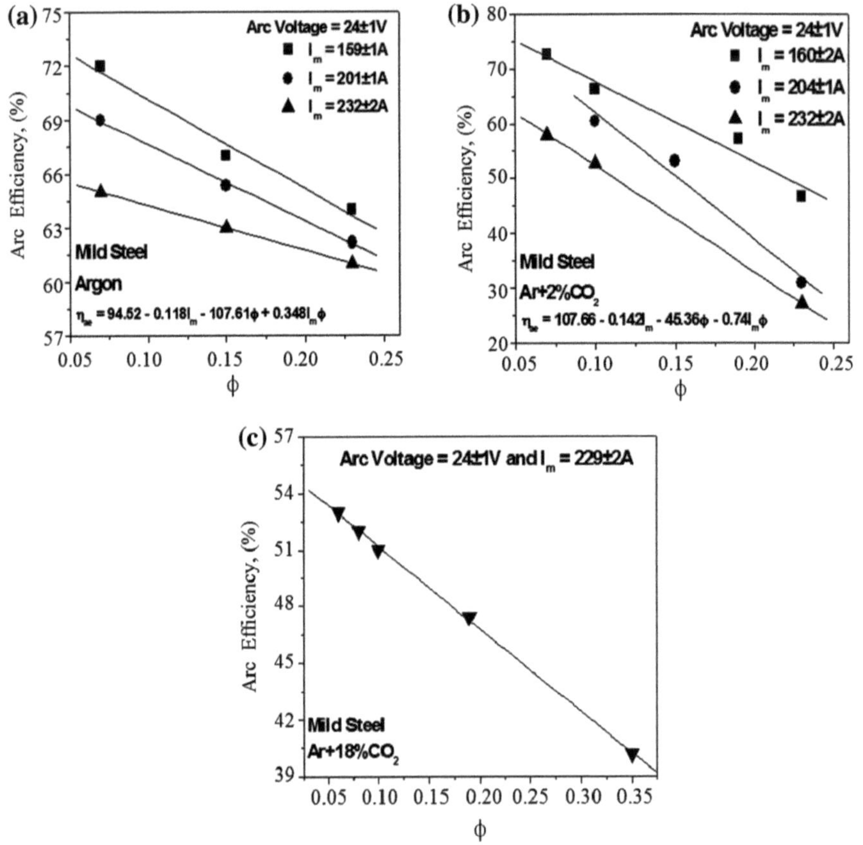

Fig. 3.45 At a given arc voltage effect of ϕ and I_m on arc efficiency under different gas shielding of **a** argon, **b** Ar + 2% CO_2 and **c** Ar + 18% CO_2 during bead on plate weld deposition of mild steel

through arc column as well as the cathodic surface heating. Here the former one depends upon arc efficiency but the latter one is directly influenced by the energy introduced to the system. At a given I_m an increase of ϕ marginally increases the T_{cath} and Q_{cath}, whereas at given ϕ the increase of I_m enhances them considerably (Fig. 3.48). Thus, it may be realized that at a given ϕ in spite of the decrease in arc efficiency increase of base plate fusion with the increase of I_m is predominantly governed by the Q_{cath}. However, at a given I_m the decrease of base plate fusion with the increase of ϕ principally dictated by the lowering of arc efficiency under the marginally incremental influence of Q_{cath}. It justifies the validity of the estimation of η_{ae} which may be used as an effective tool to characterize the P-GMAW process for controlling weld quality with due consideration of the cathodic heating rate at different I_m.

Fig. 3.46 Comparison of estimated and evaluated arc efficiency in case of weld deposition of mild steel

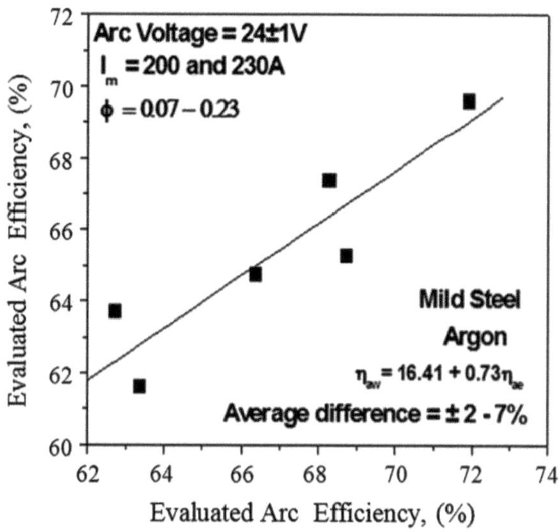

Fig. 3.47 At a given heat input (Ω) effect of ϕ on area of fusion of base plate at different Im of 160, 200 and 230 A during bead on plate weld deposition of mild steel

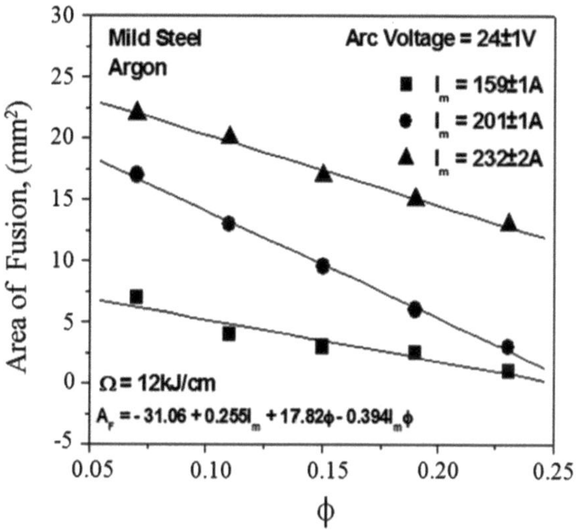

3.2.4.3 Stainless Steel

The observations on aluminium alloy and mild steel are further clarified on a different class of high alloyed comparatively low electrical and thermal conductive material as stainless steel. Here also the I_m and ϕ similarly influences the surface area of arc column and arc efficiency (Fig. 3.49a, b respectively) under argon shielding. However, considering the Figs. 3.41a, 3.44a and 3.49a it may be realised that at given gas (argon) shielding and similar process parameters of I_m and ϕ, the surface area of arc column significantly increases in order of using the type of filler

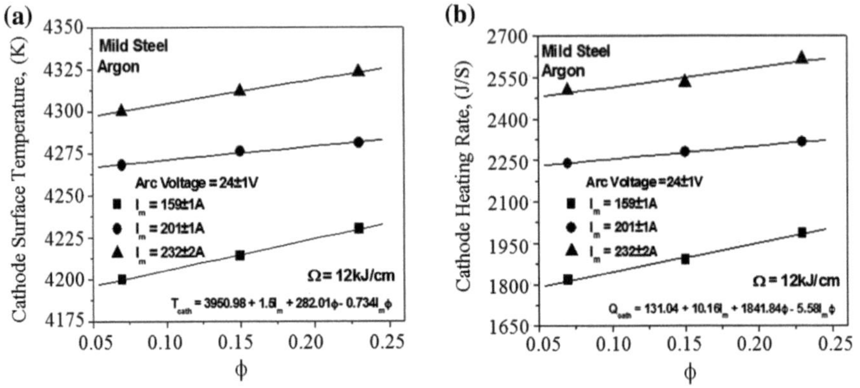

Fig. 3.48 At a given heat input (Ω) effect of ϕ on **a** cathode surface temperature and **b** cathode heating rate during bead on plate weld deposition of mild steel at different Im of 160, 200 and 230 A

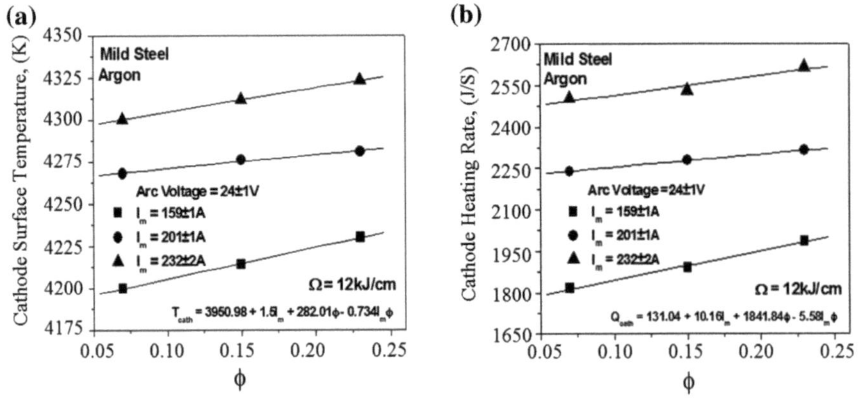

Fig. 3.49 At a given arc voltage effect of ϕ on **a** surface area of arc column and **b** arc efficiency during bead on plate weld deposition of austenitic stainless steel at different Im

wire as Al–Mg alloy, mild steel and stainless steel. Accordingly the arc efficiency decreases in the same sequence of increase of surface area of arc column in case of corresponding types of weld deposition.

Variation in geometrical characteristics of plasmatic shielded arc column primarily happens due to differences in thermal conductivity and degree of ionisation of filler material influencing the metal vapour jet created at the electrode tip. The ambient thermal conductivity (although it is a function of temperature) decreases from 222 to 51.6 and 16.6 Wm^{-1} K^{-1} for the Al–Mg alloy, mild steel and stainless steel respectively [32, 44] which reduces the anode tip energy density causing reduction in kinetic energy of vapour jet. It may also be further influenced by the degree of ionization of nonferrous and ferrous metal, which is considered as about

0.1 and 0.001 for the aluminium and iron respectively at about 5000 °K [32, 51]. The relatively low degree of ionization may also produce weaker vapour jet of steel than aluminium. Considering the well known Saha ionization equation [32, 51] it may be further realised that the degree of ionization of plasma of a single atomic gas depends upon temperature, density and ionization energies of the atoms. In case of argon gas shielding the degree of ionization of its plasma reduces largely as a direct function of its temperature depending upon lowering of anode energy density with the use of filler wire of relatively lower electrical conductivity. Thus, it may enhance a lateral diffusion of shielding environment of the arc. In view of these phenomena the increase of surface area of arc column and consequent decrease of arc efficiency with the use of filler wire in order of Al–Mg alloy, mild steel and stainless steel at a similar I_m and ϕ, can be understood, which should be studied further in detail.

3.2.5 Arc Stability

P-GMAW is a widely employed highly efficient process for welding of thin to thick sections of ferrous and non ferrous materials [3, 16, 52] due to its ability to maintain smooth control of metal transfer under pulse current [53, 54, 43]. But, the interruption of welding arc under certain pulse parameters often causes irregular behaviour of pulse waveform resulting in non-regular transfer of considerably large amount of liquid mass and heat from electrode to weld pool. Instability of pulse waveform in P-GMAW primarily happens due to short circuiting of electrode to weld pool, which largely depends upon pulse parameters and the type of shielding gas [55–57]. The nature of instability defined by its N_{PA}, I_{DC}, T_{IDC} and T_{DDC} primarily diagnoses the irregularities of pulse waveform influencing weld quality. The instability in welding arc interrupts its gaseous shield and consequently affects the weld quality depending upon overall performance of power source. In general, such a disturbance of welding arc is largely studied by measuring the mean value and standard deviation of welding current and arc voltage [47, 58–60].

Stability of arc within a standoff distance of 12 mm between the tip of 1.2 mm diameter mild steel filler wire and similar base plate under Ar + 2% CO_2 and Ar + 18% CO_2 gas shielding at a flow rate of 18 l/min was studied. It was studied on simultaneously recorded pulse characteristics (Fig. 3.50a, b) with the help of a transient recorder fitted with the electrical circuit of P-GMAW process. The instability of pulse waveform is arbitrarily defined by a deviation of shooting beyond a current (I_S) marked as 30 A more than I_P. The degree of instability is estimated [61] by counting the average number of peaks (N_{PA}) shooting above the I_S at a given interval of arcing of 200 ms. Numerical expressions for estimation of these aspects of pulse current can be stated as

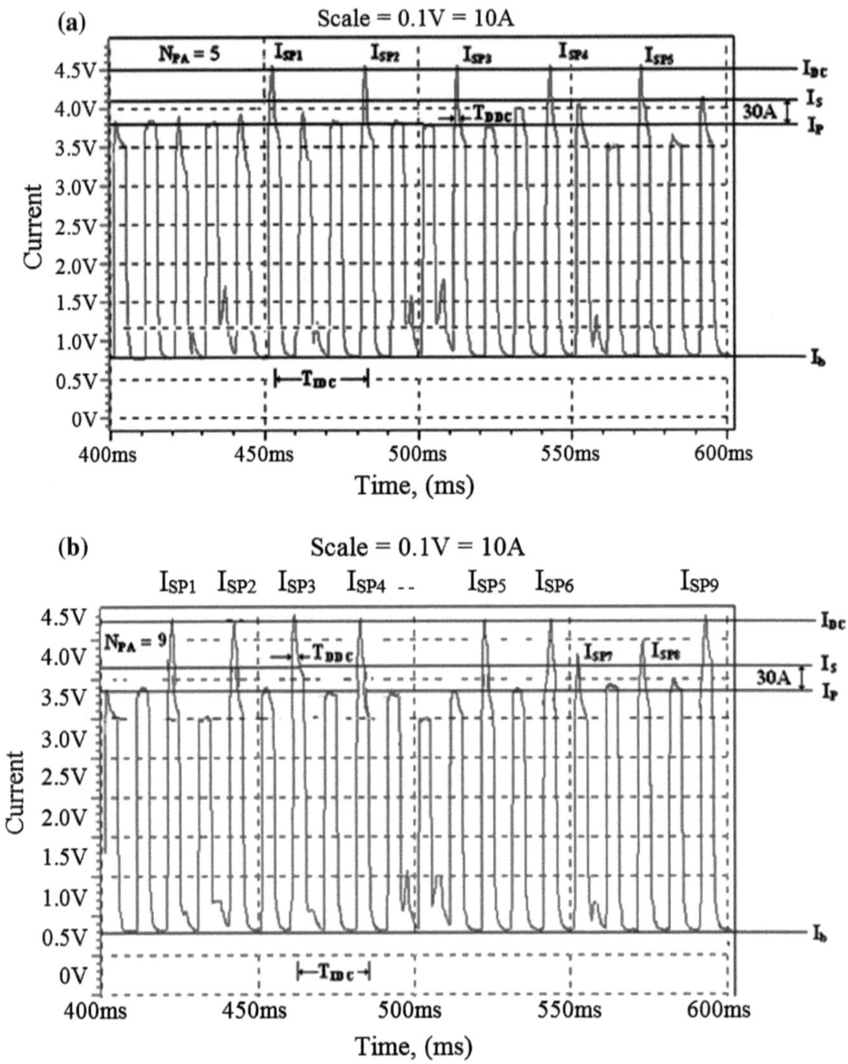

Fig. 3.50 Typical nature of instability in pulse waveform observed during welding at a given ϕ, I_m and arc voltage of 0.10, 230 A and 24 V respectively under different gas shielding of **a** Ar + 2% CO_2 and **b** Ar + 18% CO_2

$$I_S = I_P + 30 \tag{3.37}$$

$$I_{DC} = \frac{\sum_{i=1}^{N_{PA}} I_{SP}}{N_{PA}} \tag{3.38}$$

$$T_{IDC} = T_{NP2} - T_{NP1} \tag{3.39}$$

where, I_{DC} is average deviated current, I_{SP} is shooting peak current beyond the I_S per peak of instability and T_{IDC} is average time interval of deviated current. The T_{NP1} and T_{NP2} are starting time corresponding to the peaks of N_{PA1} and N_{PA2} respectively. Under different conditions of pulse current (ϕ, I_m and arc voltage) the arc stability estimated as follows has been further studied in terms of the fraction of short circuiting (SC) of metal transfer at an interval of 200 ms.

$$SC = \frac{(N_{PA} \times T_{DDC})}{200} \times 100 \tag{3.40}$$

where, T_{DDC} is average duration of deviated current. During P-GMA weld deposition of steel, the Ar + 2% CO_2 gas shielding provides better arc stability with respect to short circuiting (SC) compared to that observed during using Ar + 18% CO_2 gas shielding as shown in Fig. 3.51a, b respectively. However, at a given arc voltage and mean current the average number of instability (N_{PA}) in peaks of pulse current enhances with the change in pulse parameters giving rise to a higher value of the factor ϕ under Ar + 2% CO_2 gas shielding (Fig. 3.52a). Whereas, at any arc voltage keeping the I_m beyond transition current improves the arc stability as it is observed even under Ar + 18% CO_2 gas shielding (Fig. 3.52b). Such a change of N_{PA} as function of ϕ and I_m primarily happens due to considerable increase in velocity of droplets (V_i) and number of droplets transfer per pulse as well as decrease of arc length with the increase of ϕ and I_m [19, 20]. Validity of the observations on instability of pulse waveform with the change in pulse parameters under different gas shielding confirms the corresponding high speed (10^4 frames per second) video-graphs of the arc environment (Fig. 3.53). In the context of reducing

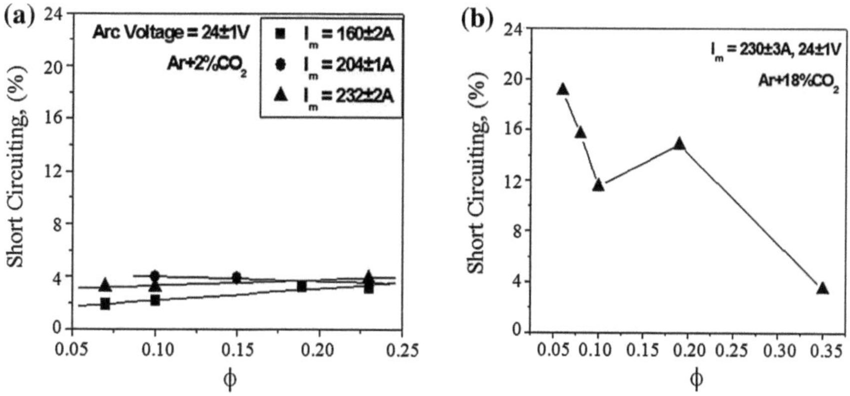

Fig. 3.51 At a given mean current and arc voltage effect of ϕ on fraction of short circuiting per period of 200 ms under different gas shielding of **a** Ar + 2% CO_2 and **b** Ar + 18% CO_2

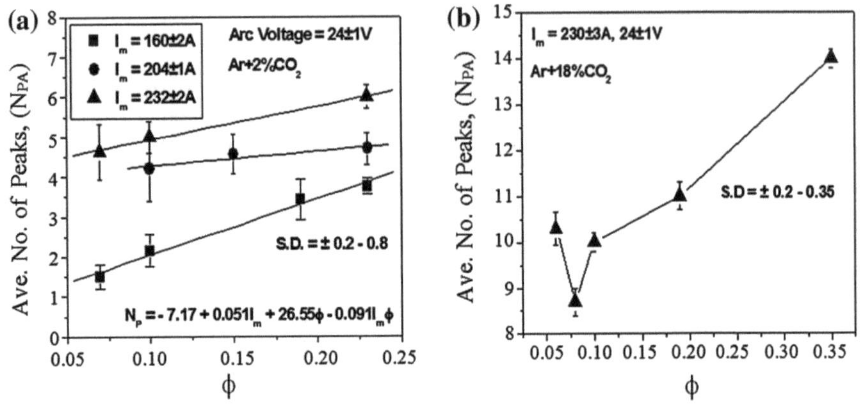

Fig. 3.52 At a given mean current and arc voltage effect of ϕ on average number of peaks shooting above average peak current per period of 200 ms under different gas shielding of **a** Ar + 2% CO_2 and **b** Ar + 18% CO_2

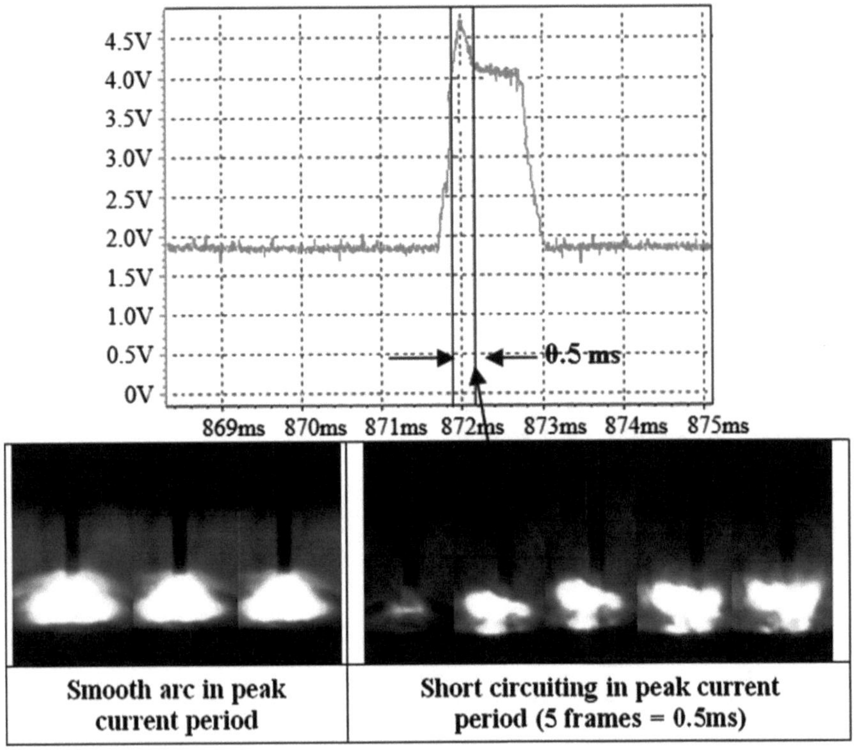

Fig. 3.53 At a given I_m, V and ϕ of 230 A, 24 V and 0.35 respectively typical sequence of short circuiting of electrode to weld pool under Ar + 18% CO_2 gas shielding

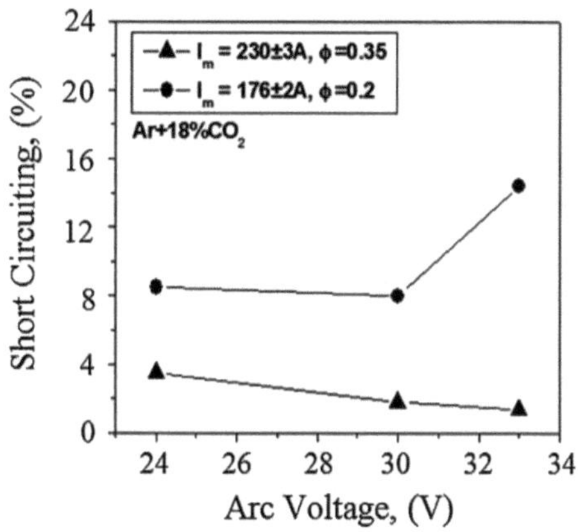

Fig. 3.54 At a given combination of ϕ and Im effect of arc voltage on percentage of short circuiting per 200 ms under Ar + 18% CO_2 gas shielding

the short circuiting it is noted that an increase of arc voltage is also appreciably useful along with keeping relatively higher I_m (230 A) beyond transition current and larger ϕ (0.35) of pulse current as shown in Fig. 3.54. But it is not very much similar at a relatively lower I_m (176 A) than the transition current where, at a given ϕ of 0.2 the increase of arc voltage from 24 to 30 V marginally reduces the short circuiting followed by a considerable increase in it with a further increase of arc voltage to 33 V. It basically happens because a relatively softer arc at higher arc voltage discourages effective spray transfer especially at low I_m. The knowledge of arc stability as a function of interruption of pulse waveform under different pulse parameters and shielding environment in P-GMAW may be beneficial for better control of P-GMAW process to produce desired weld quality. An appropriate control of critical aspects of arc stability of P-GMAW process may facilitate its more effective application in automation.

3.3 Shielding of Arc Environment

Stability of shielding of arc environment in gas metal arc welding can be analytically understood based on the basic knowledge of arc physics and its response to service in case of generally found uninterrupted arc profile in conventional process. At an appropriate gas flow rate the interruption in profile of arc and its gas shield of P-GMAW process largely depends upon magnitude and frequency in fluctuation of arc pressure at different arc stiffness and arc length. The arc length and stiffness varies as a function of ϕ, I_m and arc voltage [17, 18, 20, 62], whereas the degree of fluctuation of arc pressure significantly depends upon the ratio of (I_b/I_p) [17, 18]. An improper shielding can introduce porosity and inclusion in pulsed current GMA weld of both the ferrous and nonferrous metals.

3.3.1 Aluminium Alloy

Using 1.6 mm diameter Al–Mg filler wire with argon shielding at a flow rate of 18 l/min an increase of arc voltage from about 20 to 25 V considerably enhances the arc length of pulse off as well as pulse on periods (Figs. 3.55 and 3.56) of P-GMAW where, the I_m and ϕ practically remains constant with a low I_b of the order of 72–98 A. Welding at an arc voltage within this range by keeping (I_b/I_p) ratio of about 0.345 may lead to formation of vortex in shielding jacket penetrating the arc environment (Fig. 3.57a) even at a stiff arc of high I_b of 144 A. But no such behaviour occurs (Fig. 3.57b) at further stiffer arc during pulse on time at I_p of about 417 A. However, at a comparatively longer arc length of higher arc voltage of 25 V the fluctuation in arc pressure due to large variation in I_p and I_b resulting an (I_b/I_p) ratio of 0.253 at I_b of 98 A creates instability in shielding gas jacket during both the pulse off (Fig. 3.55) and pulse on periods (Fig. 3.56b) of the arc. Stability of arc environment is primarily a function of arc length and arc stiffness, where welding at a relatively lower arc voltage below about 22 V with appropriately matching higher current beyond about 122 A may be beneficial. The relatively

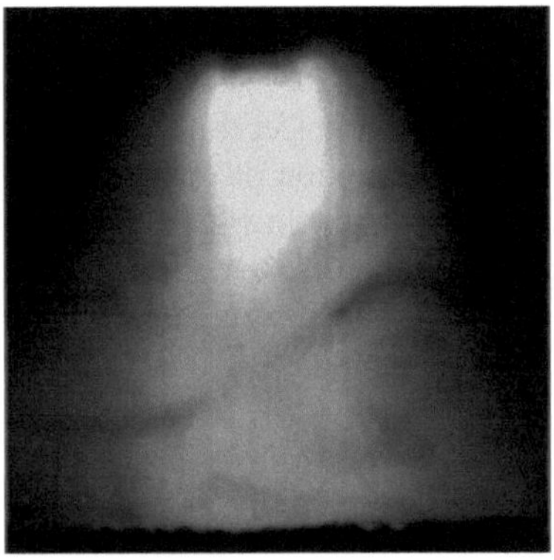

Fig. 3.55 Arc characteristic of pulse off time at a given arc voltage of 25 V, I_m = 263 A, I_p = 387 A, I_b = 98 A and ϕ = 0.1

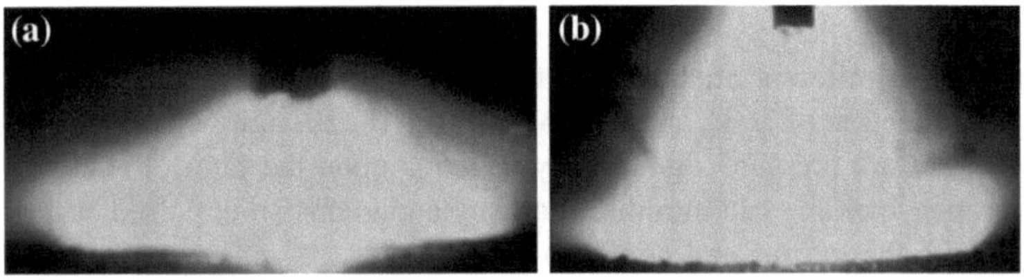

Fig. 3.56 Arc characteristics of pulse on time at **a** arc voltage of 20 V, I_m = 258 A, I_p = 332 A, I_b = 72A, ϕ = 0.059 and **b** arc voltage of 25 V, I_m = 263 A, I_p = 387 A, I_b = 98 A, ϕ = 0.1

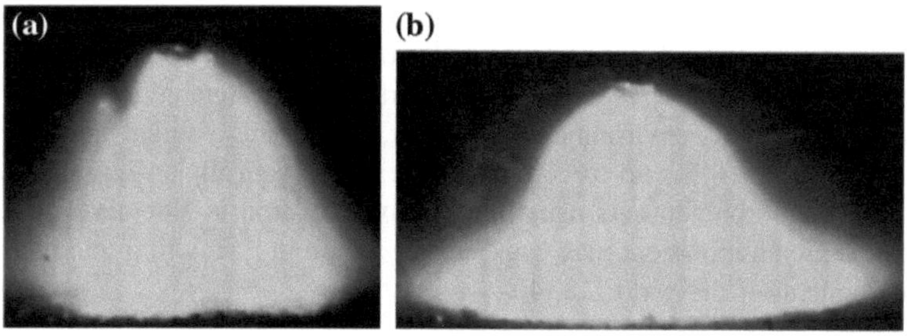

Fig. 3.57 At arc voltage of 23 V, I_m = 212 A, I_p = 417 A, I_b = 144 A and ϕ = 0.258 **a** vortex formation at pulse off time **b** no vortex formation at pulse on time in shielding jacket

softer arc of pulse off time is more prone to severity of vortex formation penetrating deep inside the arc environment than that of the pulse on time of significantly high current [17]. The vortex formation in shielding jacket penetrating the arc environment may cause air aspiration in it that enhances the chance of porosity formation in weld.

3.3.2 Stainless Steel

During welding of stainless steel using 1.6 mm diameter filler wire at given I_m and arc voltage of about 249 ± 5 A and 24 ± 1 V respectively, an increase of ϕ from 0.18 to 0.27 considerably disturbs the environment of argon shielding and natural bell shape of the arc in pulse on period (Figs. 3.58 and 3.15b). Such disturbance in arc profile with increase of ϕ also occurs at a relatively low I_m and arc voltage of 193 ± 6 A and 20 ± 1 V respectively. But, it happens at a practically insignificant intensity [18]. Increase of ϕ enhances the ratio of (I_b/I_p) and reduces arc stiffness (Fig. 3.33), when both of them favours adverse influence of fluctuation of arc pressure on stability of its environment under pulsed current. As it occurs in case of welding of aluminium alloy [17], here also an increase in ratio of (I_b/I_p) beyond about 0.289 considerably enhances turbulence in shielding environment and disturbs the arc profile. It happens due to significant enhancement in fluctuation of arc

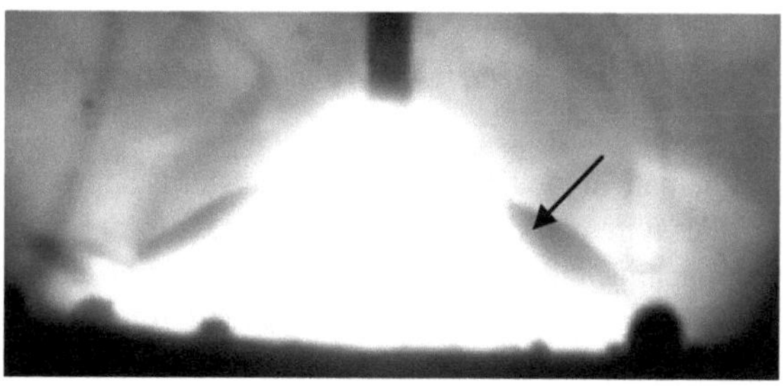

Fig. 3.58 During SS welding at a given I_m = 243 A and arc voltage of 24 V typical turbulence in arc shielding at ϕ = 0.18

pressure at low arc stiffness, especially at high arc voltage of about 24 ± 1 V. A low ϕ, (I_b/I_p), I_m and arc voltage in the range of about 0.05–0.06, 0.1–0.11, 181–199 A and 18–20 V respectively is generally found suitable for maintaining a practically stable arc environment [18]. But, a low arc voltage may allow an increase of ϕ and (I_b/I_p) up to certain extent with practically insignificant disturbance in arc environment. As usual the vortex formation in shielding jacket penetrating the arc environment may cause air aspiration in it resulting into increase in porosity content of the weld [15, 40].

3.3.3 Mild Steel

A considerable disturbance in shielding environment of Ar + 2% CO_2, leading to distortion in plasma flow around the arc from its vertical central axis, occurs in pulse on period at a low ϕ of about 0.07 (Fig. 3.18a), but it subsides with the increase of ϕ and I_m [19]. It happens also during welding of aluminium alloy but not in case of welding of stainless steel. The disturbance in arc profile increases at relatively softer arc of pulse off period. At any level of I_m the variation in ϕ, (I_b/I_p) and arc voltage significantly influences the stability of arc environment and their control at relatively higher magnitude is beneficial to maintain a stable arc environment in P-GMAW process. It may possibly happens because increase of ϕ enhances the ratio of (I_b/I_p) and increases the arc stiffness as a function of its pressure [18, 20], when both of them favour the adverse influence of fluctuation of arc pressure on stability in arc environment under pulsed current. The increase of arc voltage significantly enhances the arc length at the pulse on as well as pulse off periods, when the pulse off period (Fig. 3.31) shows comparatively more disturbance in arc profile than that of the pulse on time under Ar + 18% CO_2 shielding. A low arc voltage of about 24 V creates instability in shielding jacket due to relatively low arc pressure [20]. A higher combination of ϕ, I_m and arc voltage of about 0.35, 230 A and 33 V respectively maintains a relatively stable arc of welding by 1.2 mm diameter filler wire. At a given I_m and arc voltage the variation of ϕ with a change in pulse parameters disturbs the arc profile and by distortion with respect to its vertical central axis under argon shielding in similar fashion to that observed in Ar + CO_2 shielding [19].

3.3.4 Theoretical Analysis of Metal Transfer

Mathematical modelling of metal transfer is approached through analysis of melting of filler wire and distribution of the mass to droplets and their nature of movement approaching the weld pool driven by specific conditions of current and electrode tip. The mass of filler metal, M_t, and total surface area of molten metal, A_s, transferred per pulse are estimated as

$$M_t = [(A_W V_W \rho_w)/f] \tag{3.41}$$

$$A_S = \pi D^2 N_d \tag{3.42}$$

$$N_d = 6M_t/(\pi D^3 \rho_d) \tag{3.43}$$

The electrode tapering under the concept of energy balance in detachment dynamics [63] being most active at I_p, the metal transfer from filler wire is generally assumed to take place primarily at this phase of pulse. Accordingly the characteristics of metal transfer with respect to the diameter of droplets, D, and mass of molten metal transfer per pulse in number of droplets, N_d, are estimated by Eqs. 2.19 and 2.22 respectively. The flight speed of the droplets, $V_{i,}$ at the time of detachment from electrode as a function of I_p, size of filler wire, properties of flight environment and the type of metal being transferred is understood [63] as follows.

$$V_i = (2\gamma/\rho_d r)^{1/2}\left[1 + 0.187\theta - 1.226\theta^{0.142}\right]^{1/2} \tag{3.44}$$

Velocity of droplet at the time of vertically down hand deposition (V_{de}) is expressed as

$$V_{de} = V_i + (a_{cc} + g)\tau \tag{3.45}$$

$$\tau = \left[-V_i \pm \sqrt{(V_i^2 + 2(a_{cc} + g)L)}\right]/(a_{cc} + g) \tag{3.46}$$

where, L is the arc length and τ is the flight time of a drop from the electrode tip to weld pool. Velocity of droplet during its travel (V_t) from the electrode tip to weld pool can also be estimated by measuring its position in succeeding frames of video-graphs.

$$V_t = \left[V_p^2 + 2(a_{cc} + g)\,h\right]^{0.5} \tag{3.47}$$

where, V_p is the estimated velocity in preceding frame, h is distance of travel in between two consecutive frames and g is gravitational acceleration. The acceleration, a_{cc}, of droplet due to plasma aerodynamics drag force is estimated as [3, 15, 64, 65].

$$a_{cc} = \left(3V_{eff}^2 \rho_g C_d\right)/(4\,D\,\rho_d) \tag{3.48}$$

$$C_d = 17.251\,R_e^{-0.5902} \tag{3.49}$$

$$R_e = V_{eff} D/n \tag{3.50}$$

$$V_{eff} = 1/2\left[\mu_o I_p j_g / (2\pi\, \rho_g)\right]^{1/2} \quad (3.51)$$

where, V_{eff} is the effective velocity of plasma, ρ_g is mass density of plasma in arc column, C_d is drag coefficient, ρ_d is density of molten filler metal at temperature of its detachment from the wire electrode, R_e is Reynold's number, ν is kinematic viscosity of ionized shielding gas and j_g is current density of plasma in arc column. The j_g is estimated as division of effective current (I_{eff}) by cross sectional area of electrode, assuming that it truly represents the most active area of plasma in an arc column. The I_{eff} is estimated as

$$I_{eff} = \left[k_p I_p^2 + (1 - k_p) I_b^2\right]^{0.5} \quad (3.52)$$

where, k is the pulse duty cycle defined as,

$$k_p = t_p / t_{pul} \quad (3.53)$$

where, t_{pul} is pulse cycle time period (t_p + t_b). The measured behaviour of metal transfer studied as function of the factor ϕ and the mean current (I_m) is compared with their theoretical values estimated at same welding parameter.

3.3.5 Transfer Behaviour of Different Materials

The basic characteristics of metal transfer in P-GMAW process are considered as size (diameter), number and velocity of droplets. These characteristics estimated by the mathematical expressions are verified by experimental observations under different welding process parameters. The mode of metal transfer in P-GMAW process is broadly understood by single drop transfer and multiple drop transfer per pulse with or without transfer at base current. The entire mode of transfer is governed by the pulse parameters. However, at the end of any current pulse a residual transfer of mass may exist resulting into a comparatively bigger size of droplet transfer at a relatively lower current. The mechanism of relatively simpler single drop transfer per pulse is fairly well understood. But, paucity of knowledge still exists about the process of multiple drop transfer per pulse. High speed video-graphs at a speed of 10^4 frames per second during welding by 1.6 mm diameter Al–Mg filler wire and 1.2 mm diameter stainless steel and mild steel filler wires at the pulse parameters giving multiple drop transfer per pulse with and without transfer at base current throws some light on characteristics of droplet diameter and its velocities at different stages of travel on the way of transfer to weld pool. The physically observed behaviours of metal transfer agree to the theoretical understandings and gives satisfactory empirical correlation as a function of pulse parameters. Accordingly the knowledge as a whole gives significant opportunity to operate and control the P-GMAW process for desired nature of metal transfer determining the weld quality.

Based on the model expressions as stated earlier some thorough theoretical studies on behaviour of metal transfer, during weld deposition of Al–Mg alloy using large variety of pulse parameters from different power sources, gives good opportunity to study the effect of pulse parameters on it [66]. In spite of wide variety of welding conditions the behaviour of metal transfer, using 1.6 mm diameter filler wire at electrode extension in the range of 15–18 mm with commercially pure argon shielding, maintains good correlation with the pulse parameters through their summarised influence as the factor ϕ. The theoretical understandings of the behaviour of metal transfer are largely justified through high speed videography of the deposition of Al–Mg alloy at various conditions of welding as stated below. In the light of well justified knowledge of high speed videography of metal transfer at various pulse parameters, various aspects of its behaviour under high deposition multiple drop transfer per pulse are analysed also for stainless steel and mild steel weld depositions. The characteristics of metal deposition are also predicted by tracking the signature of arc sound and weld temperature [67]. The viscous flow and inertial force of molten metal generated by peak current have significant effect on dynamic characteristics of metal transfer in pulsed current gas metal arc welding. The detachment characteristic is analyzed by numerical computation of the situation in a volume of fluid using the electromagnetic force, surface profile and velocity distribution of fluid in a drop [68].

3.3.6 Aluminium

At different I_m the increase of ϕ exponentially reduces D but, enhances N_d by maintaining a good correlation as shown in Figs. 3.59 and 3.60 respectively. The empirical correlations of D and N_d with the ϕ and I_m reveal that at given ϕ the increase of mean current also decreases D and increases N_d but, this effect becomes significant at ϕ beyond about 0.2. The change in diameter of droplet and number of droplets transfer per pulse with a variation in ϕ affects the total surface area (A_s) of molten metal transfer per pulse negotiating their corresponding influence on it. At a given I_m the increase of ϕ beyond about 0.2 enhances A_s at a comparatively faster rate especially at higher I_m. At a given ϕ the A_s becomes larger also at higher I_m with more prominence at higher ϕ (Fig. 3.61). The theoretical flight speed or velocity (V_i) of droplet estimated by the Eq. (3.44) enhances exponentially with the increase of ϕ at different I_m as shown in Fig. 3.62. The effect of ϕ on enhancement of V_i remains significant up to about 0.2 with practically insignificant influence of I_m on it. However, at a higher ϕ beyond about 0.2 the V_i becomes less sensitive to ϕ with lowering of I_m showing a relatively higher V_i at larger I_m of a given ϕ. Theoretical values of D of droplet of aluminium alloy of P-GMAW with its experimentally measured values at comparable pulse conditions, especially with respect to the I_p, studied by high speed videography [66] and some other

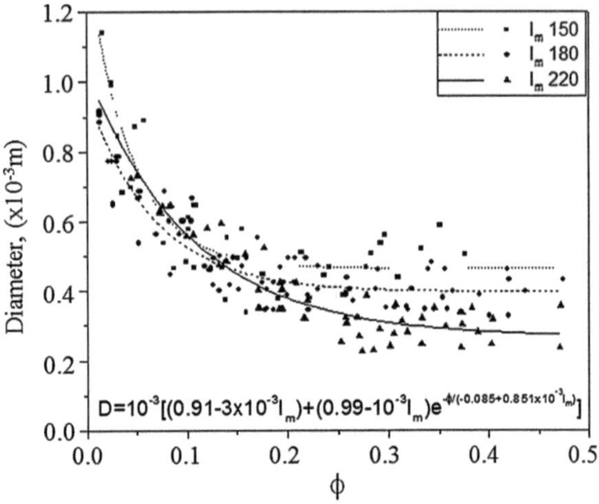

Fig. 3.59 At different Im of 150, 180 and 220 A effect of ϕ on estimated diameter of droplets (D) of Al–Mg filler metal

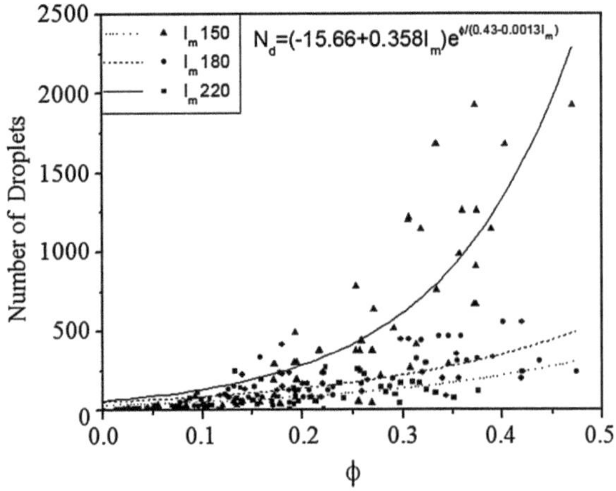

Fig. 3.60 At different Im of 150, 180 and 220 A the effect of ϕ on estimated number of droplets (Nd) transfer per pulse from Al–Mg electrode

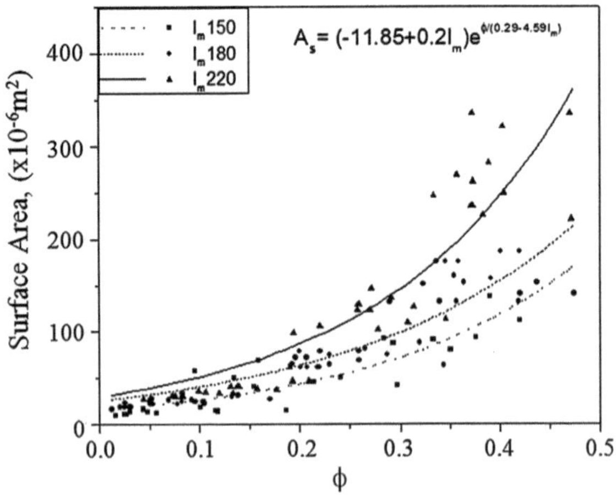

Fig. 3.61 At different Im of 150, 180 and 220 A effect of ϕ on estimated total surface area (As) of molten Al–Mg filler metal transferred per pulse

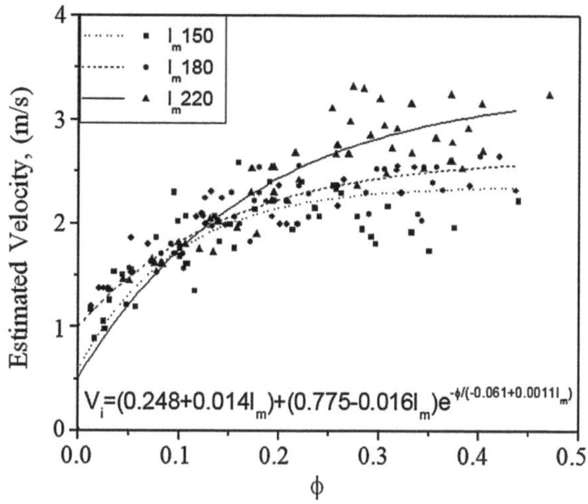

Fig. 3.62 At different I_m of 150, 180 and 220 A effect of ϕ on estimated velocity (V_i) of droplet of Al–Mg filler wire

experimental method [69] are found closely in agreement with a little underestimation [17]. The diameter of droplet logarithmically reduces with the increase of I_p but, enhances with the increase of ϕ during weld deposition of Al–Mg alloy.

At an I_m (160 A) lower than the transition current of 1.6 mm diameter Al–Mg filler wire the metal transfer from electrode tip takes place in lump (Fig. 3.63a–b) or its fragmentation (Fig. 3.63c) or in relatively smaller droplets (Fig. 3.63d) with heterogeneous nature. But, in certain cases of higher I_m (254 A) above transition current it appears that metal transfer takes place in tiny droplets from multiple locations of electrode surface without any appreciable tapering (Fig. 3.64a–b) in it. At a long duration (14.8 ms) of high base current (191 A) along with large ϕ of about 0.4 significant melting takes place at electrode tip (Fig. 3.65), which may cause metal transfer also at the pulse off period. It is in agreement to general understanding of current and duration required for metal transfer from 1.6 mm diameter aluminium alloy filler wire under argon shielding in GMAW process. However, the behaviour of metal transfer may be primarily dictated by the I_p which enhances with increase of I_m.

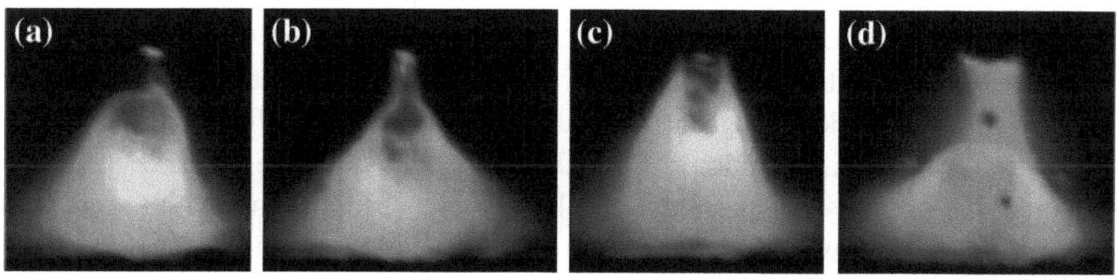

Fig. 3.63 Photographs of heterogeneous nature of metal transfer at arc voltage of 21 V, I_m = 160 A, I_p = 328 A, I_b = 92 A, t_p = 4 ms, t_b = 10 ms and ϕ = 0.2 depicting; **a–b** lump, **c** fragmentation and **d** small droplets

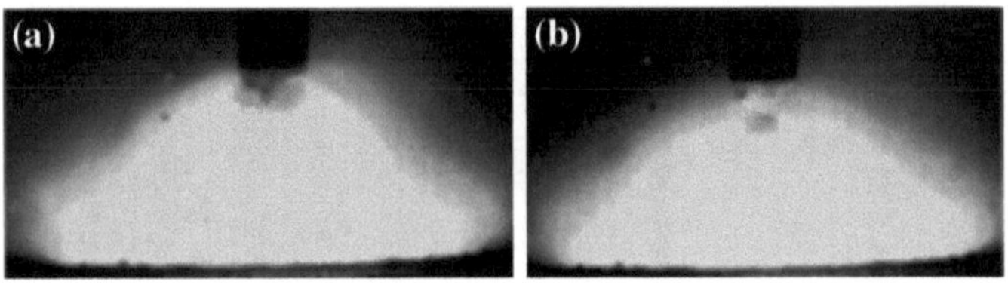

Fig. 3.64 Photographs showing transfer of tiny droplets from multiple locations of Al–Mg electrode during peak current at arc voltage of 23 V, $I_m = 254$ A, $I_p = 444$ A, $I_b = 141$ A, $t_p = 4.4$ ms, $t_b = 7.6$ ms and $\phi = 0.181$

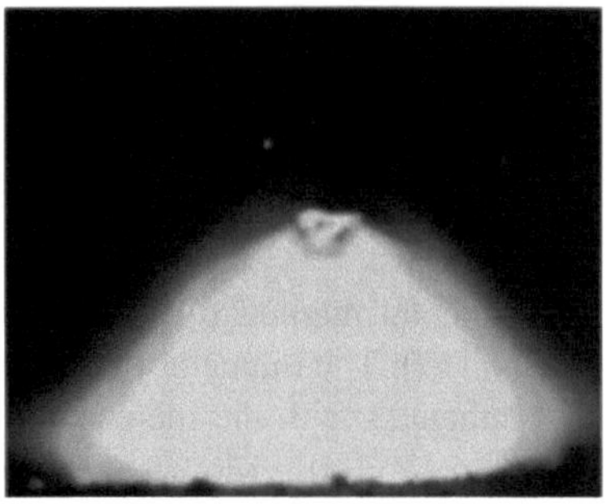

Fig. 3.65 Typical arc characteristic observed during pulse off time at a given arc voltage of 20.5 V, $I_m = 212$ A, $I_p = 411$ A, $I_b = 191$ A, $t_p = 1.2$ ms, $t_b = 14.8$ ms and $\phi = 0.426$

In pulsed current GMAW the average diameter (D) of droplet transfers per pulse is a strong function of peak current (I_P), and naturally all other parameters as I_m and ϕ also significantly influence the D due to their influence on I_P. By its predominance the I_P considerably reduces the D with appreciable consistency which is well in agreement to their corresponding theoretically estimated values (Fig. 3.66) at wide range of I_m and ϕ. At a given I_m the increase of ϕ up to certain extent initially reduces the droplet diameter followed by an insignificant influence on it with further increase of ϕ beyond a critical value. However, it appears that the critical value of ϕ enhances with the increase of I_m from about 160–260 A (Fig. 3.67). The measured values of droplet diameter at I_P of different pulse parameters are almost found in agreement (Fig. 3.68) to their empirically estimated values.

The velocity (V_i) of metal transfer from electrode tip to weld pool in lump or tiny droplet, measured by video-graphy (Fig. 3.69), primarily depends upon magnitude of current in pulse on and off times (Table 3.4). The velocity of droplet at any parameter lies close to its estimated (Eq. 3.44) value and maintains a steady linear growth within a critical range of current as shown in Fig. 3.70a. At high current beyond about 417 A, a predominant influence of considerable increase in plasma velocity [70] possibly enhances the droplet velocity at higher rate with the increase of current. At too low and high currents in reference to the limit beyond the linear part of measured velocity, the gravitational transfer and plasma driven transfer respectively predominate the velocity of droplet. However, during transfer of metal in lump or a relatively bigger drop at low I_m and as a small drop at high I_m the arc

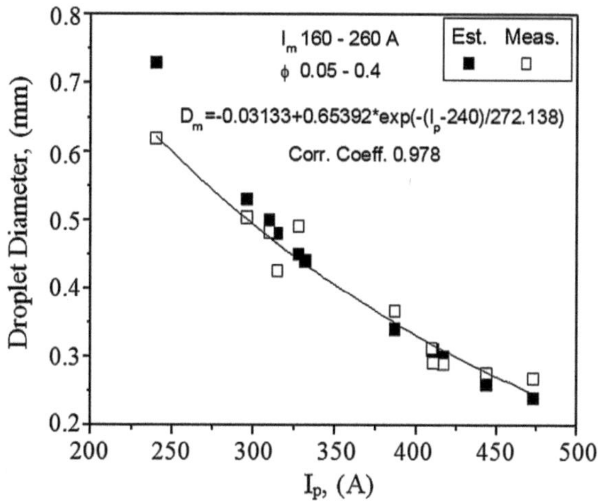

Fig. 3.66 At a given range of I_m (160–260 A) and ϕ (0.05–0.4) effect of IP on measured and theoretically estimated droplet diameter

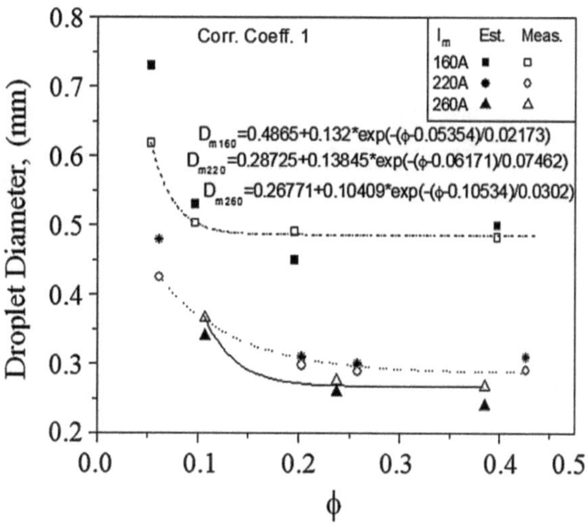

Fig. 3.67 At a given I_m of 160, 220 and 260 A effect of ϕ on measured and theoretically estimated droplet diameter

root becomes significantly narrow as shown in (Fig. 3.63a–b). It primarily happens due to tapering of electrode tip, which is measured [31] to have coefficient, δ, of about 0.838 ± 0.05 irrespective of pulse parameters. This value of δ for Al–Mg filler wire is also similar to that considered in case of mild steel filler wire [71]. At given pulse parameters of I_m and ϕ of 160 A and 0.196 respectively the typical measured velocity (6.865 m.s^{-1}) of a metal drop at the time of deposition (V_{de}) during pulse on time lies in close approximation to its theoretical value of 6.728 m.s^{-1} estimated by the Eq. 3.45. But, the velocity of plasma accelerates the movement of droplet within the arc environment up to certain extent resulting the V_{de} relatively higher than its V_i of detachment (2.46 m.s^{-1}) at the same welding parameters.

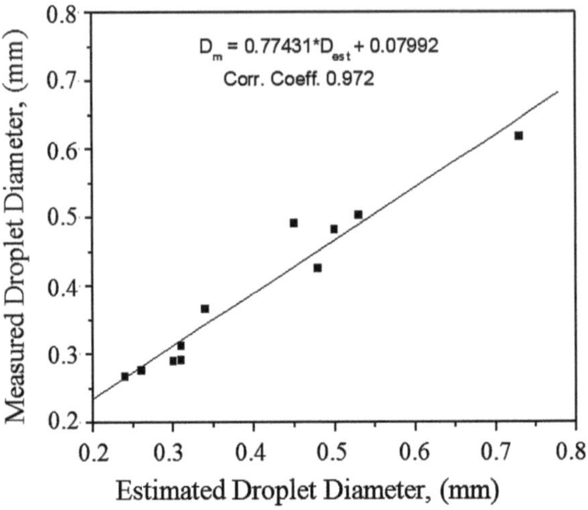

Fig. 3.68 Comparison between estimated and measured droplet diameter at a given range of I_m (160–260 A) and ϕ (0.05–0.4)

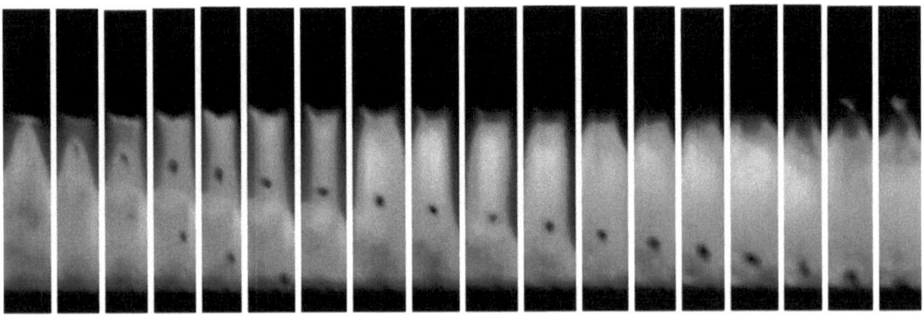

Fig. 3.69 Series of consecutive video-graphs showing transfer of tiny droplets from Al–Mg electrode tip to weld pool at arc voltage of 21 V, I_m = 160 A, I_p = 328 A, I_b = 92 A and ϕ = 0.2

3.3.7 Stainless Steel

During welding of stainless steel the detachment of droplet from electrode followed by its movement towards weld pool as captured in consecutive frames of video-graphs facilitates the measurement of droplet diameter and its detachment velocity, as typically shown in Fig. 3.71. The measured D and V_i are well in agreement to their theoretical values estimated at the same welding parameters with the help of the Eqs. 2.19 and 3.44 respectively. The empirical expressions are developed based on the experimentally measured values of effective radius (r) of tapered electrode (Fig. 3.4a) as 0.77 $\pm$ 0.16 mm and arc length (L) along with the reported [72, 23, 73] values of coefficient of surface tension (γ), density of molten

Table 3.4 Estimated and measured velocity of droplet at the time of detachment in pulse on and pulse off periods

ϕ	I_m (A)	I_{eff} (A)	I_p (A)	I_b (A)	V_w (m/s)	Pulse on period (t_p)			Pulse off period (t_b)			Average difference ± SD
						V_i (m/s)		Difference (%)	V_i (m/s)		Difference (%)	
						Estimated	Measured		Estimated	Measured		
0.196	160	190	328	92	0.0867	2.402	2.46	−2.41	–	–	–	6.58 ± 3.22
0.397	168	162	310	145	0.08	2.178	2.41	−10.65	–	–	–	
0.204	213	248	410	122	0.11167	3.138	2.9	7.58	–	–	–	
0.258	212	241	417	144	0.11167	3.159	2.97	5.98	–	–	–	
0.181	254	280	444	141	0.13	3.399	3.52	−3.56	–	–	–	
0.426	212	215	411	191	0.104	–	–	–	1.213	1.326	−9.32	
0.385	262	278	473	220	–	–	–	–	–	1.86	–	–

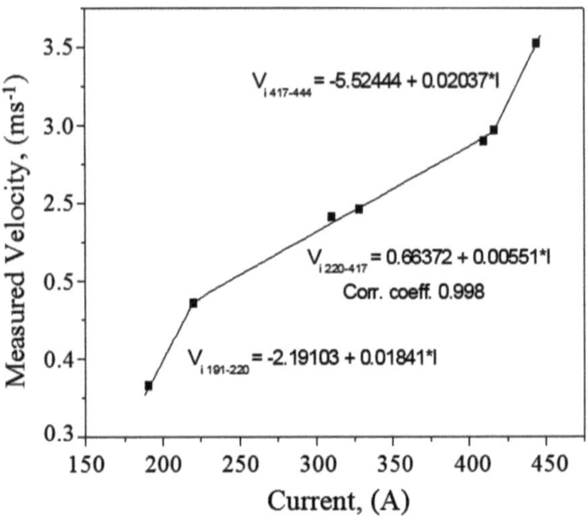

Fig. 3.70 Effect of current on measured velocity (V_i) of droplet at the time of detachment from 1.6 mm diameter Al–Mg electrode

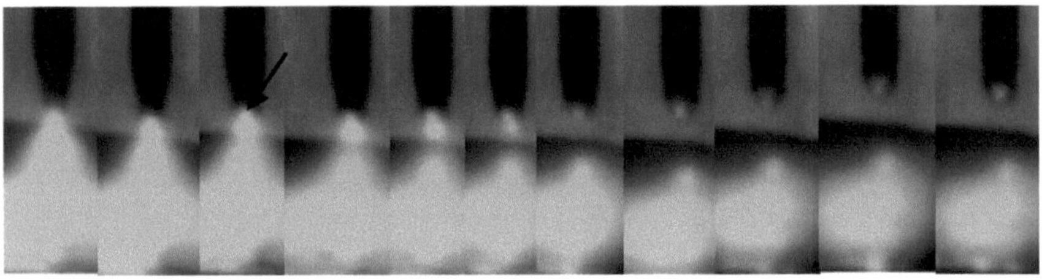

Fig. 3.71 Typical tapering of ASS electrode followed by transfer of droplet at a pulse parameter of $\phi = 0.08$, $I_m = 206$ A and arc voltage of 23 V

filler metal (ρ_d) and permeability of free space (μ_o) as 1.2 N m^{-1}, 7507 kg m^{-3} and $4\pi \times 10^{-7}$ N A^{-2} respectively. In spite of inherited heterogeneity of welding process the measured values of V_i and D mostly lies in close approximation to their theoretical values (Table 3.5) with an average difference of about 7.25 and 20.7% respectively. The D and V_i of metal transfer in P-GMAW process predominantly depends upon I_P irrespective of ϕ and arc voltage. However, as it happens in case of aluminium [74] here also at a given arc voltage increase of ϕ increases D but reduces V_i, whereas at a given ϕ increase of arc voltage generally increases both the D and V_i relevantly with respect to their corresponding I_P [18, 75, 76].

Table 3.5 Estimated and measured diameter (D) and detachment velocity (V_i) of droplet at different ϕ

Wire feed rate (mm/s)	ϕ	I_m (A)	V_i Est. (m/s	V_i Obs. (m/s)	Difference (%)	D Est. (mm)	D Obs. (mm)	Difference (%)
158.3	0.20	248	2.21	2.57	16.3	0.20	0.25	25
158.3	0.24	255	1.22	1.29	5.7	0.53	0.50	5.7
92	0.27	248	1.98	1.72	13.1	0.25	0.32	28
125	0.21	220	2.09	2.14	2.4	0.23	0.28	21.7
141.7	0.06	211	1.90	1.93	1.6	0.27	0.33	22.2
125	0.17	214	1.61	1.54	4.4	0.23	0.28	21.7
			Average difference (%)		7.25	Average difference (%)		20.7

3.3.8 *Mild Steel*

The composition of shielding gas significantly affects the burn-off rate [77, 23, 73] due to its influence on thermal efficiency and thus consequently influences behaviour of metal transfer of P-GMAW process [20] as typically shown in Figs. 3.72a, b and 3.73. This phenomenon controls diameter and velocity of droplets transferred from mild steel filler wire at different pulse parameters under commonly used (Ar + 2% CO_2 and Ar + 18% CO_2) shielding gases. The D and V_i of droplet transfer almost linearly enhances with the increase of I_P at relatively wide variation of I_m and ϕ as shown in Fig. 3.74a, b respectively. Droplet diameter also significantly decreases with the increase of ϕ in case of Ar + 18% CO_2 shielding. However, an increase of I_m, arc voltage and ϕ as well as the use of Ar + 18% CO_2 instead of Ar + 2% CO_2 reduces the D more significantly [20]. This behaviour of

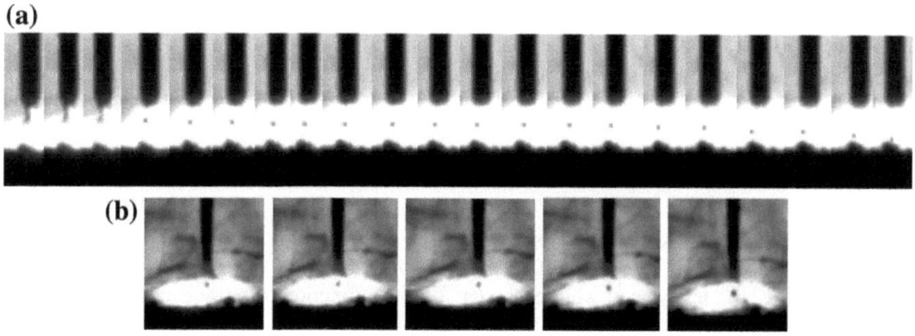

Fig. 3.72 At a given arc voltage of 24 ± 1 V the typical droplet transfer from 1.2 mm diameter mild steel electrode tip to weld pool under different I_m and ϕ respectively of **a** 176 A and 0.11 and **b** 205 A and 0.23 in Ar + 2% CO_2 gas shielding

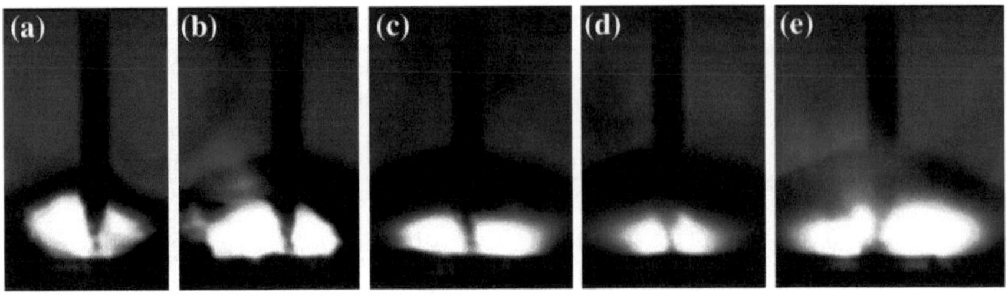

Fig. 3.73 Typical appearance of droplet after detachment from 1.2 mm diameter mild steel electrode tip at different I_m, ϕ and arc voltage respectively of **a** 179 A, 0.19 and 25 V, **b** 180 A, 0.19 and 33 V, **c** 232 A, 0.10 and 25 V, **d** 227 A, 0.35 and 24 V and **e** 227 A, 0.35 and 33 V in Ar + 18% CO_2 gas shielding

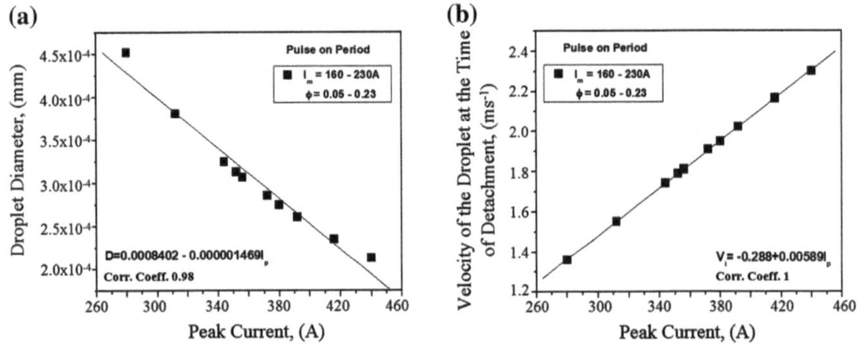

Fig. 3.74 At a given arc voltage of 24 ± 1 V effect of I_p on **a** droplet diameter and **b** detachment velocity of droplet under varying I_m and ϕ in the range of 160–230 A and 0.06–0.23 respectively in Ar + 2% CO_2 gas shielding

metal transfer is similar to that happens in case of aluminium and stainless steel where, at given I_m and arc voltage increase of ϕ reduces D but enhances V_i primarily due to increase of I_P.

References

1. Ghosh, P.K., Rai, B.K.: Characteristics of pulsed current bead on plate deposit in flux cored GMAW process. ISIJ Int. **36**(8), 1036–1045 (1996)
2. Ghosh, P.K.: An analysis of weld characteristics as a function of pulse current MIG welding parameters. Int. J. Join. Mater. **8**(4), 157–161 (1996)
3. Quintino, L., Allum, C.J.: Pulsed GMAW: Interactions between process parameters. Part I Weld. Metal Fab. **16**(4), 126–129 (1984)
4. Ghosh, P.K., Gupta, S.R., Gupta, P.C., Rathi, R.: Fatigue characteristics of pulsed MIG welded Al–Zn–Mg alloy. J. Mat. Sci. **26**, 6161–6170 (1991)

5. Ghosh, P.K., Dorn, L., Issler, L.: Fatigue crack growth behaviour of pulsed current MIG weld of Al–Zn–Mg alloy. Int. J. Join. Mater. **6**(4), 163–168 (1994)
6. Potluri, N.B., Ghosh, P.K., Gupta, P.C., Reddy, Y.S.: Studies on weld metal characteristics and their influence on tensile and fatigue properties of pulse current GMA welded Al–Zn–Mg alloy. Weld. J. **75**(2), 62s–70s (1996)
7. Matsuda, F., Nakata, K., Tsukamoto, K., Tohgan, S.: Combined effect of current pulsation and Zr addition on improvement of solidification cracking of AI–Zn–Mg alloy weld metal. Trans. IWRI **14**(2), 99–104 (1985)
8. Ghosh, P.K., Gupta, P.C.: Influence of pulsed current MIG welding on the characteristics of Al–Zn–Mg alloy weldments. Trans. Indian Inst. Met. **44**(4), 317–326 (1991)
9. Ghosh, P.K., Dorn, L., Goecke, S.F.: Universality of correlationships among pulse parameters for different MIG welding power sources. Int. J. Join. Mater. **13**(2), 40–47 (2001)
10. Ghosh, P.K.: Decide pulse parameters for desired properties of pulsed current GMAW weld. In: International Welding Conference (IWC 99) on Welding and Allied Technology—Challenges in 21st century, pp. 18–28. New Delhi, Feb. 15–17 (1999)
11. Randhawa, H.S., Ghosh, P.K., Gupta, S.R.: Geometric characteristics of pulsed current positional GMA weld. ISIJ Int. **38**(4), 276–284 (1998)
12. Ghosh, P.K., Rai, B.K.: Correlations of pulse parameters and bead characteristics in pulsed current flux cored GMAW process. Ind. Weld. J. **31**(4), 30–39 (1998)
13. Ghosh, P.K., Gupta, S.R., Randhawa, H.S.: Characteristics and criticality of pulsed current vertical up GMA weld in steel. Int. J. Join. Mater. **11**(4), 99–110 (1999)
14. Randhawa, H.S., Ghosh, P.K., Gupta, S.R.: Some basic aspects of geometrical characteristics of pulsed current vertical up GMA weld. ISIJ Int. **40**(1), 71–76 (2000)
15. Ghosh, P.K., Hussain, H.M.: Morphology and porosity content of multipass pulsed current GMA weld of Al–Zn–Mg alloy. Int. J. Join. Mater. **14**(1/2), 16–27 (2002)
16. Ghosh, P.K., Randhawa, H.S., Gupta, S.R.: Characteristics of a pulsed-current, vertical-up gas metal arc weld in steel. Met. Mater. Trans. A **31A**(9), 2247–2259 (2000)
17. Ghosh, P.K., Dorn, L., Hübner, M., Goyal, V.K.: Arc characteristics and behaviour of metal transfer in pulsed current GMA welding of aluminium alloy. J. Mater. Process. Tech. **194**(4), 163–175 (2007)
18. Ghosh, P.K., Dorn, L., Kulkarni, S.G., Hofman, F.: Arc characteristics and behaviour of metal transfer in pulsed current GMA welding of stainless steel. J. Mater. Process. Tech. **209**, 1262–1274 (2009)
19. Ghosh, P.K., Dorn, L., Devakumaran, K., Hofmann, F.: Pulsed current gas metal arc welding under different shielding and pulse parameters; Part-1: Arc characteristics. ISIJ Int. **49**(2), 251–260 (2008)
20. Ghosh, P.K., Dorn, L., Devakumaran, K., Hofmann, F.: Pulsed current gas metal arc welding under different shielding and pulse parameters; Part-2: behaviour of metal transfer. ISIJ Int. **49** (2), 261–269 (2009)
21. Ghosh, P.K., Agrawal, B.: Extra narrow gap gas metal arc welding of thick high strength low alloy steel. Proceedings of the 2nd South East European IIW International Conference, Pipeline welding current topic of the region, Sofia, Bulgaria, 21–24 Oct., 168–173 (2010)
22. Choi, S.K., Yoo, C.D., Kim, Y.S.: The dynamic analysis of metal transfer in pulsed current gas metal arc welding. J. Phys. D Appl. Phys. **31**, 207–215 (1998)
23. Wang, G., Huang, P.G., Zhang, Y.M.: Numerical analysis of metal transfer in gas metal arc welding under modified pulsed current conditions. Met. Mater. Trans. B **35B**(10), 857–866 (2004)
24. Liu, Anhua, Tang, Xinhua, Fenggui, Lu: Arc profile characteristics of Al alloy in double-pulsed GMAW. Int. J. Adv. Manuf. Technol. **65**, 1–7 (2013)
25. Welding Alcoa Aluminum: Aluminum Company of America. Pittsburgh, Pennsylvania (1972)
26. Ghosh, P.K., Devakumaran, K., Goyal, V.K., Kulkarni, S.G., Ghosh, Aritra K.: Proceedings of the 14th ISME International Conference. Mechanical Engineering in knowledge and age, New Delhi, India, December, Delhi college of engineering, Paper 338 (2005)

27. Goyal, V.K., Ghosh, P.K., Saini, J.S.: Analytical studies on thermal behaviour and geometry of weld pool in pulsed current gas metal arc welding. J. Mater. Process. Tech. **209**, 1318–1336 (2009)
28. Pires, I., Quintino, L., Miranda, R.M.: Analysis of the influence of shielding gas mixtures on the gas metal arc welding metal transfer modes and fume formation rate. Mater. Des. **28**, 1623–1631 (2007)
29. Dillenbeck, V.R., Castagno, L.: The effects of various shielding gases and associated mixtures in GMA welding of mild steel. Weld. J. **66**(9), 45–49 (1987)
30. Goyal, V.K., Ghosh, P.K., Saini, J.S.: Process-controlled microstructure and cast morphology of dendrite in pulsed-current gas-metal arc weld deposits of aluminum and Al-Mg alloy. Met. Mater. Trans. A **38**(8), 1794–1805 (2007)
31. Goyal, V.K.: Effect of thermal and solidification behaviour on characteristics of pulsed current GMA weld. Ph.D. thesis, Indian Institute of Technology Roorkee, India (2007)
32. Lancaster, J.F.: The Physics of Welding, 1st Ed., pp. 134–230. IIW, Pergamon Press, New York (1984)
33. Lin, M.L., Eagar, W.: Influence of arc pressure on weld pool geometry. Weld. J. **64**(6), 163s–169s (1985)
34. Kodama, S., Ikuno, Y.: Process modeling of short-circuiting GMA welding and its application to arc sensor control, Nippon Steel Technical Report No. 95 January, 71–75 (2007)
35. Murphy, A.B., Tanaka, M., Yamamoto, K., Tashiro, S., Sato, T., Lowke, J.J.: Modelling of thermal plasmas for arc welding: the role of the shielding gas properties and of metal vapour, J. Phys. D Appl. Phys. **42**(19), 194006 (2009)
36. Jenney, C.L., O'Brien, A.: AWS welding handbook', 9th Ed. **1**, 110–154 (2001)
37. Radaj, D.: Heat Effects of Welding', 1st Ed. Springer, Berlin (1992)
38. Rhee, S., Asibu, E.K.: Analysis of arc pressure effect on metal transfer in gas metal arc welding. J. Appl. Phys. **70**(9), 5068–5075 (1991)
39. Kulkarni, S.G.: Effect of narrow gap welding on characteristics of weld joint of austenitic stainless steel. Ph.D. Thesis, Indian Institute of Technology Roorkee, India (2009)
40. Heat Flow in Welding, Suranaree University of Technology, Sep-Dec (2007)
41. Praveen, P., Kang, M.J., Yarlagadda, P.K.D.V.: Arc voltage behavior in GMAW-P under different drop transfer modes. J. Achievements Mater. Manuf. Eng. **32**(2), 196–202 (2009)
42. Subramanium, S., White, D.R., Lyons, D.W.: Experimental approach to selection of pulsing parameters in pulsed GMAW, AWS. Weld. J. **78**, 166s–172s (1999)
43. Praveen, P., Yarlagadda, P.K.D.V., Kang, M.J.: Advancements in pulse gas metal arc welding. J. Mater. Process. Technol. **164–165**, 1113–1119 (2005)
44. Lienhard, J.H. IV, Lienhard, J.H. V: A Heat Transfer Text Book, 3rd Ed. Phlogiston Press, Cambridge, Massachusetts, USA (2008)
45. Redding, C.J.: Fume model for gas metal arc welding. Weld. J. **81**(6), 95s–103s (2002)
46. Gery, D., Long, H., Maropoulos, P.: Effects of welding speed, energy input and heat source distribution on temperature variations in butt joint welding. J. Mater. Process. Technol. **167**, 393–401 (2005)
47. Joseph, A., Harwig, D., Farson, D., Richardson, R.: Measurement and calculation of arc power and heat transfer efficiency in pulsed gas metal arc welding. Sci. Technol. Weld. J. **8** (6), 400–406 (2003)
48. Hu, J., Tsai, H.L.: Heat and mass transfer in gas metal arc welding. Part II. The metal. Int. J. Heat Mass Transf. **50**(5/6), 808–820 (2007)
49. Goyal, V.K., Ghosh, P.K., Saini, J.S.: Influence of pulse parameters on characteristics of bead-on-plate weld deposits of aluminium and its alloy in the pulsed gas metal arc welding process. Met. Mater. Trans. A **39A**(12), 3260–3275 (2008)
50. Eager, T.W.: The physics of arc welding process, Advanced Joining Technologies, T. H. North Ed., London, pp. 61–68 (1990)

51. Bellan, P.M.: Fundamentals of Plasma Physics. Cambridge, pp. 1–22 (2004)
52. Nasir Ahmed: New Developments in Advanced Welding, Woodhead Publishing Ltd, Abington, pp. 9–13 (2005)
53. Choi, S.K., Ko, S.H., Yoo, C.D., Kim, Y.S.: Dynamic simulation of metal transfer, Part II: Short circuit transfer modes. Weld. J. **77**(1), 45s–51s (1998)
54. Jacobsen, N.: Monopulse investigation of drop detachment in pulsed gas metal arc welding, Phys. D. Appl. Phys. **25**, 783–797 (1992)
55. Era, T., Ueyama, T.: Spatter reduction in GMAW by current waveform control. Weld. Int. **20** (7), 496–501 (2007)
56. Rouffet, M.E., Wendt, M., Goett, G., Kozakov, R., Schoepp, H., Weltmann, K.D., Uhrlandt, D.: Spectroscopic investigation of the high-current phase of a pulsed GMAW process. J. Phys. D Appl. Phys. **43**(43), 434003 (2010)
57. Quinn, T.P., Madigan, R.B., Siewert, T.A.: An electrode extension model for gas metal arc welding. Weld. J. **73**(10), 241s–248s (1994)
58. Belinga, E.M.M.: Applictions and benefits of adaptive pulsed GMAW, Lappeenranta University of Technology, Faculty of Technology, Mechanical Engineering, Master's Thesis (2012)
59. Cantin, G.M.D., Francis, J.A.: Arc power and efficiency in gas tungsten arc welding of aluminium. Sci. Technol. Weld. Join. **10**(2), 200–210 (2005)
60. Wu, C.S., Chen, M.A., Lu, Y.F.: Effect of current waveforms on metal transfer in pulsed gas metal arc welding. Meas. Sci. Tech. **16**(12), 2459–2465 (2005)
61. Ghosh, P.K., Devakumaran, K., Piyush, M.: Arc stability of pulse current gas metal arc welding of low alloy steel under different pulse parameters and shielding gas compositions. Indian Weld. J. **44**(4), 29–42 (2011)
62. Ghosh, P.K., Dorn, L., Devakumaran, K., Hofmann, F.: Influence of welding parameters and shielding gas on arc characteristics and behavior of metal transfer in GMA welding of mild steel. Indian Weld. J., **41**(2), 23–33 (2008)
63. Allum, C.J., Quintino, L.: Control of fusion characteristics in pulsed current MIG welding, Part-II, Simple model of fusion characteristics. Metal Constr. **17**(5), 314R–317R (1985)
64. Allum, C.J., Quintino, L.: Control of fusion characteristics in pulsed current GMAW—Part II Development of model for arc welding. Tech. Report, IIW Doc. 212-589-84 (1984)
65. Dilthey, U., Killing, R.: Heat input for pulse GMAW. Weld. Des. Fab. **63**(9), 51–53 (1990)
66. Cornu, J.: Advanced Welding System, vol. 2, pp. 127–165. IFS Publication Ltd. UK (1988)
67. Pal, K., Bhattacharya, S., Surjaya, K.P.: Prediction of metal deposition from arc sound and weld temperature signatures in pulsed MIG welding. Int. J. Adv. Manuf. Technol., **45**(11/12), 1113–1130 (2009)
68. Choi, S.K., Yoo, C.D., Kim, Y,S.: The dynamic analysis of metal transfer in pulsed current gas metal arc welding. J. Phys. D Appl. Phys. **31**(2), 207–215 (1998)
69. Waszink, J.H., Piena, M.J.: Experimental investigation of drop detachment and drop velocity in GMAW. Weld. J. **65**(11), 289s–298s (1986)
70. Pintard, J.: Some experimental data on short circuit transfer, in Physics of the Welding Arc. Institute of welding, London (1966)
71. Alum, C.J.: Metal transfer in arc welding as a varicose instability: I. Varicose instabilities in a current carrying liquid cylinder with surface charge. J. Phys. D Appl. Phys. **18**, 1431–1446 (1985)
72. Lancaster, J.F., Metallurgy of Welding, pp. 144–229. Allen and Unwin (Publishers) Ltd. (1987)
73. Melton, G.B.: Proc. Euro join2, Florence, Italy, 16–18 May, Italian Institute of Welding, 249–258 (1994)
74. Subramanium, S., White, D.R., Jones, J.E., Lyons, D.W.: Droplet transfer in pulsed gas metal arc welding of aluminium. Weld. J. **77**(11), 458s–464s (1998)

75. Kim, Y.S., Eagar, T.W.: Metal transfer in pulsed current gas metal arc welding. Weld. J. **72** (7), 279s–287s (1993)
76. Rajasekharan, S., Kulkarni, S.D., Mallya, U.D., Chaturvedi, R.C.: Molten droplet detachment characteristics in steady and pulsed current GMA welding AIMg alloys. In: Proceedings of the 6th International Conference on Aluminum Weldrnents, pp. 207–214. Cleveland, Ohio (1995)
77. Kim, I.S., Basu, A.: A mathematical model of heat transfer and fluid flow in gas metal arc welding process. J. Mater. Process. Technol. **77**, 17–24 (1998)

Chapter 4
Thermal Behaviour of Pulse Current Gas Metal Arc Weld

Abstract The complex nature of heat flow behaviour due to interruption of metal deposition in pulse current gas metal arc weld which is affecting its thermal behaviour has been discussed to realise the basis of selection of pulse parameters to produce a sound weld. The thermal characteristics of P-GMA weld are critically understood through model analysis considering the influence of arc heating and super heating of filler metal under the pulse current parameters of simultaneously interactive nature. The behaviour of metal transfer under pulse current that is governing the cavity formation in weld pool and temperature distribution in weld has been described to express the mechanism of establishing the isotherm and thermal cycle of weld pool and heat affected zone of weld joint. The thermal nature of P-GMAW process has been correlated to the factor ϕ as the summarized influence of pulse parameters for better understanding of more precise control of the welding process.

Keywords Thermal behaviour · P-GMA weld · Pulse parameters · HAZ · Analytical model · Arc heating · Superheated filler metal · Metal transfer · Weld pool · Cavity formation · Nature of welding process · Control

The mechanism of variation in thermal behaviour of weld as a function of the characteristics of depositing weld metal followed by heat flow behaviour governing thermal cycle of pulsed current GMA welding is quite complex. Several theoretical and experimental attempts have been made [1–11] to understand the influence of various pulse parameters on nature of metal transfer aiming at one-drop transfer per pulse. But, in all these efforts hardly any basic understanding has been developed about the effect of interruption of metal deposition on thermal cycle of weld deposit, especially by considering the effect of pulse off period on it, irrespective of one or multiple drop transfer per pulse. Considering the characteristics of metal transfer in each pulse, theoretically derived analytical model for estimation of thermal behaviour of depositing weld metal under pulsed current addresses this issue up to certain extent. In pulsed current GMA welding variety of pulse parameters give stable weld deposition with different behaviour of metal transfer (Fig. 3.2) due to

P.K. Ghosh, *Pulse Current Gas Metal Arc Welding*, Materials Forming, Machining and Tribology, DOI 10.1007/978-981-10-3557-9_4

their influence on arc characteristics. The mechanism of droplet detachment from the electrode [12] significantly affects the weld thermal cycle. The variation in pulse parameters resulting stable arc for welding can be established by using different proportion of energy (I_b/I_p) introduced and distributed in terms of pulse-off-time fraction ($f \, . \, t_b$) in the system. It may be justified to assume that a control over these parameters through the factor ϕ as a function of them may be useful to produce a weld of desired quality. From Eq. 3.1 it may be realised that a given value of ϕ can be obtained at any I_m of proper arc stability by appropriate adjustment of (I_b/I_p) and ($f \, . \, t_b$) and consequently it may affect the thermal behaviour of weld. In view of this it is necessary to understand the thermal behaviours of depositing weld metal in P-GMAW as a function of the factor ϕ and I_m in order to control the welding process to introduce desired quality in the joint. By taking advantage of using the factor ϕ as summarised influence of pulse parameters the experimental observations reported in this regard by various workers could be successfully compared. It verifies the solutions of the mathematical models for estimation of thermal behaviour of P-GMA weld, which is otherwise difficult due to use of dissimilar combination of pulse parameters in different work.

4.1 Analytical Model of Thermal Behaviour

Proper selection of pulse parameters prior to carrying out pulsed current GMA welding resulting desired weld quality is critical due to simultaneous interaction of the pulse parameters affecting thermal and transfer behaviours of weld metal. Mathematical models developed for thermal analysis facilitate the prediction of temperature distribution and geometry of weld pool including isotherms and thermal cycle of a P-GMA weld. The concept of model is primarily based on basic assumptions regarding consequence of interruption in metal deposition on thermal behaviour of weld metal depositing per pulse in consideration of its transfer characteristics. In P-GMAW process, the pulses of high current, I_p, are superimposed on the continuing arc of low base current, I_b, enhancing the arc heating and pinch effect at the electrode tip causing detachment of molten drops of filler metal from it. Model reasonably assumes that the molten metal transfers from filler wire in spherical drops primarily at peak current (I_p) which largely governs their heat content [2]. The transfer of metal at peak current produces a considerable effect on interruption in metal deposition as well as heat transfer to the weld pool due to practically insignificant or no metal transfer at base current period. The model estimates temperature distribution in weld by appropriately considering two primary heat sources arising out of arc and superheated filler metal acting on it. The process of heat transfer to the weld pool simultaneously takes place by initial arc heating followed by deposition of superheated filler metal transferring heat at certain depth of cavity formed by the impact of the droplets on molten weld metal. The effect of the later one superimposed on the first one augments the size of the initial weld pool developed by arc heating. Model analysis of weld isotherms at melting point of base metal estimates the weld pool geometry and nature of heat affected zone (HAZ). The analysis becomes more realistic due to modifications of the solutions of the point [13] and distributed heat sources [14] by considering the effects of arc heating [9, 15], thermal behaviour and nature of metal transfer [16, 17] and process of heat transfer by droplets inside the weld pool [18, 19]. Due to presence of two dissimilar natures of heat sources, the temperature at any point of a weld is analysed by a combined heat source technique of superimposing the analytical solution of distributed heat source on that of the point heat source.

4.1.1 *Influence of Arc Heating*

The contribution of arc heating is considered from two aspects as amount of heat transfer toweld pool by arc heating (H_{AW}) and temperature distribution in it. The arc heat continuously acting at the surface of the base plate is considered as a distributed heat source of double ellipsoidal nature [20]. It melts the base plate and creates an initial weld pool in base metal depending upon energy input. The arc heat transferred to base metal (H_{AW}) is realised by deduction of heat consumed (H_A) for melting of filler wire per unit time from the arc heat (H_{arc}) generated by the energy input. The (H_{arc}) is expressed as

$$(H_{arc}) = VI_{eff} \tag{4.1}$$

where, V is arc voltage and effective current I_{eff} is estimated by the Eq. 3.32 as root mean square of the pulsed current waveform. In consideration of the heterogeneous nature of energy distribution in practically irregular pulse wave of P-GMAW process, the effective current instead of mean current is found more appropriate for better realistic estimation of thermal behaviour of arc heat, as proposed earlier [2, 9, 15, 17]. Assuming H_A is proportional to the effective current it is expressed as

$$(H_A) = \psi I_{eff} \tag{4.2}$$

where, ψ is effective melting potential at anode. Therefore, arc heat transferred to weld pool is expressed as

$$H_{AW} = [VI_{eff} - \psi I_{eff}]\eta_a \tag{4.3}$$

where, η_a is arc heat transfer efficiency.

The temperature distribution in a weld can be analysed by estimating temperature, T_d, of any point (x(ξ), y, z) of a weld (Fig. 1.3) as follows [14] by expressing the Eq. 1.21 as a function of arc heating H_{AW} of weld pool under pulsed current.

$$T_d = \frac{3\sqrt{3} \cdot Q_{AW}}{\rho \cdot c \cdot \pi\sqrt{\pi}} \int_0^t \left[\frac{dt'}{\sqrt{(12a(t-t') + a_h^2)} \cdot \sqrt{(12a(t-t') + b_h^2)}} \cdot \left(\frac{A'}{\sqrt{(12a(t-t') + c_{hf}^2)}} + \frac{B'}{\sqrt{(12a(t-t') + c_{hb}^2)}} \right) \right] + T_o \quad (4.4)$$

4.1.2 *Influence of Superheated Filler Metal*

In GMAW process the superheating of depositing filler metal primarily takes place by the arc and resistive heating at the electrode tip. The heat generation at electrode tip per unit time (H_o) due to arc heating (H_A) is estimated by Eq. 4.2 and the heat generation by resistive heating (H_R) per unit time due to electrical resistance, R_W, of the filler wire extension is estimated [9, 15, 17, 21] as

$$H_R = I_{eff}^2 R_W \quad (4.5)$$

A minimum electrode extension, E_W, is generally kept in GMAW process to avoid energy loss by the resistive heating. Thus, the effect of temperature on resistance of wire electrode R to appreciably affect the arc characteristics resulting into a significant change in thermal behaviour of welding may be considered practically negligible. Accordingly the H_R develops as a function of resistivity at ambient temperature R_o and cross sectional area A_w of the filler wire and joins the heat consumed (H_A) for melting of filler wire per unit time giving rise to total heat generation (H_o) at the tip of the filler wire per unit time.

$$H_R = \left[\left(I_{eff}^2 E_w R_o\right)/A_w\right] \quad (4.6)$$

$$H_o = \left[\psi I_{eff} + \left(E_w R_o I_{eff}^2\right)/A_w\right] \quad (4.7)$$

The H_o may also be expressed in terms of the heat content, H_i, per unit mass of the filler wire causing its melting and superheating from ambient temperature, T_a, to temperature of droplet, T_i, at the time of detachment.

$$H_o = \left[H_i A_w V_w \rho_w\right] \quad (4.8)$$

The V_w and ρ_w are the wire feed speed and mass density of filler wire respectively. The Eqs. 4.7 and 4.8 may be solved for H_i expressed as a function of effective current density, j_{eff}, at the electrode tip.

$$H_i = \left[\left(\psi I_{eff} + E_w R_o I_{eff}^2\right)/\left(A_w V_w \rho_w\right)\right] \quad (4.9)$$

$$H_i = \left[\left(\psi j_{eff} + E_w R_o j_{eff}^2\right)/\left(V_w \rho_w\right)\right] \quad (4.10)$$

For a smooth weld deposition, total heat generated per unit time (H_o) at tip of the filler wire should be equal to the heat required, H_w, per unit time by the filler metal for its melting and producing a superheated drop from ambient temperature, T_a, to temperature, T_i. The H_w as a function of physical properties of filler wire is expressed as

$$H_w = \left[\left\{C_{p(s)}(T_m - T_O) + L_h + C_{p(l)}(T_i - T_m)\right\}A_w V_w \rho_w\right] \quad (4.11)$$

where, $C_{p(s)}$ and $C_{p(l)}$ are specific heat of solid and molten filler metal respectively, T_m is melting temperature of filler metal and L_h is its latent heat of fusion. Thus, the temperature, T_i, of droplets at the time of their detachment from the tip of a filler wire can be expressed by solving the Eqs. 4.8 and 4.11 of H_o and H_w respectively.

$$H_i = \left[C_{p(s)}(T_m - T_O) + L_h + C_{p(l)}(T_i + T_m)\right] \quad (4.12)$$

$$T_i = \left[\left\{H_i - C_{p(s)}(T_m - T_O) - L_h\right\}/C_{p(l)}\right] + T_m \quad (4.13)$$

Considering the heat required (H_m) in converting unit mass of filler wire from its solid state at ambient temperature, T_a, to melting point, T_m, the T_i can be expressed as

$$T_i = \left[\left\{(H_i - H_m)/C_{p(l)}\right\}\right] + T_m \quad (4.14)$$

The first term of this expression denotes change in temperature (ΔT_i) which is the measure of degree of superheating of a metal drop at the time of its detachment from electrode tip.

$$\Delta T_i = T_i - T_m \quad (4.15)$$

During flight of a droplet from electrode tip to weld pool it loses heat [2, 21, 22] primarily by convection, H_{cv}, and radiation, H_r per unit mass of filler metal. However, the process of heat loss from a droplet in flight is a quite complex phenomenon dictated by several aspects of welding conditions. They are primarily considered as variation in shape and size [21] of arc column, deviation in flight path of a droplet from the vertical central axis of arc column and nature of variation in diminishing temperature of arc column [19] from the point of initiation to the weld pool at different welding parameters. All these aspects particularly affect the pattern of heat loss of droplet in the passage of its transfer through the arc column. To maintain a simplicity in assessment of heat loss with sufficient accuracy for all practical purposes in determining weld quality affecting its properties, it is determined on the basis of temperature difference between the droplet and ambient temperature. Therefore, heat content, H_{de}, per unit mass of droplets at the time of deposition can be estimated [21] as

$$H_{de} = H_i - (H_{cv} + H_r) \tag{4.16}$$

The temperature, T_{de}, of a droplet at the time of deposition in weld pool can be expressed as a function of its heat content per unit mass as well as H_m as follows.

$$H_{de} = \left[C_{p(s)}(T_m - T_O) + L_h + C_{p(l)}(T_{de} - T_m)\right] \tag{4.17}$$

$$T_{de} = \left[\left\{H_{de} - C_{p(s)}(T_m - T_O) - L_h\right\}/C_{p(l)}\right] + T_m \tag{4.18}$$

$$T_{de} = \left[(H_{de} - H_m)/C_{p(l)} + T_m\right] \tag{4.19}$$

First term of the expression of T_{de} is denoting a change in temperature (ΔT_{de}) representing the measure of the degree of superheating of droplet at the time of its deposition in weld pool.

$$\Delta T_{de} = T_{de} - T_m \tag{4.20}$$

Heat loss per unit mass of filler metal during flight of droplets from electrode tip to weld pool due to convection, H_{cv}, and radiation, H_r, are considered as

$$H_{cv} = [\alpha_h(T_i - T_O)A_S\tau]/M_t \tag{4.21}$$

and

$$H_r = \left[\varepsilon\sigma Ti^4 A_S\tau\right]/M_t \tag{4.22}$$

where, α_h is heat transfer coefficient of droplets in flight, ε is the emissivity, σ is the Stefan-Boltzmann constant and τ is the flight time of droplet. The A_s and M_t are total surface area of the droplets and mass of filler wire transferred per pulse respectively as stated in Sect. 3.4. The distance travelled by a droplet during its flight through the arc length, L, can be estimated as follows as a function of the V_i, a_{cc} and τ as also stated earlier in Sect. 3.4.

$$L = \left[V_i\tau + \left[(a_{cc} + g)\tau^2\right]/2\right] \tag{4.23}$$

The maximum axial velocity, V_{max}, of the electro-magnetically induced plasma jet at peak current of pulsation [21, 23] accelerates the droplet during its flight through the plasma stream along the arc column. At this point it is assumed that the current density, j_g, in the plasma of arc column [24] truly represents the most active area of plasma within the arc.

$$V_{mx} = \left[\left(\mu_0 I_p j_g\right)/(2\pi\rho_g)\right]^{1/2} \tag{4.24}$$

$$j_g = [I_{eff}/A_w] \tag{4.25}$$

The axial velocity of plasma falls rapidly in radial direction. Therefore effective velocity, V_{eff}, of the plasma is assumed [23] to be half of the maximum axial velocity as stated earlier in Eq. 3.51, when the aerodynamic drag force, F, on a droplet within the arc column is expressed [23] as

$$F = \frac{\left(\pi V_{eff}^2 \rho_g C_d D^2\right)}{8} \tag{4.26}$$

During transfer of droplet, the drag coefficient C_d maintains a good empirical correlation [23] with the Reynolds number, R_e, estimated by the Eqs. 3.49 and 3.50 respectively. The heat transfer coefficient, α_h, is expressed [21] as a function of Nusselt number (N_u), thermal conductivity of gas shielding (λ_a) and diameter of droplet (D).

$$\alpha_h = [(N\lambda_a)/D] \tag{4.27}$$

$$N_u = 2.0 + 0.6\ R_e^{0.5} P_r^{0.33} \tag{4.28}$$

$$P_r = [(C_p\eta)/\lambda_a] \tag{4.29}$$

C_p and η are the specific heat of argon plasma and viscosity of ionized argon respectively. The heat content of the total filler metal transferred per pulse, H_{dp}, and per unit time, H_f, is expressed as follows.

$$H_{dp} = H_{de}M_t \tag{4.30}$$

$$H_f = H_{de}M_t f \tag{4.31}$$

In spite of the assumption that metal transfer primarily takes place during peak current of P-GMAW, a possibility of metal transfer during its base current duration cannot be ignored. It may especially happen when the base current and its duration is high enough to melt and detach the molten drop from the filler wire. In case of larger duration of a base current than the minimum time required for drop transfer at it, the metal transfer in pulse off period may be considered inevitable. However, the metal transfers at base current in globular form of comparatively larger diameter with lesser in number than that observed at the peak current period. Thus, under such circumstance it is also necessary to estimate the heating of weld pool through transfer of superheated filler metal during the base current of a pulsed process. Heat transfer within the peak and base periods can be estimated by solving the Eqs. 4.5–4.31. The heat transfer by superheated filler metal within the duration of peak current, H_{fpp}, and base current, H_{fpb}, of a pulse can be expressed as follows.

$$H_{fpp} = H_{dep}M_{tp} \tag{4.32}$$

and

$$H_{fpb} = H_{deb} M_{tb} \tag{4.33}$$

where, H_{dep} is heat content per unit mass of filler metal at the time of deposition during peak current period, H_{deb} is heat content per unit mass of filler metal at the time of deposition during base current period, M_{tp} is mass of filler metal transferred during peak current period and M_{tb} is mass of filler metal transferred during base current period. As a result total heat input by superheated filler metal transferred per pulse, H_{dp}, and per unit time, H_f, can be expressed as

$$H_{dp} = H_{fpp} + H_{fpb} \tag{4.34}$$

and

$$H_f = H_{dp} f \tag{4.35}$$

The total heat, H, transferred per pulse to the weld pool in P-GMAW process can be expressed as.

$$H = H_{AW} + H_f \tag{4.36}$$

Considering the knowledge of heat flow in a weld as discussed in Sect. 1.3.1, the temperature, T_d, at any point (x(ξ), y, z) of a weld (Fig. 1.3) due to heat transfer by superheated filler metal per unit time in pulsed current GMW can be expressed [13] by considering it as a point heat source.

$$T_d = \frac{Q_f}{2.\pi.k_t} e^{-\lambda v \xi} \left[\frac{e^{-\lambda v R}}{R} + \sum_{m=1}^{m=\infty} \frac{e^{-\lambda v R_m}}{R_m} + \frac{e^{-\lambda v R'_m}}{R'_m} \right] + T_o \tag{4.37}$$

$$\lambda = 1/(2a) \tag{4.38}$$

where, k_t is the thermal conductivity of base metal and the R, R_m and R'_m are expressed as stated earlier in Eqs. 1.8, 1.11 and 1.12 respectively.

During P-GMA welding using aluminium and Al–Mg alloy filler wires the H_{aw}, H_f and total heat (H) transferred to the weld pool estimated by the Eqs. 4.3, 4.31 and 4.36 respectively reveals some interesting variation as a function of ϕ and I_m as shown in Fig. 4.1a, b respectively. At a given I_m the increase of ϕ from 0.05 to 0.1 relatively enhances the arc heat, heat of super heated filler metal and total heat transferred to the weld pool followed by a reduction in them with a further increase of it up to 0.3 or 0.4. However, at a given ϕ the increase in I_m enhances them significantly [25]. Thus, the variation in ϕ and I_m is having significant capacity to influence the thermal behaviour of weld pool governing the geometry, microstructure and properties of weld.

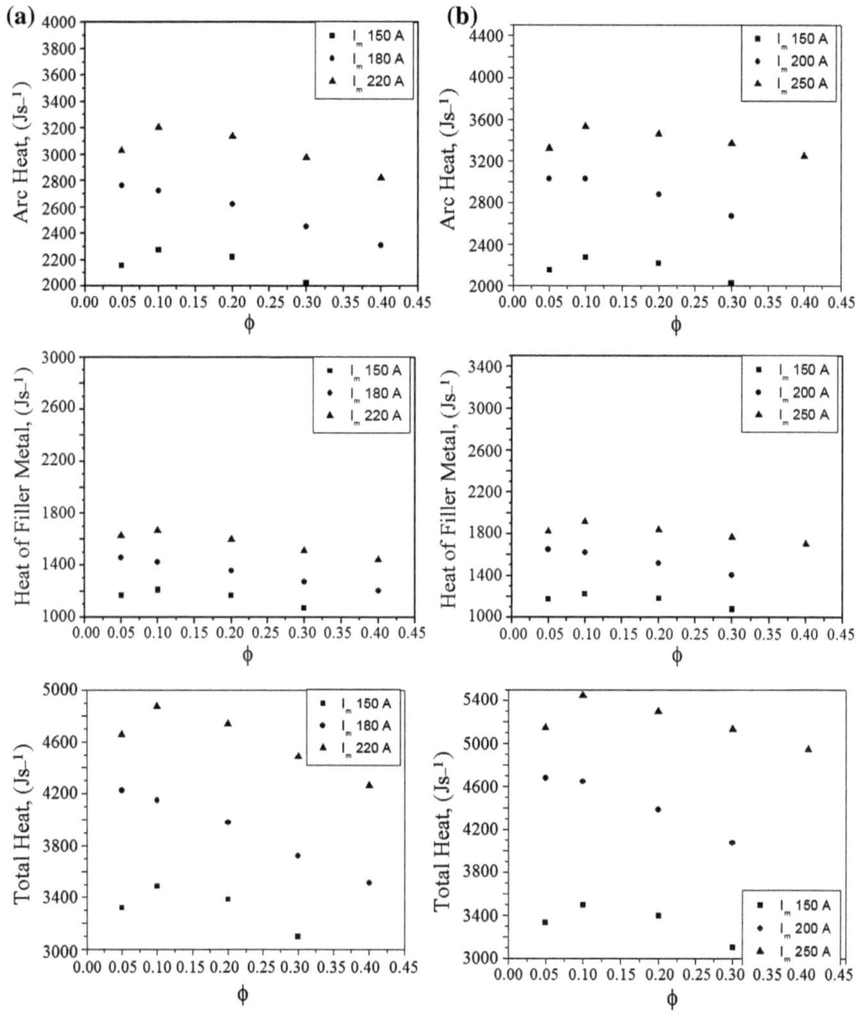

Fig. 4.1 Effect of ϕ on heat transfer to the weld pool of **a** aluminium and **b** Al–Mg alloy at different mean currents

4.2 Influence of Metal Transfer

The thermal behaviour of GMA weld to a great extent depends upon the transfer mechanism of heat by super heated droplets pushed inside the molten weld pool. The total impact of transferring droplets on weld pool determines the depth of penetration of it [26]. In spray transfer at peak current of P-GMAW process the kinetic energy of tiny droplets may not be sufficient to impart enough impact force on surface of a weld pool that can create a pressure on it to penetrate. But it can elastically stretch the surface film of molten metal apparently to a paraboloidal shape as shown in Fig. 4.2. Thus, it may be assumed that immediately after the velocity of pushing droplet comes to rest with no kinetic energy the weld pool surface starts regaining its shape. However, during movement of droplet at relatively lower velocity towards end of surface stretching of molten weld pool than its initial velocity it may consequently receive additional energy from strike of a following droplet moving freely with full energy through the depressed zone on pool surface. Under these circumstances the drop may subsequently break the surface tension of molten metal. In such cases the droplets may merge inside a cavity formed into the weld pool. Accordingly, a cavity on the molten weld pool can exist in case of a low time interval between successive droplet transfer than the time required to disperse and reform the molten pool surface.

4.2.1 Cavity Formation in Weld Pool

The kinetic energy imparts by a droplet on weld pool is largely spent on deforming the surface of the molten metal in the direction of developing a cavity (Fig. 4.2) in it [19]. The total work done, w, in creation of a paraboloidal cavity of focal length r_1 and depth h can be estimated as

$$w = \int_0^h \left(p \, . \, dV_c + 2 \, . \, \pi \, . \, r_c \, . \, \gamma \, . \, \sqrt{\left(1 + \frac{r_1}{h}\right)} \, . \, dh \right) \tag{4.39}$$

First term of the integral denotes work done against hydrostatic pressure and the second term represents work done in creating a new surface. At the projected surface of cavity, the hydrostatic pressure, volume, of the paraboloidal cavity and radius at the projected surface of the cavity denoted as p, V_c and r_c, respectively are expressed as follows

$$p = \rho_m g h \tag{4.40}$$

$$r_c^2 = 4 r_1 h \tag{4.41}$$

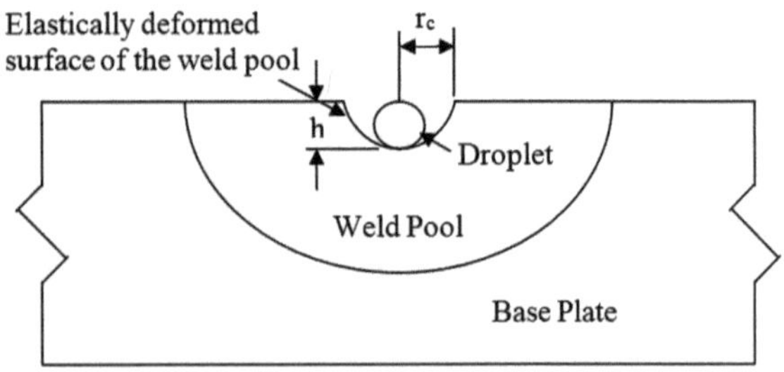

Fig. 4.2 Mechanism of cavity formation

$$V_c = 2\pi r_1 h^2 \tag{4.42}$$

where, ρ_m is density of molten metal at the weld pool. In the process of extending the cavity inside weld pool by penetration of droplet, the differentiation of Eq. 4.42 gives

$$dV_c = 4\pi r_1 h dh \tag{4.43}$$

In consideration of the above expressions the Eq. 4.39 for total work done in creating a paraboloidal cavity can be resolved as follows.

$$w = \int_0^h \left(\rho_m g h 4\pi r_1 h dh + 2\pi\gamma\sqrt{4 r_1 h}\sqrt{\left(1 + \frac{r_1}{h}\right)} dh\right) \tag{4.44}$$

$$w = \frac{4\pi r_1 \rho_m g h^3}{3} + 2\pi\gamma\sqrt{r_1}\left(\sqrt{h}\sqrt{(r_1 + h)} + r_1 \ln\left(\frac{\sqrt{h} + \sqrt{(r_1 + h)}}{\sqrt{r_1}}\right)\right) \tag{4.45}$$

The kinetic energy, E, of the droplet contributing to work done, w, can be expressed as a function of mass of a droplet, m_d, and its velocity, V_{de}, at the time of vertically down hand striking to weld pool as follows, according to the M_t, N_D and V_{de} as stated earlier in Sects. 3.4 and 2.2.2.

$$w = \frac{1}{2} m_d V_{de}^2 \tag{4.46}$$

$$m_d = \frac{m_t}{N_D} \tag{4.47}$$

Considering the kinetic energy of droplet is totally spent in work done to deform the weld pool surface into the paraboloidal cavity, the Eqs. 4.45 and 4.46 can be equated as follows. The depth of cavity, h, formed by the impact of droplet on weld pool can be determined by solving the above expressions, while the focal length of paraboloidal cavity, r_1, is assumed equal to radius of the droplet.

$$\frac{1}{2} m_d V_{de}^2 = \frac{4\pi r_1 \rho_m g h^3}{3} + 2\pi\gamma\sqrt{r_1}\left(\sqrt{h}\sqrt{(r_1 + h)} + r_1 \ln\left(\frac{\sqrt{h} + \sqrt{(r_1 + h)}}{\sqrt{r_1}}\right)\right) \tag{4.48}$$

4.2.2 *Temperature Distribution in Weld*

Temperature at any point of a weld can be estimated by superimposing the analytical solution of distributed heat source of arc heating given in Eq. 4.4, on the solution of the point heat source of superheated filler metal (Eq. 4.37) acting at the depth of cavity formed in weld pool. The temperature, T, at a distance R with respect to central axis of the welding arc is estimated as follows using the combined heat source technique.

$$T = \frac{Q_f}{2.\pi.k_t} e^{-\lambda.v.\xi}.\left[\frac{e^{-\lambda.v.R}}{R} + \sum_{n=1}^{n=\infty}\left(\frac{e^{-\lambda.v.R_n}}{R_n} + \frac{e^{-\lambda.v.R'_n}}{R'_n}\right)\right] + T_d \tag{4.49}$$

where, R, R_n and R'_n is estimated as

$$R = \sqrt{\xi^2 + y^2 + (z-h)^2} \tag{4.50}$$

$$R_n = \sqrt{(2.n.d-(z-h))^2 + \xi^2 + y^2} \tag{4.51}$$

$$R'_n = \sqrt{(2.n.d+(z-h))^2 + \xi^2 + y^2} \tag{4.52}$$

4.3 Thermal Nature of Welding Process

The thermal nature of P-GMAW process is largely controlled by the heat content and temperature of molten filler metal at the point of deposition. But, it critically depends upon heat gain at the time of detachment from the electrode and heat loss during flight of a droplet from the electrode tip to weld pool, based on the pulsed parameters. Under a given shielding of GMAW process operated at a specific gap between the electrode tip to work piece the heat loss in flight of a droplet is primarily considered as a function of its temperature at the time of detachment from the electrode tip, surface area and flight time. The increase of surface area enhances the heat loss whereas the increase of velocity reduces the heat loss of droplet during its flight. At different pulse parameters the total heat loss during flight of molten metal transfers per pulse (H_{tl}) as a summation of the heat loss due to convection (H_{cv}) and radiation (H_r) is largely dictated by the influence of a competitive process of opposite nature affecting the heat loss. The opposite nature of heat loss arises due to increase of both the A_s and V_i with the increase of ϕ at a given I_m as shown in Figs. 3.61 and 3.62 respectively, where the increase of A_s enhances the heat loss by providing more area of contact for heat transfer but the increase of V_i reduces the loss by allowing less time for heat transfer.

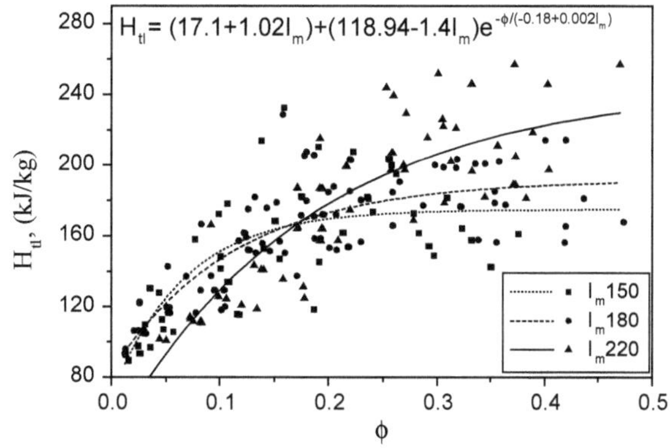

Fig. 4.3 At different I_m effect of ϕ on estimated total heat loss (H_{tl}) during flight of molten Al–Mg filler metal transferred per pulse

To understand the effect of pulse parameters on thermal behaviours of P-GMAW, the model expressions for various thermal characteristics of the process as stated above are correlated with wide range of ϕ from different power sources used in weld deposition of Al–Mg alloy at different I_m. It is observed [21] that increase of ϕ enhances the H_{tl} significantly but its rate of increase becomes comparatively slower at higher ϕ. At a given ϕ the H_{tl} enhances significantly with the increase of I_m (Fig. 4.3). However, the H_{de} and T_{de} estimated by the Eqs. 4.17 and 4.18 respectively, reduces steadily with the increase of ϕ and at a given ϕ the increase of I_m also reduces them significantly as shown in Figs. 4.4 and 4.5 respectively. The H_{tl}, H_{de} and T_{de} follows a good empirical correlations with the ϕ and I_m keeping a coefficient of more than 0.9.

$$H_{tl} = (17.1 + 1.02I_m) + (118.94 - 1.4I_m)e^{-}\phi/(-0.18 + 0.002I_m) \qquad (4.53)$$

$$H_{de} = (2917.83 - 3.76I_m) - (103.66 + 0.86I_m)\phi \qquad (4.54)$$

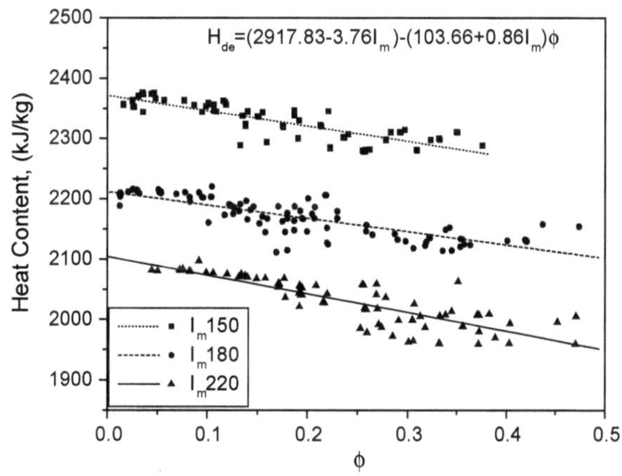

Fig. 4.4 At different I_m effect of ϕ on estimated heat content (H_{de}) per unit mass of droplet during deposition of Al–Mg filler metal

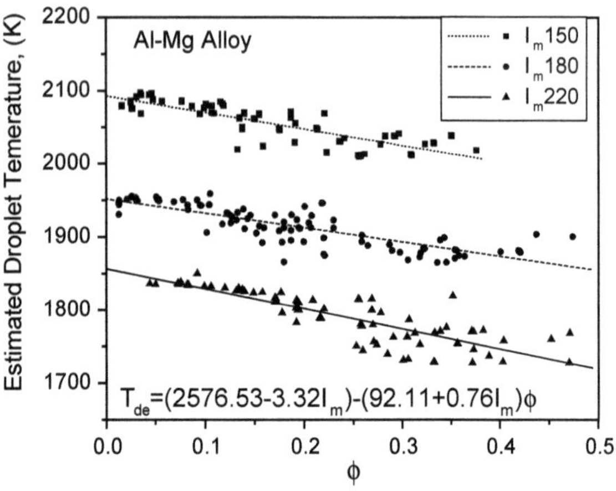

Fig. 4.5 At different I_m effect of ϕ on estimated temperature (T_{de}) of droplet during deposition of Al–Mg filler metal

$$T_{de} = (2576.53 - 3.32 I_m) - (92.11 + 0.76 I_m)\phi \qquad (4.55)$$

But, in contrast to the observations on aluminium alloy (Fig. 4.5) the estimated T_{de} of mild steel weld deposition using 1.2 mm diameter filler wire enhances [27] with the increase of ϕ in the range of about 0.05–0.25 at given mean currents of 100 and 130 A Fig. 4.6. However, in agreement to that found in case of aluminium alloy the estimated T_{de} reduces with the increase of mean current at a given ϕ. The estimated drop temperatures (T_{de}) at different pulse parameters are found to lie in the similar range as observed in other works [1, 7] on GMAW. It appears that the effect of ϕ on T_{de} at its lower range within about 0.2 is unclear. The criticality in the trend of influence of ϕ on T_{de} should be studied further with different materials but, with the present state of knowledge it is understood that at a given mean current the T_{de} reduces with the increase of ϕ from its maximum at around a ϕ of 0.2.

The heat content (Eq. 4.30) of total weld metal deposited per pulse, H_{dp} enhances significantly with decreasing pulse frequency (f) at a given I_m as shown in Fig. 4.7 due to eventual increase of ϕ at longer t_b. At a given ϕ and f the increase of

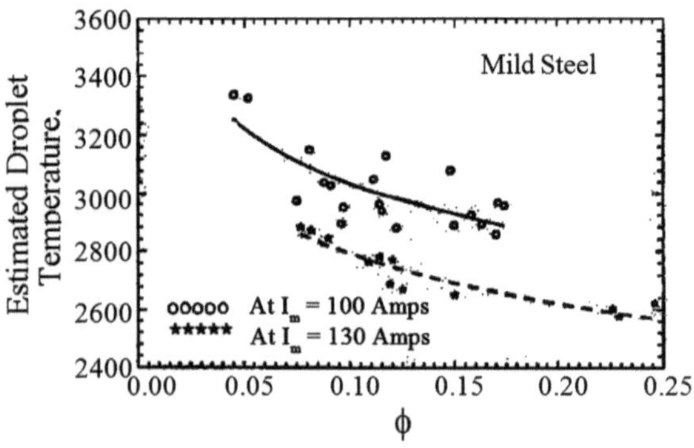

Fig. 4.6 At different I_m influence of ϕ on estimated temperature of mild steel droplet deposited at different mean currents

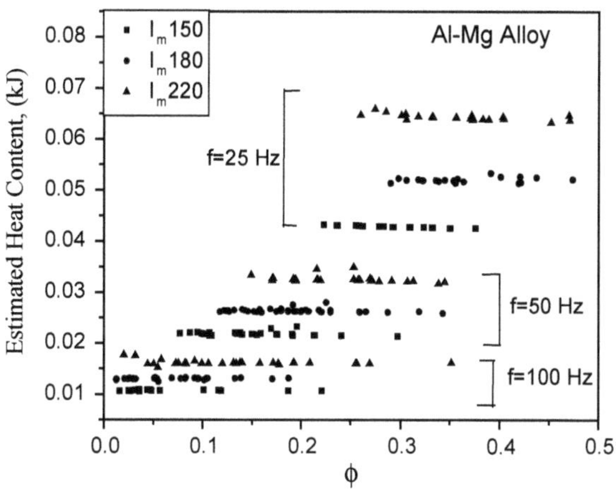

Fig. 4.7 At different I_m effect of ϕ on estimated heat content (H_{dp}) of total Al–Mg filler metal deposited per pulse

I_m enhances the H_{dp}, especially at lower f. But at a given I_m and f the increase of ϕ practically does not influence H_{dp}. In spite of the decrease of H_{de} with the increase of ϕ the significant enhancement of H_{dp} with ϕ possibly happens due to considerable increase of mass transfer per pulse (M_t), estimated by Eq. 3.41, as shown in Fig. 4.8.

The variation in temperature (T_i) of mild steel droplet at the time of detachment with a change in pulse parameter is estimated by using model expressions as a function of measured melting rate as stated in Sect. 1.2.2. It shows a fairly good agreement with its theoretical estimation by Eq. 4.13 maintaining a difference of 4.19 ± 3.52% as shown in (Table 4.1). The estimated (Eq. 4.17) heat content (H_{de}) per unit mass of weld metal at the time of deposition in weld pool also goes well in agreement with the reported [2, 9, 10] calorimetric measurements of its heat

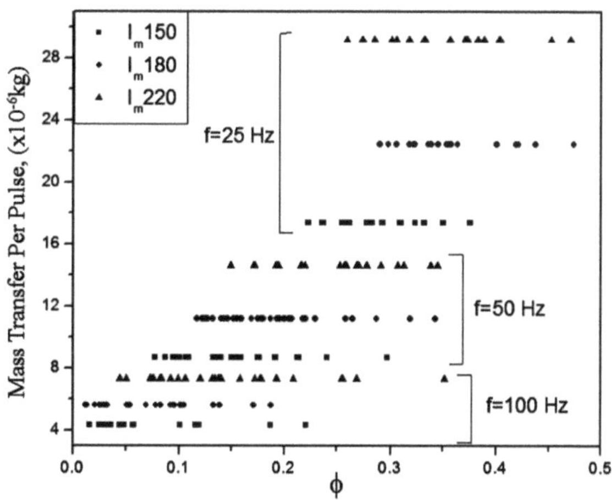

Fig. 4.8 At different I_m effect of ϕ on estimated mass (M_t) of Al–Mg filler metal transferred per pulse

Table 4.1 Temperature of droplet estimated by different models in case of P-GMAW with 1.2 mm diameter mild steel filler wire

Mean current, I_m (A)	Effective current, I_{eff} (A)	Estimated temperature of droplet, T_i (K)		
		Model (Eq. 4.13)	Model [3, 37, 38]	Difference (%)
130	135	2687	2700	−0.51
130	138	2752	2700	1.86
130	143	2861	2700	5.61
130	143	3097	2843	8.19
130	147	2949	2700	8.43
130	152	3182	2971	6.63
130	156	2535	2572	−1.46
170	187	3108	2944	5.26
180	183	2676	2635	1.51
180	183	2676	2635	1.51
180	186	2725	2635	3.31
180	189	3011	2962	1.63
180	190	2866	2658	7.27
180	197	2910	2635	9.45

Table 4.2 Estimated and measured heat content of droplet during deposition of different materials in P-GMAW

Filler wire	Mean current, I_m (A)	Peak current, I_p (A)	Base current, I_b (A)	Frequency, f (Hz)	Heat content of droplet, H_{de} (kJ/kg)			Reference
					Estimated	Measured	Diff. (%)	
M.S.	166	380	50	125	1596	1800	12.84	[9]
M.S.	300	350	50	100	1957	1804	7.82	[2]
M.S.	200	350	50	100	1835	1785	2.74	
M.S.	100	350	50	100	1750	1700	2.86	
S.S.	300	350	50	100	1891	1722	8.93	
S.S.	250	350	50	100	1765	1638	7.21	
S.S.	200	350	50	100	1639	1578	3.74	

content. This is observed (Table 4.2) during using 1.2 mm diameter mild steel and stainless steel filler wires in pulsed current GMAW. The difference between the estimated and measured heat content shows a minor variation of about $6.56 \pm 4.8\%$ and $6.63 \pm 2.64\%$ in case of using the mild steel and stainless steel filler wires respectively.

4.4 Thermal Behaviour of Weld

The thermal behaviour of weld is generally understood by the conditions of weld and heat affected zone (HAZ). This is because different nature of heating and cooling exists there in molten weld pool and solid metal adjacent to it. The basic understanding about the behaviour of molten weld pool and HAZ at different pulse parameters becomes clear to a great extent from their measured temperature and cooling rate by introducing thermocouple at desired locations (Fig. 4.9) of weld bead on plate deposition of various ferrous and non ferrous materials. The measured temperature assures the true characteristics of weld pool by avoiding direct contact of welding arc with the thermocouple through its placement at proper depth (D_T) in fused weld pool. The appropriate location of placement of the tip of thermocouple in fused weld pool was predetermined by model analysis of weld isotherm for melting point of base plate as described in previous chapter. The observed [25] weld isotherms of aluminium and Al–Mg alloy as shown in Fig. 4.10a, b respectively are considered largely realistic in view of the evaluation by considering their temperature dependent thermo-physical properties [28–32].

Thermal characteristics of weld deposit primarily depend upon energy input, but the P-GMAW can influence energy distribution in the process by controlling the pulse parameters affecting the heat buildup in weld pool [33]. Thus at a given energy input it may have ability to vary temperature of weld pool up to certain extent. The thermal characteristics of weld deposit has been found [25] to vary with a change in ϕ as a function of pulse parameters of I_p, I_b, f and t_b (Eq. 3.8) when it is considered at a given conventional energy input (E_c) per unit length.

$$E_c = [(VI_m/v)] \tag{4.56}$$

At a given I_m, arc voltage (V) and energy input controlled by welding speed (v) the change in thermal characteristics of the weld and HAZ with variation in pulse parameters justifies to assume that it happens exclusively by the influence of the latter.

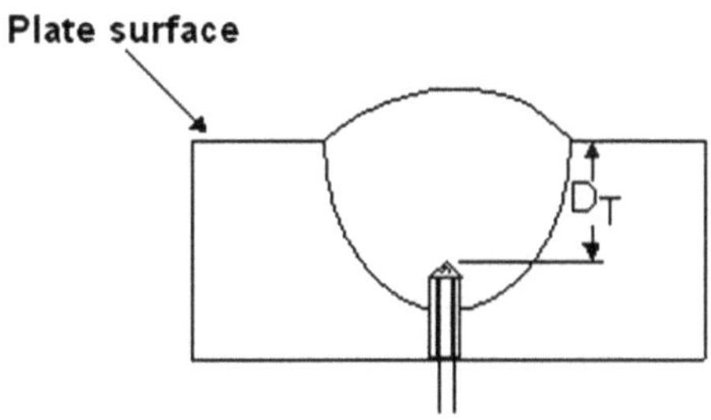

Fig. 4.9 Schematic diagram of placing of thermocouple in weld pool

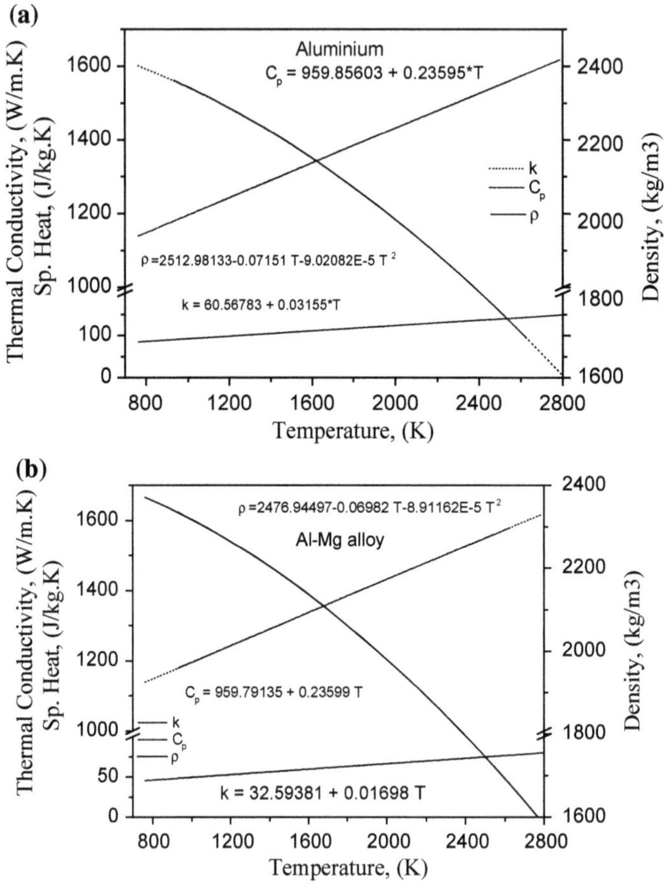

Fig. 4.10 Temperature dependent thermo-physical properties of **a** aluminium and **b** Al–Mg alloy

4.4.1 Temperature of Weld Pool

Measured temperature of geometrically different weld pool of aluminium and Al–Mg alloy at various welding parameters has been found satisfactorily in agreement to the theoretical temperatures of weld pool at the same welding parameters. The theoretical temperatures are evaluated on the basis of the double ellipsoidal heat source parameters such as a_h, b_h, c_{hf} and c_{hb} as stated earlier. The heat source parameters a_h and b_h are selected based on measured width and penetration of weld pool to find out the other two heat source parameters as $c_{hf} = a_h$ and $c_{hb} = 2c_{hf}$. Width and penetration of weld pool are estimated by employing different arbitrary values selected on the basis of the reported values [14, 34] as 6 and 8 mm for a_h and 1, 2 and 3 mm for b_h of Eq. 4.49. The measured and estimated values of width and penetration (depth of fusion) of weld pool of Al–Mg alloy at different heat source parameters, corroborating a given mean current of 200 A at varying ϕ (0.05–0.31) and heat input (6.3–10.2 kJ/cm), have been shown in Table 4.3. In view of the lowest (0.152%) average difference between the estimated and correspondingly measured width and penetration at a_h = 8 mm and b_h = 2 mm they are found [25] appropriate to use for estimation of temperature and geometry of weld pool at different welding parameters. Satisfactory applicability of these heat source parameters for prediction of temperature and weld pool geometry also of aluminium at different pulse parameters indicates that they are largely independent of welding parameters at high mean current for a given class (e.g. Al-base) of material.

The estimated (Eq. 4.49) and measured temperatures of weld pool of aluminium (ER1100) and Al–Mg alloy (ER5183) at its different depths in the range of 2.5–5.5 mm from the weld (plate) surface in 10 mm thick plates of similar compositions has been found practically well in agreement as shown in Fig. 4.11a, b respectively. At a given energy input the measured and estimated temperatures of weld poll of the aluminium and Al–Mg alloy as given in Tables 4.4 and 4.5 respectively varies significantly with pulse parameters but they lie in close approximation with a maximum deviation of about ±10%. The increase of I_m and energy input enhances the temperature of weld pool, but increase of ϕ significantly reduces the same in the line of similar observation on temperature (T_{de}) of droplet at the time of deposition as shown in Fig. 4.5. It amply justifies the understanding that the thermal characteristics of weld pool in P-GMAW significantly depend upon energy distribution in pulse system governed by the pulse parameters. The control of ϕ as a function of two terms of energy and time fraction, irrespective of the levels of I_p, I_b, f and t_b, can regulate the energy distribution in pulse system affecting the thermal characteristics of weld pool. The comparatively higher temperature of weld pool at location closer to its surface also justifies appreciable contribution of arc heating on temperature of weld pool.

An appropriate selection of double ellipsoidal heat source parameters of P-GMAW process can estimate temperature of weld pool with a reasonable accuracy. However, the Eq. 4.49 gives a comparatively poor estimation of weld pool temperature with considerable scattering with respect to its corresponding measured value at shallow depth within about 2 mm from its surface as shown in Fig. 4.12a, b. This has happened possibly because the consideration of superheated filler metal as a point heat source does not allow correct estimation of temperature at locations too close to the heat source [14, 20, 29, 35] as well as heterogeneous nature of arc heating.

The change in heat content and temperature of weld pool with respect to pulse parameters is liable to affect its cooling rate which has significant role in dictating the metallurgical characteristics of weld and its many other features including residual stresses. In the line of the earlier studies the measured cooling rate of aluminium and Al–Mg alloy weld metal (Fig. 4.9) has also been studied as a function of heat input, mean current and ϕ. At a given conventional energy input in the range of about 6.3–10.6 kJ/cm the effect of ϕ on cooling rate of weld at different mean currents measured in case of aluminium and Al–Mg alloy welds has been

Table 4.3 At a given mean current of 200 A width and depth of fusion at P-GMA weld deposition on Al–Mg alloy plate measured and estimated by considering different heat source parameters

Width/depth	ϕ	HI (kJ/cm)	Measured width and depth of fusion (mm)	Estimated width and depth of fusion at different $[a_h/b_h]$ (mm)				Difference (%) between measured and estimated width and depth at different $[a_h/b_h]$ (mm)			
				[6/2]	[8/1]	[8/2]	[8/3]	[6/2]	[8/1]	[8/2]	[8/3]
Width	0.05	6.3	11.3	11.23	11.55	11.34	11.15	0.62	−2.21	−0.35	1.33
	0.05	8.3	13.3	12.69	13.04	12.88	12.72	4.58	1.95	3.16	4.36
	0.05	10.2	15.7	13.95	14.3	14.18	14.02	11.14	8.92	9.68	10.70
	0.11	6.3	12	11.18	11.5	11.3	11.09	6.83	4.17	5.83	7.58
	0.11	8.3	13.15	12.63	12.97	12.82	12.64	3.95	1.37	2.51	3.88
	0.11	10.2	15.5	13.88	14.22	14.08	13.94	10.45	8.26	9.16	10.06
	0.21	6.3	11.3	10.85	11.14	10.92	10.72	3.982	1.41	3.36	5.13
	0.21	8.3	13.3	12.24	12.56	12.38	12.21	7.970	5.56	6.92	8.19
	0.21	10.2	14.55	13.43	13.76	13.6	13.46	7.70	5.43	6.53	7.49
	0.31	6.3	9.5	10.46	10.7	10.48	10.28	−10.10	−12.63	−10.31	−8.21
	0.31	8.3	12.15	11.79	12.07	11.88	11.71	2.96	0.66	2.22	3.62
	0.31	10.2	14.1	12.9	13.2	13.06	12.89	8.51	6.38	7.37	8.58
Depth	0.05	6.3	4.6	5.055	4.87	4.89	4.93	−9.89	−5.87	−6.30	−7.17
	0.05	8.3	5.7	5.935	5.73	5.74	5.78	−4.12	−0.52	−0.70	−1.49
	0.05	10.2	6.6	6.79	6.55	6.58	6.61	−2.88	0.76	0.30	−0.15
	0.11	6.3	4.4	5.02	4.84	4.86	4.89	−14.09	−10.00	−10.45	−11.25
	0.11	8.3	5.2	5.89	5.68	5.7	5.73	−13.27	−9.23	−9.61	−10.29
	0.11	10.2	6.4	6.73	6.49	6.51	6.55	−5.15	−1.40	−1.72	−2.34
	0.21	6.3	4.3	4.85	4.665	4.68	4.72	−12.79	−8.49	−8.84	−9.77
	0.21	8.3	5.4	5.665	5.45	5.47	5.51	−4.91	−0.92	−1.29	−2.04
	0.21	10.2	6.9	6.44	6.2	6.22	6.26	6.67	10.14	9.85	9.27
	0.31	6.3	4.15	4.65	4.46	4.49	4.52	−12.05	−7.59	−8.19	−8.91
	0.31	8.3	4.8	5.41	5.2	5.22	5.25	−12.71	−8.33	−8.75	−9.37
	0.31	10.2	6.1	6.11	5.88	5.9	5.93	−0.16	3.61	3.28	2.78
Average difference (%)								−1.11	−0.36	0.15	0.50

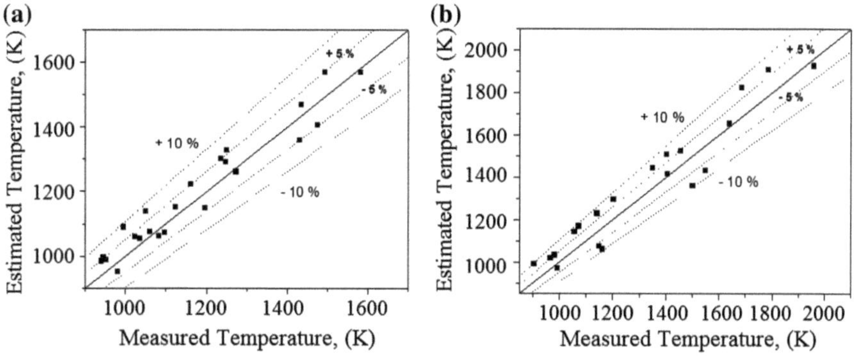

Fig. 4.11 Comparison of measured and estimated temperatures of bead on plate weld pool at its different depths within 2–5.5 mm from the weld surface in (**a**) aluminium (**b**) Al–Mg alloy at varied I_m = 150–250 A, ϕ = 0.05–0.4 and energy input = 6.2–11.2 kJ/cm

Table 4.4 Estimated and measured temperatures of weld pool of aluminium in different pulse current conditions

ϕ	I_m (A)	Energy input (kJ/cm)	Depth from plate surface (mm)	Temperature (K)			Mean difference ± SD
				Estimated	Measured	Difference (%)	
0.05	180	6.5	4	985	939	−4.90	4.28 ± 2.42%
0.1	220	6.9	2.5	1469	1434	−2.44	
			3	1260	1272	0.94	
0.2	220	6.9	2.5	1405	1475	4.75	
			3.5	1063	1083	1.85	
0.3	220	6.9	3	1148	1195	3.93	
0.05	180	8.7	4.5	988	952	−3.78	
0.4	180	8.7	4	953	981	2.85	
0.1	220	9.1	2.5	1568	1494	−4.95	
			3	1358	1430	5.03	
0.2	220	9.1	3	1301	1234	−5.43	
0.3	220	9.1	4	996	945	−5.40	
0.05	180	10.8	4	1139	1051	−8.37	
			5	981	922	−6.40	
0.2	180	10.8	4	1090	995	−9.55	
			5	941	862	−9.16	
0.05	220	10.6	4	1151	1124	−2.40	
0.2	220	10.6	4.5	1055	1036	−1.83	
0.4	220	10.6	3	1326	1249	−6.16	
			4	1074	1097	2.10	
0.05	220	11.2	3.5	1291	1246	−3.61	
			4.5	1075	1060	−1.42	
0.2	220	11.2	2.5	1568	1581	0.82	
			3.5	1223	1161	−5.34	
0.3	220	11.2	4	1059	1023	−3.52	

Table 4.5 Estimated and measured temperatures of the weld pool of Al–Mg alloy in different pulse current conditions

ϕ	I_m (A)	Energy input (kJ/cm)	Depth from plate surface (mm)	Temperature (K)			Mean difference ± SD
				Estimated	Measured	Difference (%)	
0.05	200	6.3	4	1144	1058	−8.13	6.22 ± 2.81%
0.3	200	6.3	4	1034	983	−5.19	
0.05	250	6.3	3	1508	1403	−7.48	
			4	1062	1162	8.61	
0.2	250	6.3	3	1524	1455	−4.74	
			4	1075	1150	6.52	
0.4	250	6.3	3	1445	1350	−7.04	
0.4	250	8.2	3	1654	1640	−0.85	
0.05	150	10.4	5.5	992	905	−9.61	
0.05	200	10.2	4.5	1297	1205	−7.63	
0.2	200	10.2	4.5	1230	1142	−7.71	
			5.5	1020	968	−5.37	
0.3	200	10.2	4.5	1170	1074	−8.94	
			5.5	972	994	2.21	
0.05	250	10.3	3	1908	1786	−6.83	
			4	1417	1406	−0.78	
0.2	250	10.3	3	1927	1960	1.68	
			4	1432	1549	7.55	
0.4	250	10.3	3	1824	1687	−8.12	
			4	1360	1501	9.39	

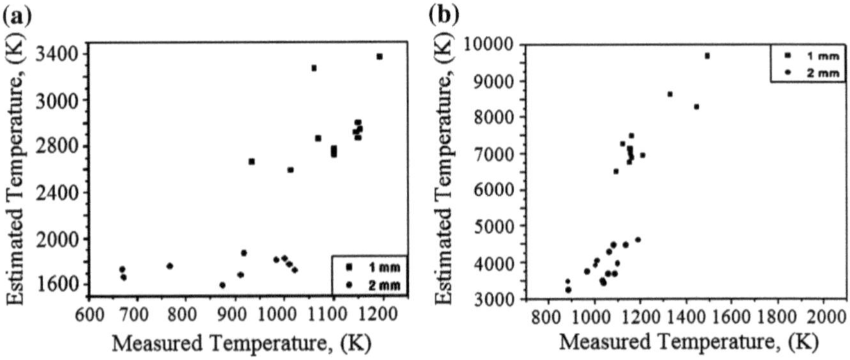

Fig. 4.12 Comparison of measured and estimated temperatures of bead on plate weld pool at different depths within 1 and 2 mm from the weld pool surface in (**a**) aluminium (**b**) Al–Mg alloy at varied I_m = 150–250 A, = 0.05–0.4 and energy input = 6.2–11.2 kJ/cm

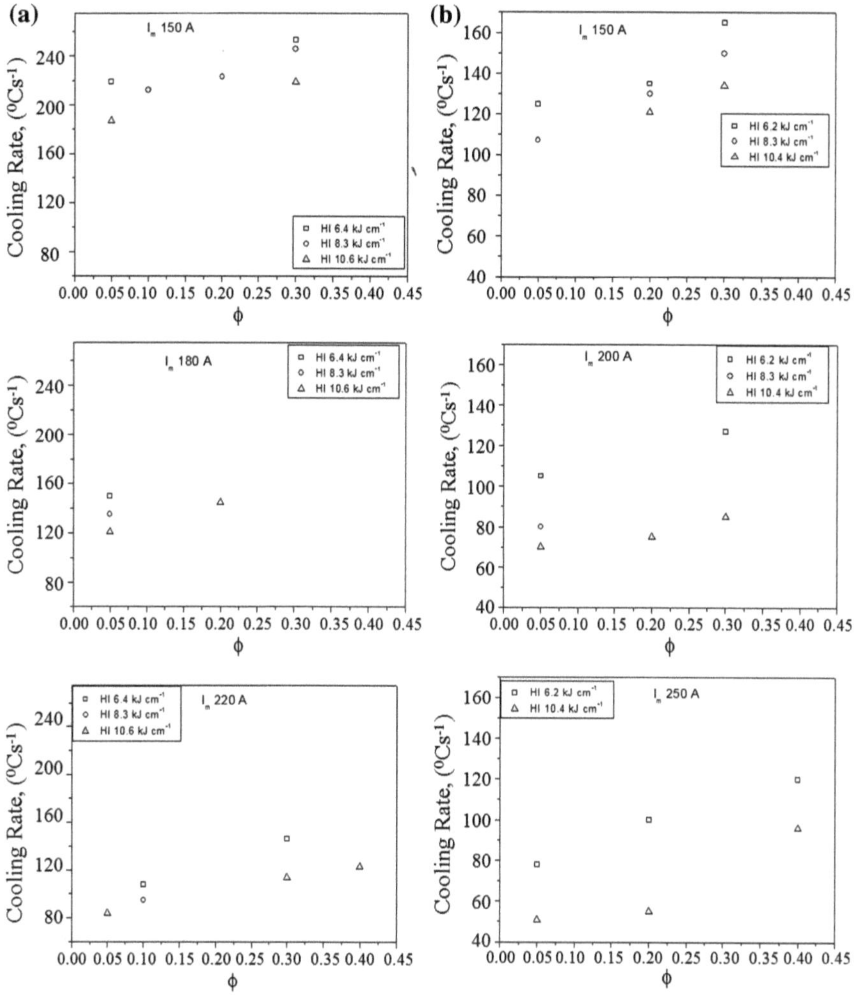

Fig. 4.13 Effect of ϕ on cooling rate of bead on plate weld of **a** aluminium and **b** Al–Mg alloy with varying energy inputs at different mean currents

shown in Fig. 4.13a, b respectively. At a given mean current and energy input cooling rate of weld metal relatively enhances with the increase of ϕ but, it reduces with an increase of energy input at a given mean current and ϕ and also with the increase of mean current at a given energy input and ϕ.

4.4.2 Temperature of HAZ

In bead on plate weld of aluminium deposited at varied pulse parameters, the measured temperatures of different locations of HAZ within 0.2–2.9 mm away from the theoretically predicted fusion line are found [25] in close approximation to their corresponding values estimated by the Eq. 4.49 as shown in Fig. 4.14. The Table 4.6 shows that the average difference of estimated values from their corresponding measured values is of the order of 4.35 ± 2.95%. Increase of energy

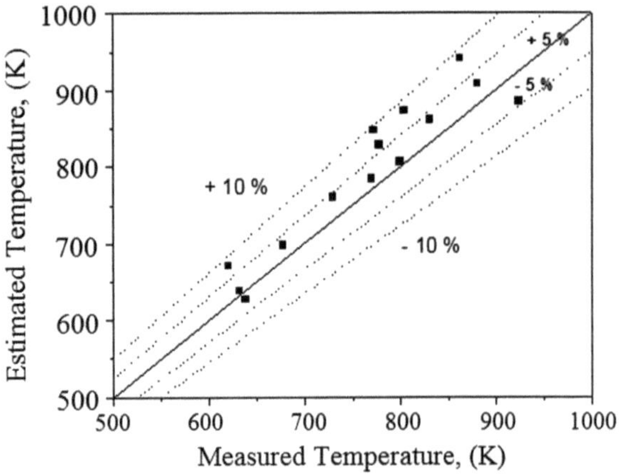

Fig. 4.14 Comparison of measured and estimated temperatures of HAZ at different distances within 0.2–2.9 mm from fusion line in bead on plate weld deposition of aluminium at varied I_m = 150–250 A, ϕ = 0.05–0.4 and energy input = 6.2–11.2 kJ/cm

Table 4.6 Estimated and measured temperatures of HAZ in aluminium at different welding parameters

ϕ	I_m (A)	Energy input (kJ/cm)	Distance from predicted fusion line (mm)	Temperature (K)			Mean difference ± SD
				Estimated	Measured	Difference (%)	
0.2	180	6.5	2.9	638	632	−0.95	4.35 ± 2.95%
0.4	180	6.5	1.8	698	678	−2.95	
			2.8	627	639	1.88	
0.3	220	6.9	2.1	671	621	−8.05	
0.2	220	6.4	0.8	806	799	−0.88	
0.4	220	6.4	1.1	760	730	−4.11	
0.05	180	8.7	0.6	861	830	−3.73	
0.2	180	8.7	0.9	828	778	−6.43	
0.2	220	9.1	0.6	885	924	4.22	
0.3	220	9.1	0.6	847	772	−9.72	
0.4	180	10.8	0.5	873	804	−8.58	
			1.5	784	770	−1.82	
0.3	220	11.2	0.2	908	880	−3.18	

input enhances the temperature of any location of HAZ adjacent to fusion line, but at a given energy input the temperature of a given location of HAZ significantly decreases with an increase of ϕ especially at higher mean current. This is also in the line of the effect of ϕ on decreasing the T_{de} and temperature of weld pool as discussed earlier. Thus, the possibility of theoretical estimation of the temperature of HAZ with respect to pulse parameters, in order to draw weld isotherm, is examined and found useful. In some other work [36] on P-GMAW of Al–Zn–Mg alloy at similar order of energy input (8.9 kJ/cm) and ϕ, the measured temperature of HAZ at 1.0 mm away from fusion line as shown in Table 4.7 is found almost

Table 4.7 Measured temperature of HAZ at 1.0 mm from fusion line in Al–Zn–Mg alloy at different ϕ of various welding parameters

ϕ	Energy input (kJ/cm)	Temperature (K)
0.04	8.9	823
0.187	8.9	815
0.33	8.9	812

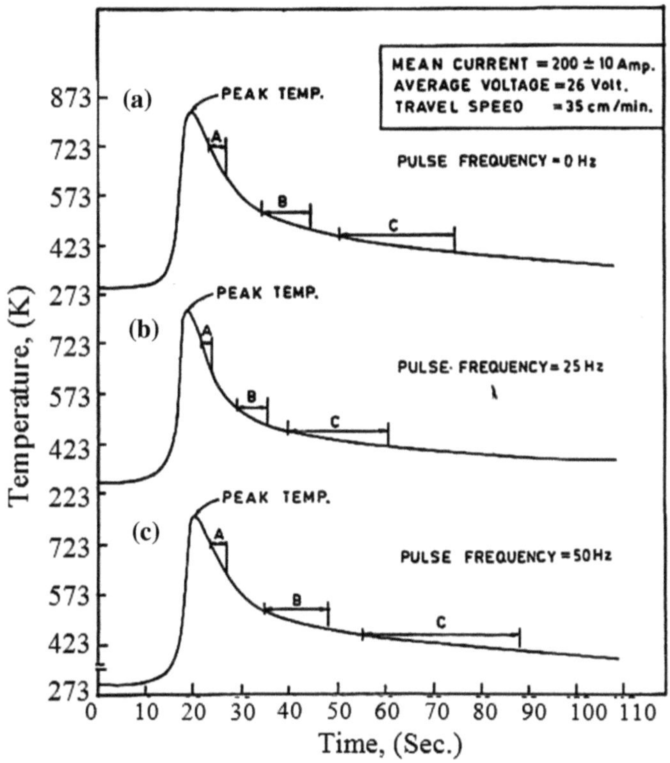

Fig. 4.15 At different pulse frequencies typical weld thermal cycle at different range of temperatures of HAZ at 1 mm away from the fusion line in 6 mm thick butt weld of Al–Zn–Mg alloy; *A* 723–623 K, *B* 523–473 K and *C* 453–403 K

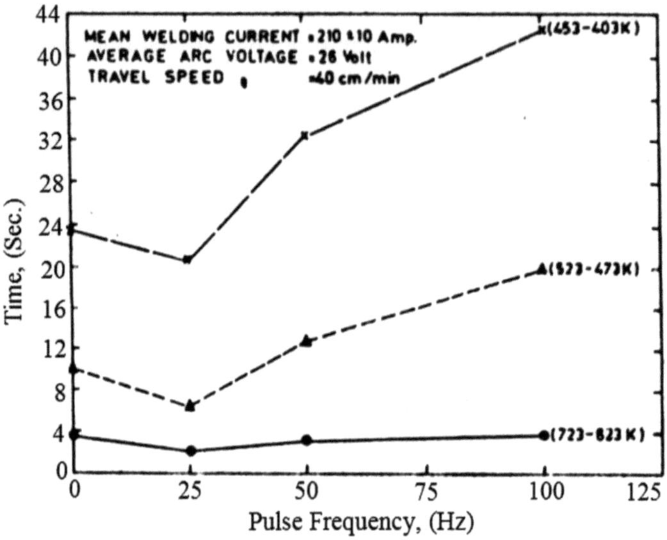

Fig. 4.16 Effect of pulse frequency on cooling time at different range of temperatures in HAZ of 1 mm from fusion line

similar to the temperature of HAZ given in Table 4.6. The thermal cycle of HAZ varies with a change of pulse parameters as it is reflected in cooling characteristics through different range of temperatures relevant to the governance of grain coarsening and phase transformation of base material Fig. 4.15. The increase of pulse frequency, which generally lowers the value of factor ϕ due to increase of (ft_b), retains the HAZ of Al–Zn–Mg alloy in its over ageing and conventional ageing temperature ranges of 523–473 K and 453–403 K respectively for longer time as shown in Fig. 4.16. This kind of behaviours primarily happens due to variation in size and shape of weld pool which results a change in width of HAZ at different welding parameters.

References

1. Allum, C.J., Quintino, L.: Control of fusion characteristics in pulsed current MIG welding, Part-II, Simple model of fusion characteristics. Metal Constr **17**(5), 314R–317R (1985)
2. Maruo, H., Hirata, Y.B.: Study of pulsed MIG welding. IIW Doc. SG 212-585-84, Welding Department, Osaka University, Japan, July 1984
3. Waszink, J.H., Piena, M.J.: Experimental investigation of drop detachment and drop velocity in GMAW. Weld. J. **65**(11), 289s–298s (1986)
4. Amin, M.: Pulse current parameters for arc stability and controlled metal transfer in arc welding. Metal Constr **15**(5), 272–287 (1983)
5. Jilong, Ma., Apps, R.L.: MIG transfer discovery of importance to industry. Weld. Metal Fab. **14**(9), 307–316 (1982)

6. Christensen, N., de Davies, V.L., Gjermundsen, K.: Distribution of temperatures in arc welding. Br. Weld. J. **12**(2), 54–75 (1965)
7. Kim, Y.S., Eagar, T.W.: Metal transfer in pulsed current gas metal arc welding. Weld. J. **72** (7), 279s–287s (1993)
8. Choi, S.K., Yoo, C.D., Kim, Y.S.: Dynamic simulation of metal transfer, Part I: Globular and spray transfer modes. Weld. J. **77**(1), 38s–44s (1998)
9. Ueguri, S., Hara, K., Komura, H.: Study of metal transfer in pulsed GMA welding. Weld. J. **64**(8), 242–250s (1985)
10. Jones, L.A.: Dynamic electrode forces in gas metal arc welding. Ph.D. dissertation, Massachusetts Institute of Technology, Cambridge, Mass (1996)
11. Matsuda, F., Ushio, M., Itonaga, K., Yokoo, T.: Pulsed CO_2 welding, Tech. Report, IIW Doc. 212-586-84 (1984)
12. Jacobsen, N.: Monopulse investigation of drop detachment in pulsed gas metal arc welding. Phys. D, Appl. Phys. **25**, 783–797 (1992)
13. Rosenthal, D.: The theory of moving sources of heat and its application to metal treatments. ASME Trans. **68**, 849–866 (1946)
14. Nguyen, N.T., Ohta, A., Matsuoka, K., Suzuki, N., Maeda, Y.: Analytical solutions for transient temperature of semi infinite body subjected to 3 – D moving heat sources. Weld. J. **78**, 265s–274s (1999)
15. Smati, Z.: Automatic pulsed MIG welding. Weld. J. **65**(1), 38s–44s (1986)
16. Needham, J.C.: Pulse controlled consumable electrode welding arcs. Br. Weld. J. **12**(4), 191–197 (1965)
17. Randhawa, H.S.: Investigation into positional welding of structural steel using pulse current GMAW process. Ph.D. Dissertation, UOR, Roorkee (1999)
18. Kumar, S., Bhaduri, S.C.: Three dimensional finite element modeling of gas metal arc welding. Met. Mater. Trans. B **25B**, 435–441 (1994)
19. Lancaster, J.F.: The physics of welding, 1st edn, pp. 134–230. IIW, Pergamon Press, New York (1984)
20. Goldak, J., Chakravarti, A., Bibby, M.: A new finite element model for welding heat sources. Met. Mater. Trans. **15B**, 299–305 (1984)
21. Quintino, L., Allum, C.J.: Pulsed GMAW: interactions between process parameters, Part I. Weld. Met. Fab. **16**(4), 126–129 (1984)
22. Ghosh, P.K., Gupta, S.R., Randhawa, H.S.: Analytical studies on characteristics of vertical up bead on plate weld deposition using pulsed current GMAW. Int. J. Joining Mater. **12**(3), 76–85 (2000)
23. Alum, C.J.: Metal transfer in arc welding as a varicose instability: II. Development of model for arc welding. J. Phys. D Appl. Phys. **18**, 1447–1468 (1985)
24. Ghosh, P.K., Dorn, L., Hübner, M., Goyal, V.K.: Arc characteristics and behaviour of metal transfer in pulsed current GMA welding of aluminium alloy. J. Mater. Process. Tech. **194**(4), 163–175 (2007)
25. Goyal, V.K., Ghosh, P.K., Saini, J.S.: Analytical studies on thermal behaviour and geometry of weld pool in pulsed current gas metal arc welding. J. Mater. Process. Tech. **209**, 1318–1336 (2009)
26. Essers, W.G., Walter, R.: Heat transfer and penetration mechanisms with GMA and plasma-GMA welding. Weld. J. **60**(2), 37–42s (1981)
27. Randhawa, H.S., Ghosh, P.K., Gupta, S.R.: Geometric characteristics of pulsed current positional GMA weld. ISIJ Int. **38**(4), 276–284 (1998)
28. Metals Handbook: Properties and Selection: Nonferrous Alloys and Pure Metals, 10th edn, vol. 2, p. 1099. ASM International, Metal Park, Ohio (1979)
29. Jeong, S.K., Cho, H.S.: An analytical solution to predict the transient temperature distribution in fillet arc welds. Weld. J. **76**, 223s–232s (1997)
30. Von Horn, K.R.: Aluminium, vol. 1, Physical Metallurgy and Phase Diagrams, vol. 4. ASM, Metal Park, Ohio (1968)
31. Mondolfo, L.F.: Aluminium Alloys: Structure and Properties. Butterworths, London (1979)

32. Smith, C.S., Palmer, E.W.: Thermal and electric conductivities of copper alloys. Trans. AIME **117**, 225–243 (1935)
33. Welding Alcoa Aluminum, Aluminum Company of America, Pittsburgh, Pennsylvania: Weld. J. **67**, 18s–27s (1988). (1988)
34. Saha, S.C., Ghosh, A.K., Malhotra, S.L.: Heat transfer in welding—a numerical approach. Indian Weld. J. **26**(4), 8–12 (1993)
35. Komanduri, R., Hou, Z.B.: Thermal analysis of the arc welding process: Part I. General solutions. Met. Mater. Trans. **31B**, 1353–1370 (2000)
36. Gupta, P.C., Ghosh, P.K., Visa, S.: Influence of pulse frequency on the properties of HAZ in pulsed MIG welded Al-Zn-Mg alloy. In: Proceedings of the International Conference on Welding Technology in Developing Countries Present Status and Future Needs, 1-71-77, 26–28 Sept 1988
37. Kim, Y.S., Mceligot, D.M., Eagar, T.W.: Analysis of electrode heat transfer in gas metal arc welding. Weld. J. **70**(1), 20s–31s (1991)
38. Jilong, Ma., Apps, R.L.: Analysing metal transfer during MIG welding. Weld. Met. Fab. **15** (4), 119–128 (1983)

Chapter 5
Geometry of Pulse Current GMA Weld

Abstract Variation in geometry of P-GMA weld as a function of pulse parameters has been discussed in order to understand the control of welding process to produce desired weld geometry of required chemistry for an acceptable weld quality. The mechanism of the control of weld geometry governed by its basic aspects as penetration, width, reinforcement, form factor, toe angle and area along with the base metal fusion and dilution has been explained as a function of pulse parameters especially with respect to ϕ and energy input. The matter has been discussed considering the cases of bead on plate weld deposition and weld cladding in flat position using solid filler wire. The cases of bead on plate weld deposition have also been considered for the flux cored filler wire and also for the out of position deposition. The geometrical aspect of weld joint as a function of pulse parameters has been addressed for flat and positional butt welding of ferrous and non ferrous metals.

Keywords Geometry of weld deposit · Bead on plate deposition · Pulse parameters · Cladding · Solid electrode · Flux cored wire · Fusion · Dilution · Flat and out of position deposition · Weld joint

The weld geometry often qualifies GMA welding process for an application through its characteristics of producing required fusion in base metal, amount of weld deposition and regulation of the flow of liquid metal to control its shape at various parameters. The shape, size or extent of fusion of base plate is primarily governed by the energy input as well as energy distribution in weld depending upon thermal and transfer behaviours of depositing metal as discussed earlier. Whereas the weld bead geometry is governed further by the amount and rate of metal deposition and its fluidity. In case of conventional continuous current GMAW it is generally understood that the bead geometry is primarily dictated by the welding current as a function of wire feed speed, arc voltage and welding speed due to their influence on deposition per unit length, area of deposition per unit length and fluidity of weld pool. But during P-GMAW at a given wire feed rate or mean current, arc voltage and welding speed the bead geometry varies with pulse parameters. For example a change in pulse frequency (f) or pulse duration (t_p) may affect the weld geometry

P.K. Ghosh, *Pulse Current Gas Metal Arc Welding*, Materials Forming, Machining and Tribology, DOI 10.1007/978-981-10-3557-9_5

primarily due to their significant influence on the peak (I_p) and base (I_b) of the pulse current. Thus, it regulates the characteristics of metal transfer at the peak current as well as the possibility of metal transfer at the base current and influences energy distribution in the welding process. Due to its simplicity, in most of the cases the basic understanding on the influence of welding parameters on weld geometry is popularly acquired by studying the features of weld bead on plate deposition. However, because of relatively complex nature of influence of pulse parameters on arc physics, behaviour of metal transfer and fluid flow in weld pool, computer simulation of these aspects has become of considerable interest to predict the weld shape in bead on plate deposition [1]. In the light of this knowledge a welding process control is suitably applied for preparation of weld joint of desired quality with respect to its geometry, chemistry and properties.

5.1 Weld Bead on Plate Deposition

The welding of aluminium alloy using P-GMAW produces practically spatter free weld which is commonly observed in case of conventional (0 Hz) GMAW, especially during welding at a current lower than transition current of drop to spray transfer. But due to significant influence of pulse parameters on temperature and behaviour of metal transfer in P-GMAW, a change in pulse parameters varies interval of weld deposition and forms fine to highly coarse ripples on the weld beads. However, the welding with a current pulse of higher frequency (≥ 100 Hz) and longer duration preferably with larger mean current significantly improves the situation. For example using mean current beyond 180 A with 1.6 mm diameter Al–Mg filler wire under argon shielding gives rise to appreciably smooth weld deposition [2] The influence of individual pulse parameters on bead geometry is difficult to understand because of appearance of significant irregularities in their response (Fig. 5.1). It happens due to interactive nature of pulse parameters posing different nature of influence on weld pool as it is typically reported [2] in case of the deposition of Al–Mg alloy on 10 mm thick Al–Zn–Mg alloy plate. Under such conditions the role of different pulse parameters on geometrical characteristics of weld deposition cannot be thoroughly realised. In this regard, consideration of summarised influence of pulse parameters defined by the factor ϕ along with the theoretical knowledge of various aspects of P-GMAW process as stated earlier may provide a more clear understanding of the process.

The performance of P-GMAW in fusion of base material at various pulse parameters can be basically understood [3–5] from a theoretically estimated shape and size of plate fusion as stated earlier and its superimposition on digitally traced geometry of fusion on transverse section of bead on plate deposition at a given parameter. A comparison of the theoretically estimated and actual shape and size of plate fusion in aluminium at a given welding parameter is typically shown in Fig. 5.2. The real and analytically predicted geometries of fusion especially about the width and depth of fusion are to a great extent of complementary in nature,

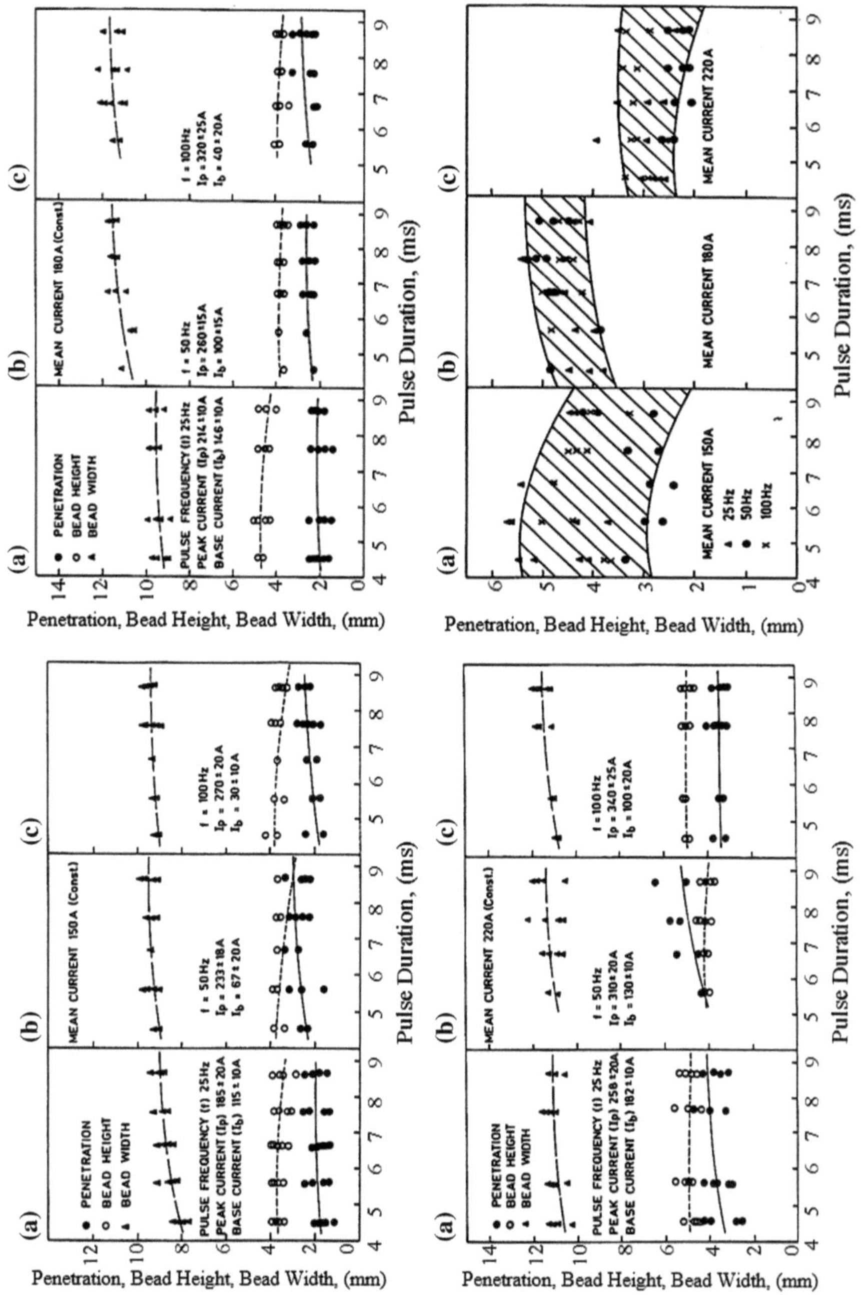

Fig. 5.1 Influence of pulse parameters on geometry of bead on plate deposition of Al–Mg alloy on 10 mm thick plate of Al–Zn–Mg alloy at different mean currents of **a** 150 A, **b** 180 A and **c** 220 A

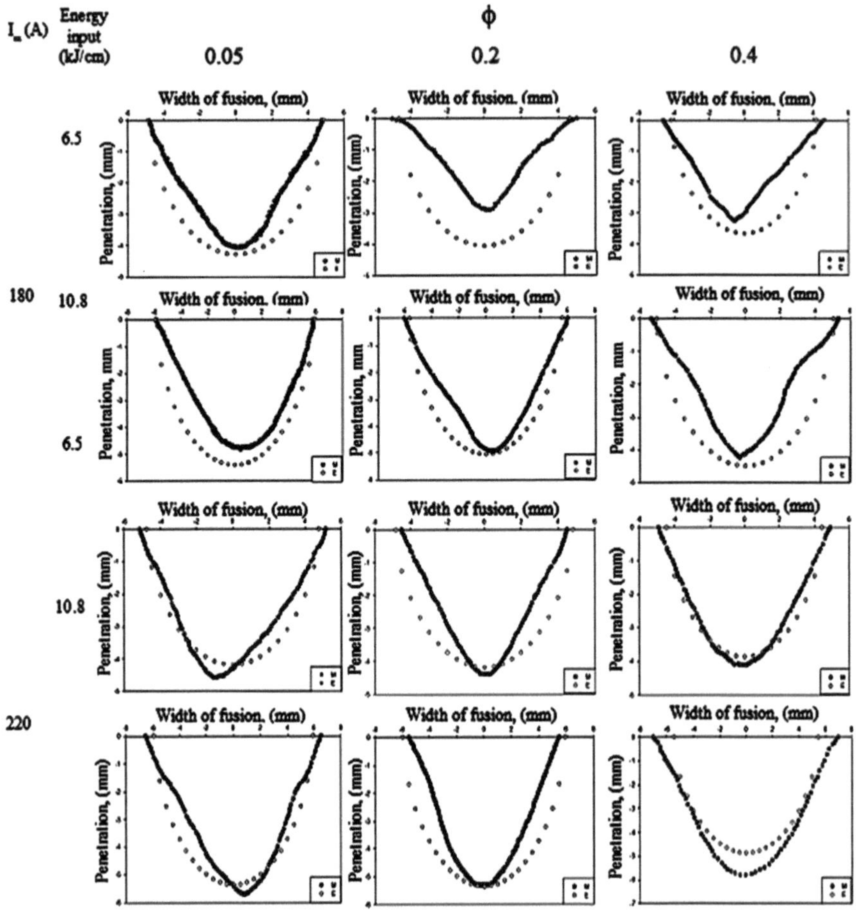

Fig. 5.2 Predicted and measured geometry of aluminium weld pool at different levels of ϕ, energy input and mean current

where the conventional energy input and the factor ϕ lies within 6.2–10.4 kJ/cm and 0.05–0.4 respectively. The similarity of real shape and size of the fused weld pool deviates from their theoretically predicted features with lowering of I_m and energy input below about 180 A and 8.0 kJ/cm respectively. At higher I_m beyond 200 A the estimated depth of fusion in case of both the aluminium and Al–Mg alloy broadly agrees to the measured one within a difference of $\pm 10\%$. In bead on plate weld deposition the weld geometry is generally studied by the depth of fusion or penetration (P_n), bead width (W_b), bead height (H_b) or reinforcement, weld toe angle (θ), area of plate fusion (A_f) and area of weld deposit (A_d) as schematically shown in Fig. 5.3.

Influence of energy input and the factor ϕ on the geometry of weld bead is quite significant. It reveals in transverse section as shown in Figs. 5.4a, b and 5.5a, b for

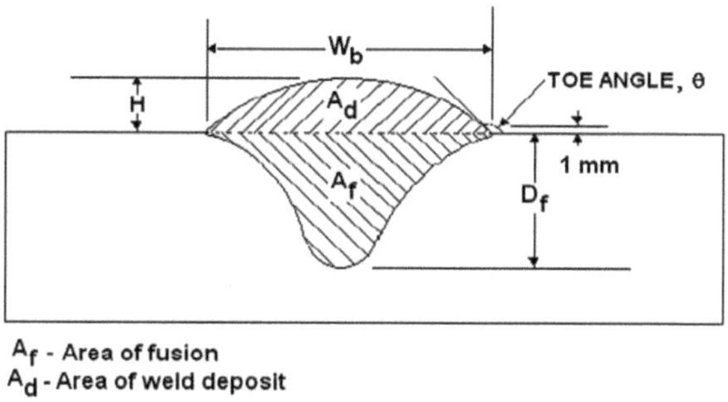

Fig. 5.3 Schematic diagram of bead on plate weld deposit

bead on plate deposition of aluminium and Al–Mg alloy at relatively low as well as high mean currents respectively. It is understood that depending upon parameters the primary aspects of weld geometry as weld penetration, bead width, bead height and weld toe angle changes appreciably within the commonly used range of parameters for the factor ϕ, mean current and heat input as of 0.05–0.4, 150–250 A and 6.3–10.6 kJ/cm respectively. In case of P-GMA welding a due knowledge about their correlations is necessary for controlling the weld appropriately to produce acceptable geometry.

5.1.1 Depth of Penetration

The depth of fusion as a measure of penetration is a function of conventional energy input and mean current as well as the pulse parameters involved in factor ϕ. The fusion vertically penetrating the base plate in bead on plate deposition significantly enhances with the increase of heat input or men current at a given condition of the remaining two out of the three conventional functional parameters in this regard. But, at a given energy input and mean current a relative enhancement of the depth of fusion in base plate takes place with the increase of ϕ from 0.05 to about 0.1 followed by a comparative reduction in it with a further increase of ϕ up to 0.3 or 0.4. These behaviours are clearly revealed in Fig. 5.6a, b for the aluminium and Al–Mg alloy respectively. As per Eq. 3.8, increase of pulse duration at a given peak current, base current and pulse frequency reduces ϕ. Accordingly as per summarized influence of pulse parameters the relative increase of penetration with the increase of pulse duration (Fig. 5.1) also justifies the observations of Fig. 5.6. It primarily happens due to change in arc heat (H_{aw}) and heat of superheated molten metal (H_f) affecting the total heat (H) transferred per pulse to the weld pool with the change of ϕ as stated earlier in Eqs. 4.3, 4.31 and 4.36 respectively as well as with the change in mean current. In view of the discussions of earlier chapters on the thermal and transfer behaviours of molten metal and cavity formation in weld pool

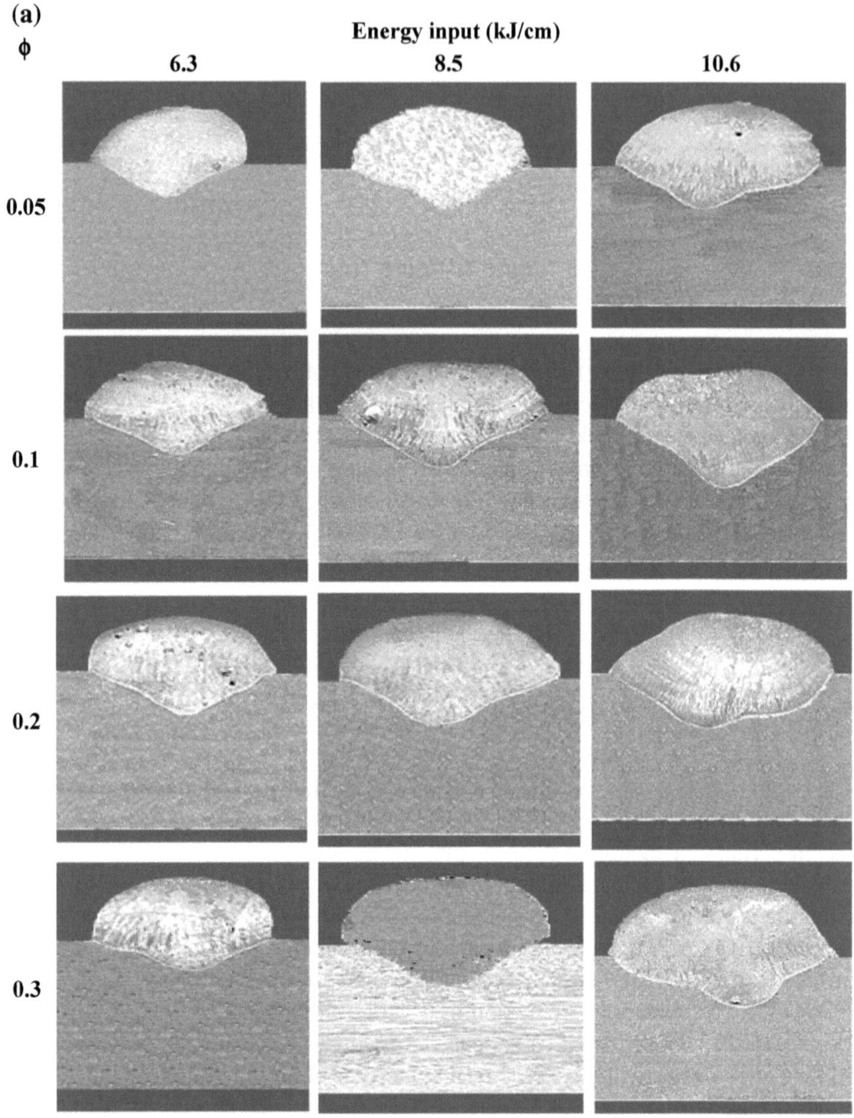

Fig. 5.4 **a** At a given mean current of 150 A typical geometry of bead on plate weld deposition of aluminium at different energy input and ϕ. **b** At a given mean current of 220 A typical geometry of bead on plate weld deposition of aluminium at different energy input and ϕ

it may be realized that the influence of pulse parameters on weld penetration is not very straight forward in nature. The complication arises due to interactive nature of varied degree of different kind of influence of parameters on it.

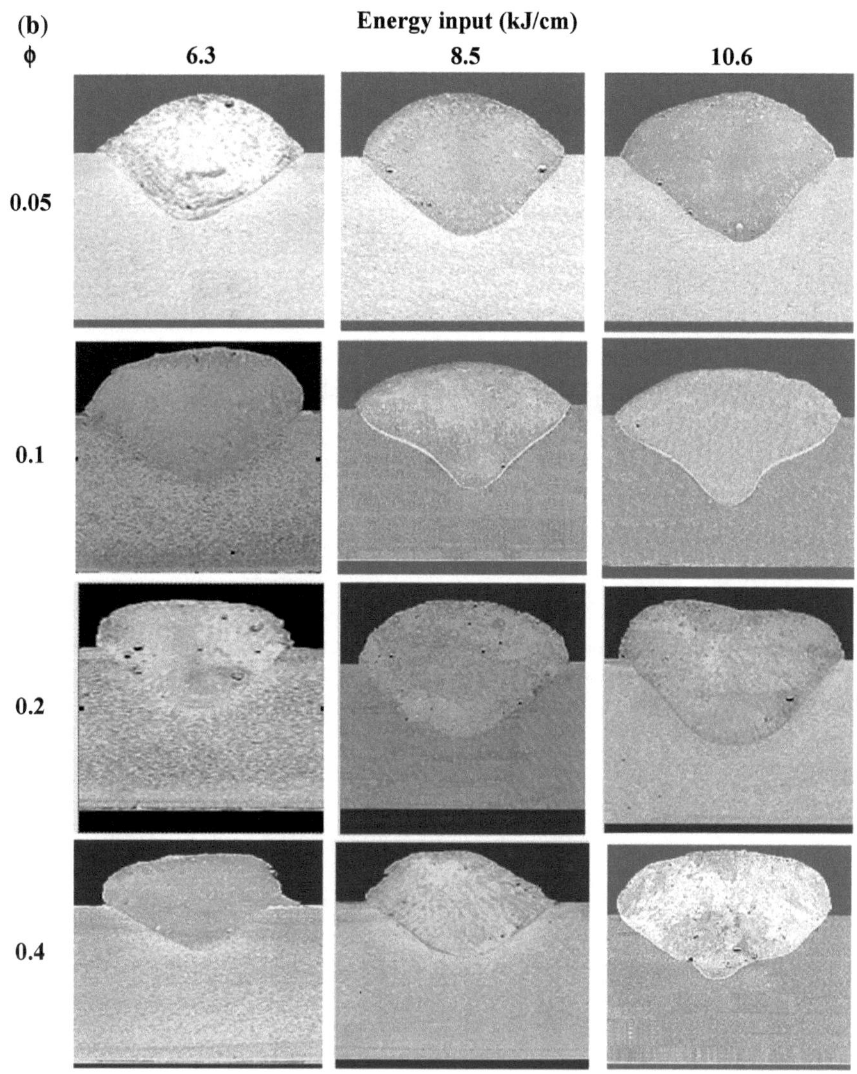

Fig. 5.4 (continued)

The increase of energy input enhances fluidity of molten weld pool facilitating deep penetration of transferring droplets depending upon more energetic V_{de} with respect to V_i and thus increases weld penetration. The general response of the increase of ϕ on reduction of penetration can be understood from its identical influence (Fig. 4.5) on temperature of the depositing molten metal which forms a

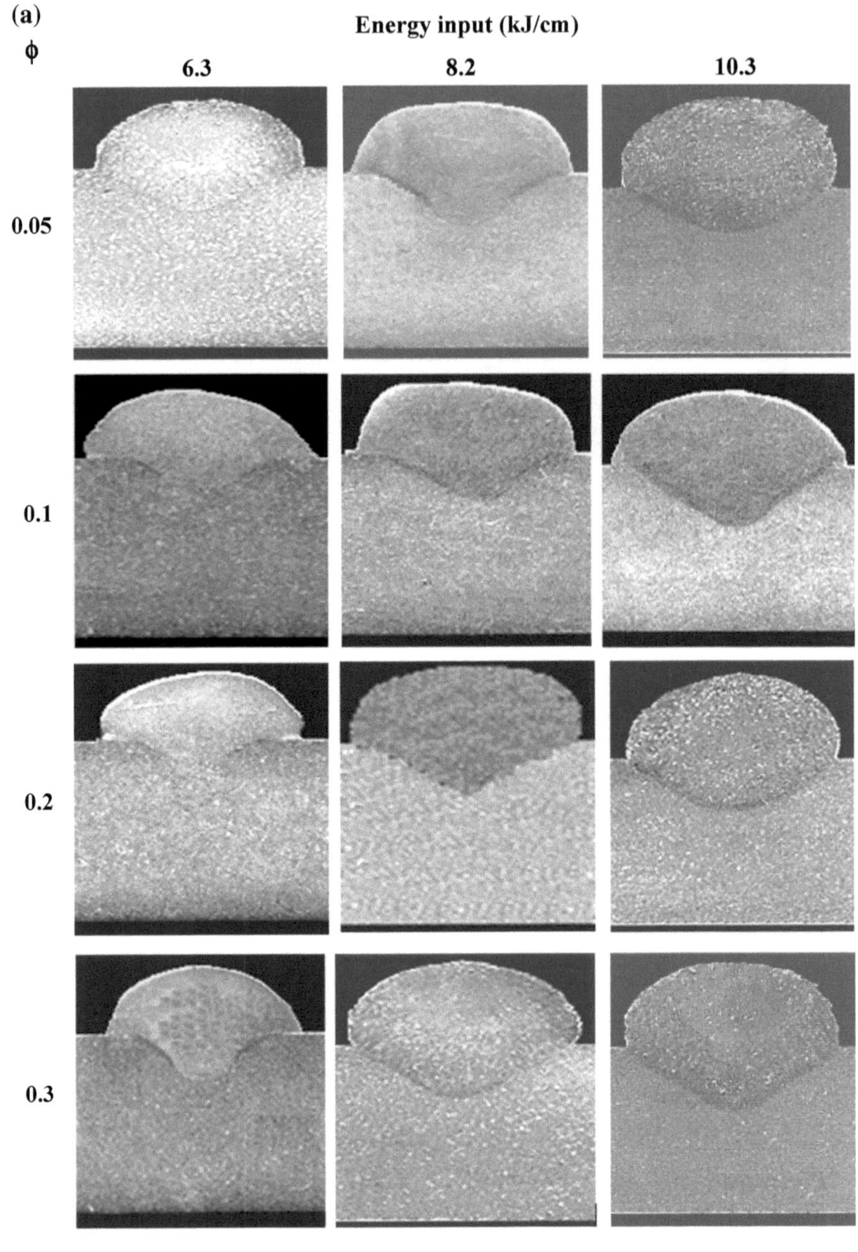

Fig. 5.5 **a** At a given mean current of 150 A typical geometry of bead on plate weld deposition of Al–Mg alloy at different energy input and ϕ. **b** At a given mean current of 250 A typical geometry of bead on plate weld deposition of Al–Mg alloy at different energy input and ϕ

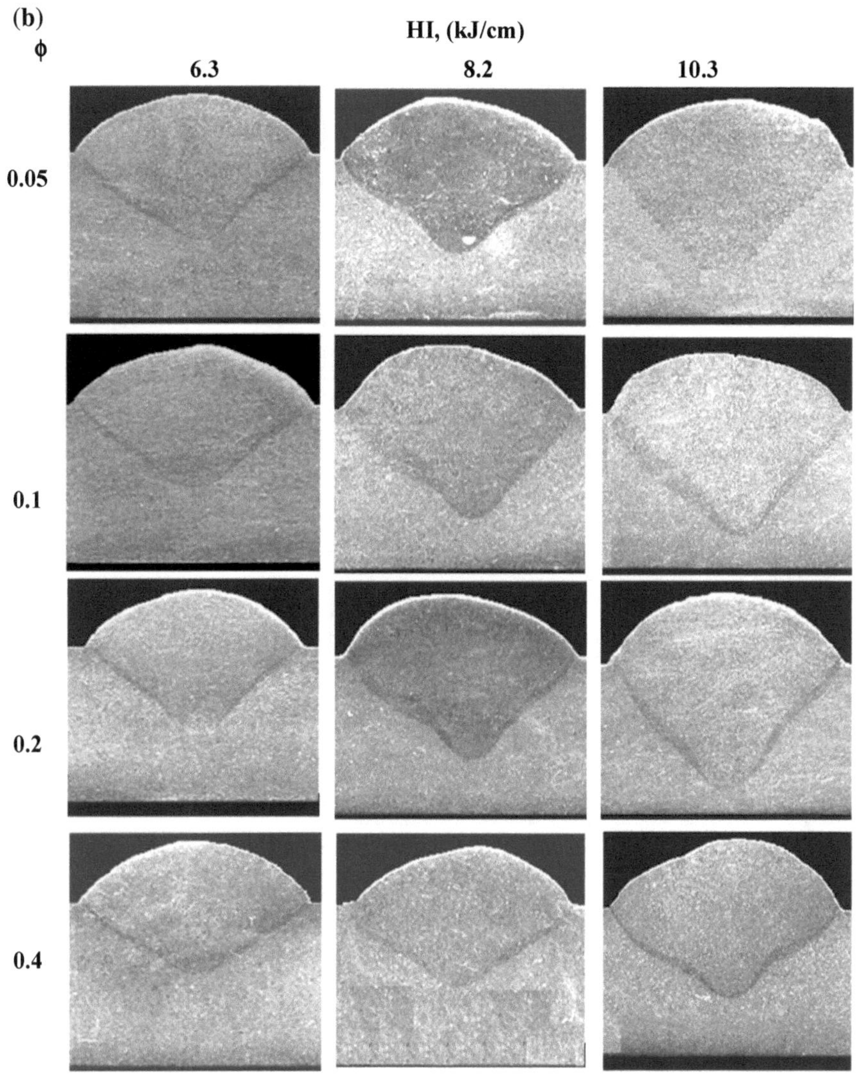

Fig. 5.5 (continued)

more viscous weld pool. At a given I_m the increase of ϕ exponentially enhances the V_i as shown in Fig. 3.62, which is suppose to increase penetration. But the influence of I_m on V_i is of opposite nature at the low and high values of ϕ holding a criticality at 0.1. Such empirical observations along with basic understanding of the

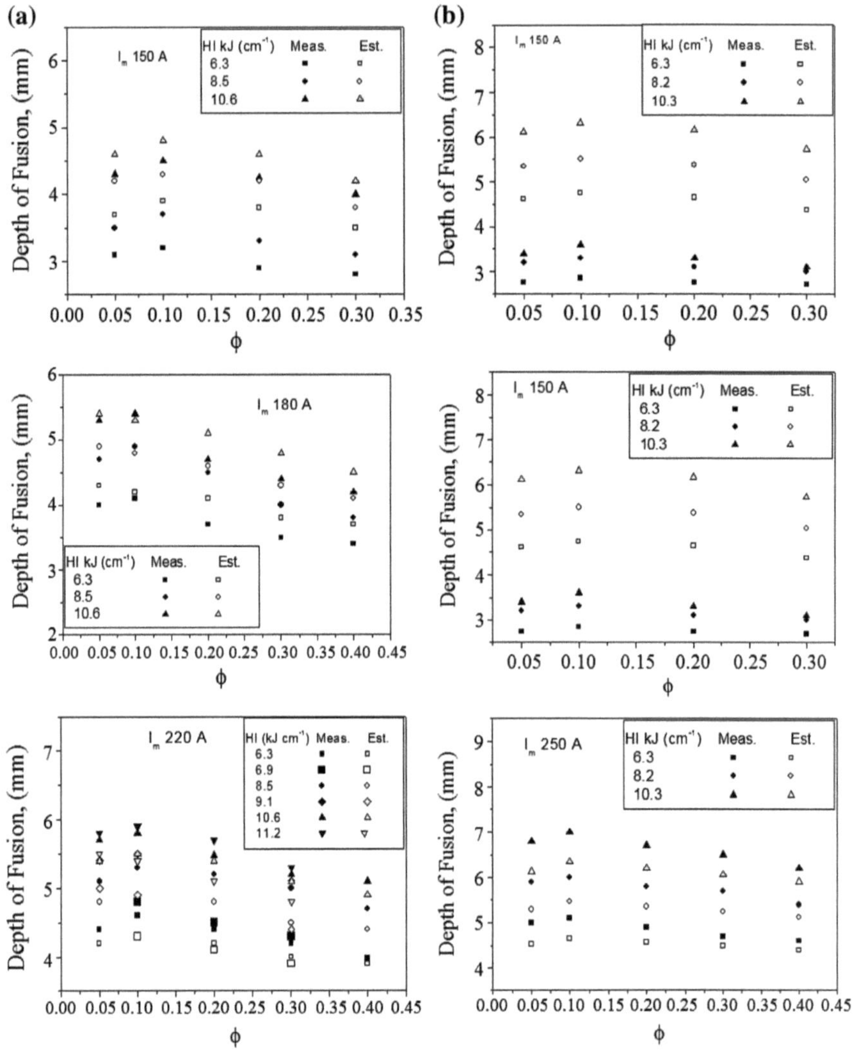

Fig. 5.6 Effect of ϕ on measured and estimated depth of fusion (penetration) in bead on plate deposition of **a** aluminium and **b** Al–Mg alloy with varying energy input at different mean current

V_i and V_{de} as stated in Eqs. 3.44 and 3.45 govern the penetration in P-GMAW process.

At low mean current the insufficient arc heat cannot raise an initial weld pool in base metal enabling the heat of the deposited weld metal to reach at the liquid-solid interface by cavity formation in the weld pool. Moreover, at low mean current a comparatively less number of bigger drop transfer at a relatively lower rate and velocity may not be active enough to penetrate deep inside the weld pool approaching closer to its interface with the solid base metal. It affects the size and shape of the weld pool significantly. This process becomes relatively slower in case of Al–Mg alloy due to formation of comparatively bigger weld pool through more melting of base metal arising out of relatively sluggish heat dissipation at lower thermal conductivity than aluminium. But, the situation becomes reverse at higher mean currents due to creation of larger initial weld pool by increased arc heating and more kinetically energetic metal transfer as observed at I_m of 200 and 250 A. An increased temperature and fluidity of molten base metal in relatively less conductive Al–Mg alloy facilitates the high velocity metal droplets to penetrate deeply inside the weld pool that results more depth of fusion as compared to that occurs in aluminium. On this matter the role of heat build-up in weld pool influencing the viscosity and fluid dynamics of molten metal significantly under the interrupted impact of super heated droplets on it cannot be ignored.

5.1.2 Width of Weld Bead

The width (W_b) of weld bead varies significantly under the influence of conventional energy input, mean current and the pulse parameters involved in factor ϕ as shown in Fig. 5.7a, b for the bead deposits of aluminium and Al–Mg alloy respectively. With their comparatively more predominant effect the energy input as well as I_m enhances the bead width at a given ϕ primarily due to increase in metal deposition per unit length of weld with its enhanced fluidity. But, at a given energy input and I_m laying in the range of 6.3–11.2 kJ/cm and 150–250 A respectively the increase of ϕ from 0.05 to 0.1 relatively enhances the bead width followed by a significant reduction in it with a further increase of ϕ to about 0.3–0.4. The general tendency of reduction in bead width with the increase of ϕ arises out of reduction in metal deposition per pulse [4] and decrease in temperature of depositing droplet (Fig. 4.5) which lowers the fluidity of weld pool. The effects of ϕ, analogous to that of pulse duration as stated above, and I_m on the increase of bead width are also in agreement to the observations (Fig. 5.1) on Al–Zn–Mg alloy. At a given energy input the reduction in bead width (W_b) at lower mean current (Fig. 5.7) possibly predominantly happens due to comparatively poorer heat content of weld deposit. However, at a given mean current a comparatively lower penetration of droplets may cause more lateral displacement of liquid in weld pool. Thus, it results a relatively larger width of fusion as it happens in case of comparatively less conductive Al–Mg alloy than aluminium.

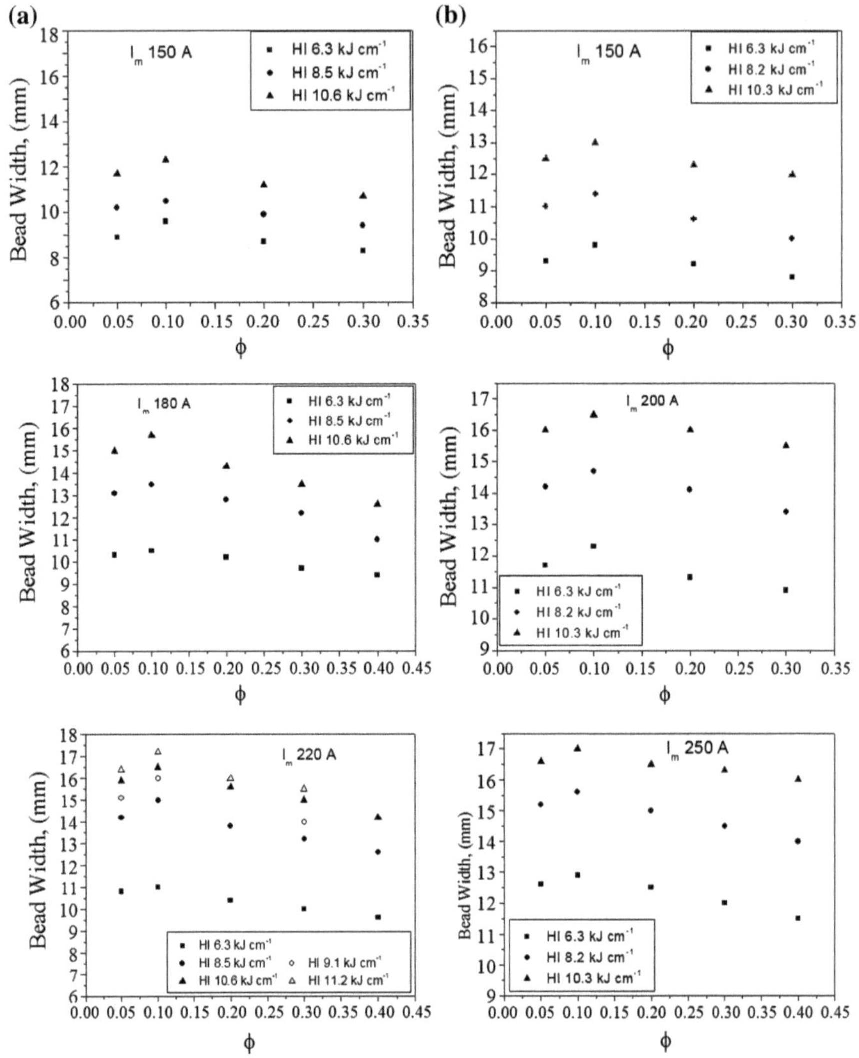

Fig. 5.7 Effect of ϕ on bead width in bead on plate deposition of **a** aluminium and **b** Al–Mg alloy with varying energy input at different mean current

5.1.3 Height of Bead Reinforcement

Similar to those observed in the cases of penetration and bead width here also at a given energy input and I_m, the ϕ apparently maintains a criticality at 0.1 and the increase of ϕ beyond this point up to 0.3 or 0.4 reduces the bead height (H_b) in both the cases of aluminium and Al–Mg alloy as shown in Fig. 5.8a, b respectively. But, at a given ϕ and I_m the increase of energy input and at a given ϕ and energy input

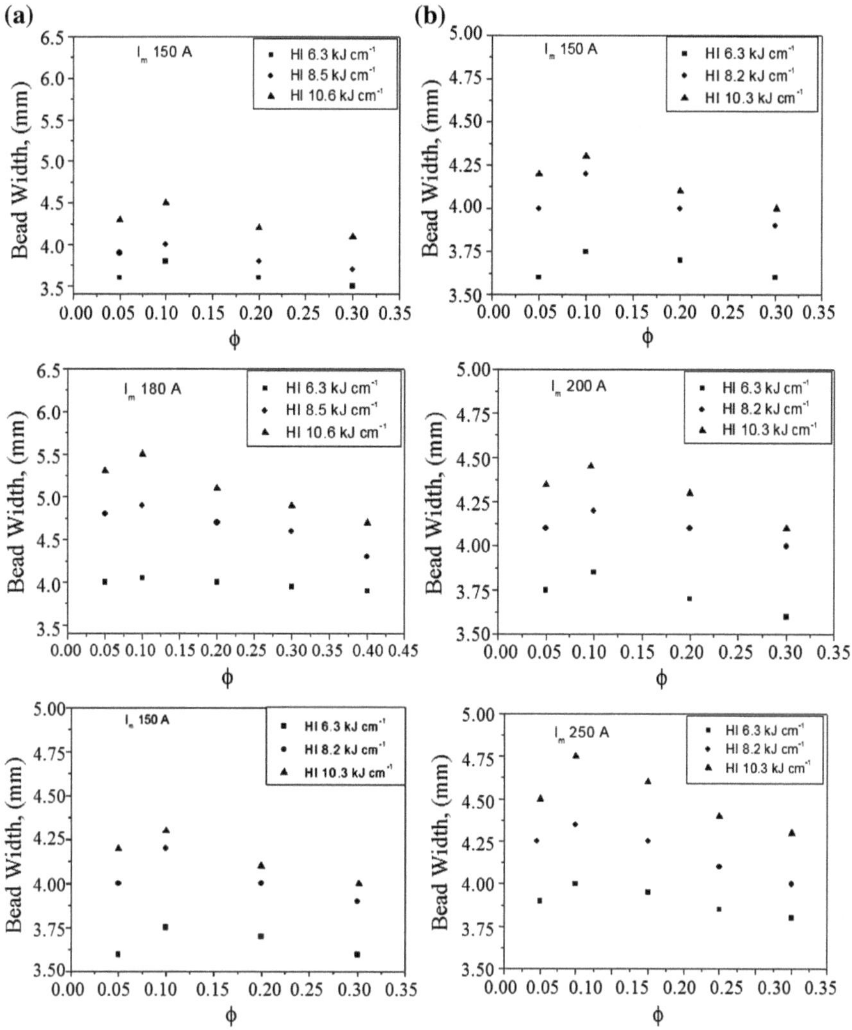

Fig. 5.8 Effect of ϕ on bead height in bead on plate deposition of **a** aluminium and **b** Al–Mg alloy with varying energy input at different mean current

the increase of I_m enhances bead height. The ϕ dictates bead height primarily by its similar decreasing trend of influence on rate of metal deposition per pulse [3–5]. The enhancement of bead height with the increase of mean current at a given ϕ and energy input and also with the increase of energy input at a given ϕ and mean current may be attributed to increase in the rate of weld metal Metal depositionper unit time and per unit length of weld. It happens due to increase in wire feed speed and decrease in welding speed respectively. A similar response of I_m on bead height is also confirmed in other studies on aluminium alloy (Fig. 5.1).

5.1.4 Area of Bead Deposit

In view of the significant response of height and width of weld bead to the pulse parameters it is interesting to understand their effect on area of weld bead on plate deposition. At a given heat input and mean current increase of ϕ from 0.05 to 0.1 relatively enhances the area of bead deposit followed by a reduction in it with a further increase of ϕ up to 0.3 or 0.4. It possibly happens due to variation of M_t in similar fashion in both the cases of aluminium and Al–Mg alloy as shown in Fig. 5.9a, b respectively. Area of bead deposit enhances appreciably with the

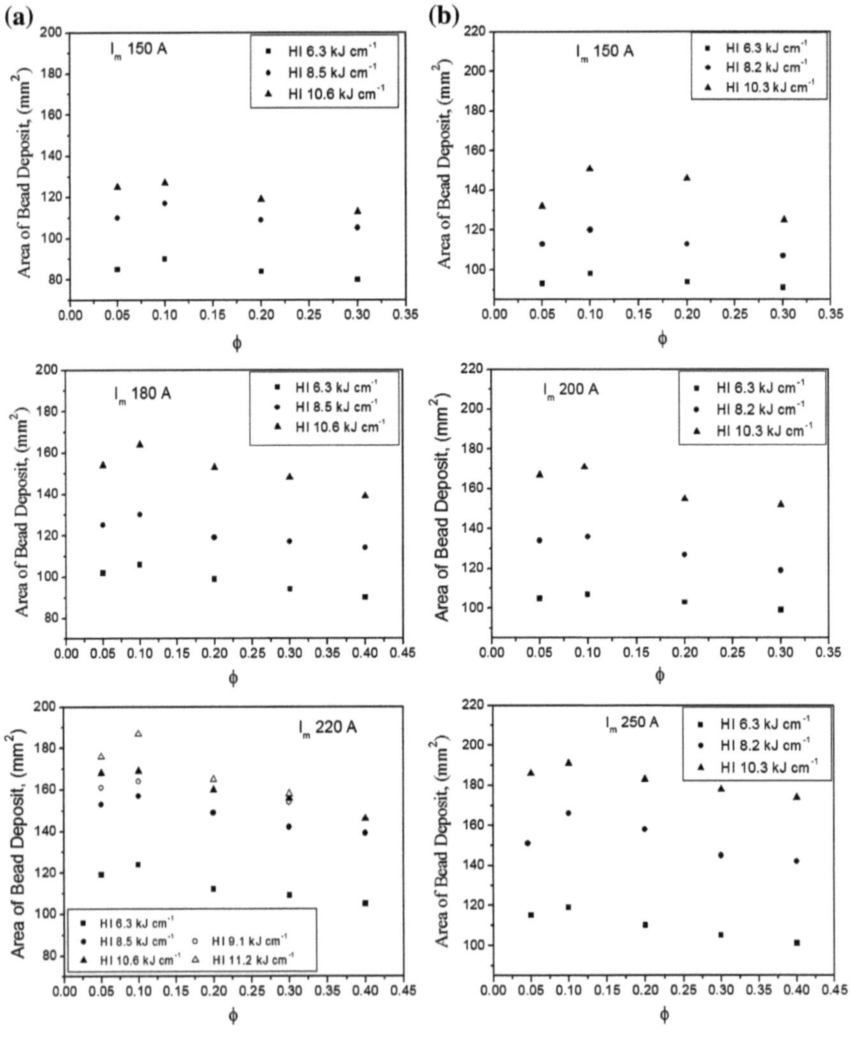

Fig. 5.9 Effect of ϕ on area of bead on plate deposit of **a** aluminium and **b** Al–Mg alloy with varying energy input at different mean current

increase of energy input at a given ϕ and mean current as well as with the increase of mean current at a given ϕ and energy input.

5.1.5 Weld Bead Toe Angle

Variation in height and width of reinforcement of weld deposit with a change in welding parameters shows a similar trend. But, the effect of welding parameters on bead toe angle of weld deposit indicates that the proportional change in height and width of reinforcement is somehow disturbed by predominance of either of these two factors. It may be either rate of metal deposition influencing the bead height or the fluidity of molten metal affecting the bead width by regulating its lateral spread after fusion of the plate surface. Unlike to that occurs in case of the penetration, width and height of weld bead the excluded bead toe angle, θ, (Fig. 5.3) enhances with the increase of ϕ up to 0.3 or 0.4 beyond a critical point of 0.1 in both the cases of aluminium and Al–Mg alloy as shown in Figs. 5.10a, b respectively. In other side of this point it apparently shows an opposite trend with the decrease of ϕ. However, out of the three basic criterion as the ϕ, I_m and energy input affecting the bead geometry the increase of I_m or energy input at a given other two parameters enhances the bead toe angle significantly. At a given ϕ, I_m and energy input the bead toe angle becomes comparatively lower in low conductive Al–Mg alloy than in aluminium.

Increase in rate of metal deposition may reduce θ by increasing the bead height, whereas increase of heat transfer to weld pool may enhance θ by increasing its fluidity. So the variation of θ can be considered as a function of a competitive process of the effects of rate of weld deposition and heat transfer to it. Variation in bead toe angle as a function of ϕ, mean current and energy input may have primarily caused by change in amount of metal deposition per unit time and length of deposit reflected in area of bead deposit (Fig. 5.9) as well as variation in heat transfer (Fig. 4.1) to weld pool affecting fluidity of deposited weld metal as stated earlier.

5.1.6 Deposition form factor

The geometrical form factor (F_f) of weld deposition defined by its ratio of bead width (W_b) to bead height (H_b) as a measure of acceptance of weld reinforcement with respect to its toe angle often becomes a matter of interest to control the weld quality. The significant influence of welding parameters on width and height of weld bead makes it imperative to know their effect on geometrical form factor of weld bead in P-GMAW process. At a given energy input and mean current increase of ϕ from 0.05 to 0.1 moderately enhances the form factor followed by a reduction in it with a further increase of ϕ up to 0.3 or 0.4 in both the cases of aluminium and

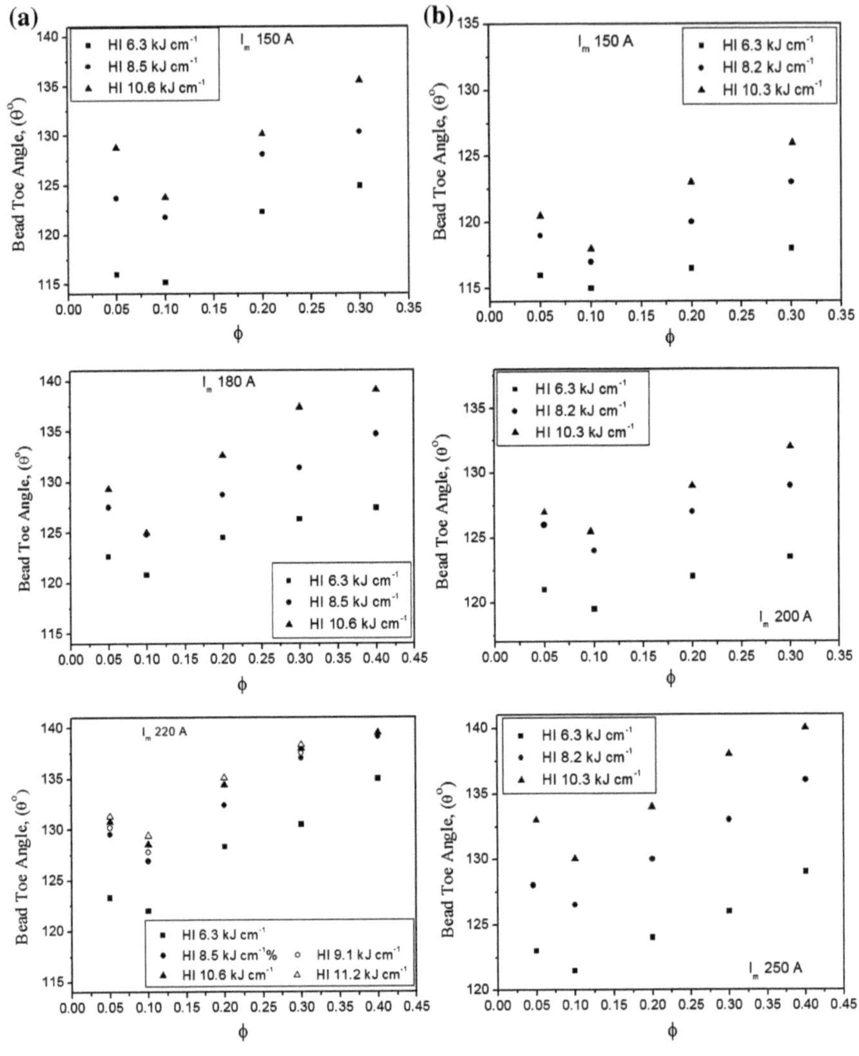

Fig. 5.10 Effect of ϕ on bead toe angle of weld deposition of **a** aluminium and **b** Al–Mg alloy with varying energy input at different mean current

Al–Mg alloy as shown in Fig. 5.11a, b respectively. However, at a given ϕ and mean current increase of energy input and at a given ϕ and energy input increase of mean current enhances the form factor of weld deposition. Form factor of both the aluminium and Al–Mg alloy deposits generally lies within an acceptable range (1.2–3.5) [3, 5, 6] of sound deposit in down hand welding. But, in case of Al–Mg alloy deposit produced at a relatively higher mean current and heat input of

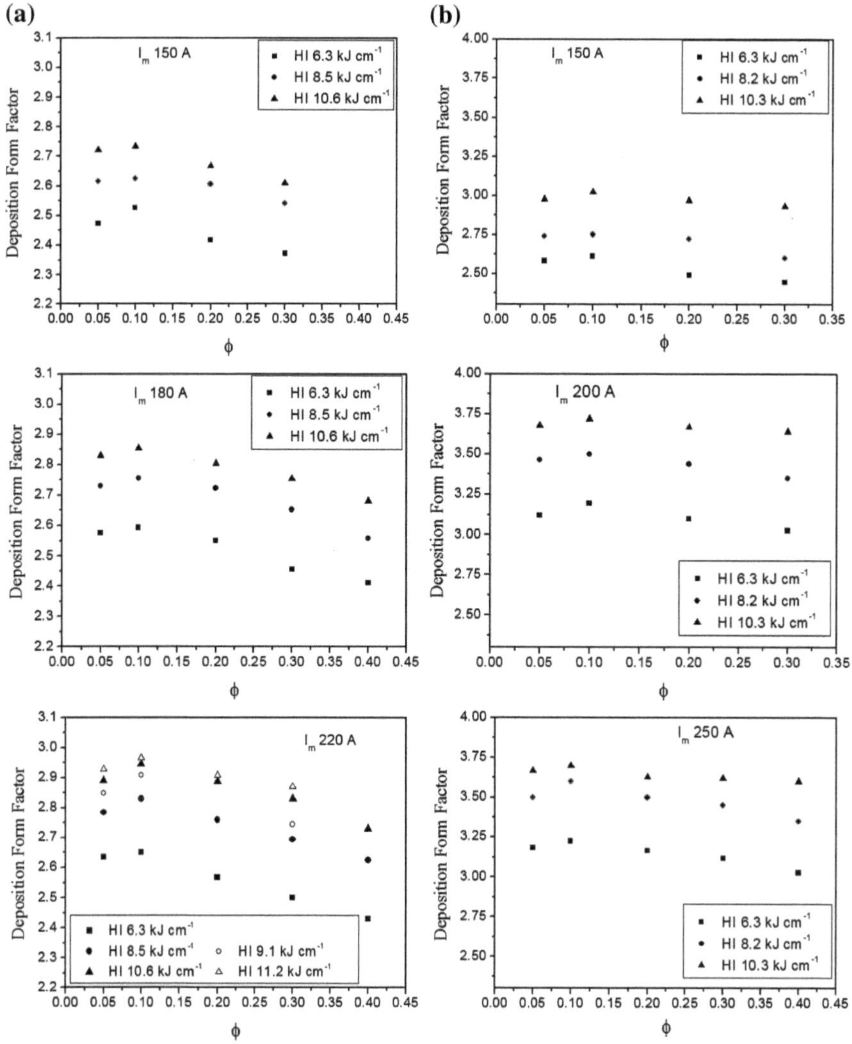

Fig. 5.11 Effect of ϕ on form factor of weld deposition of **a** aluminium and **b** Al–Mg alloy with varying energy input at different mean current

≥ 200 A and 10.3 kJ/cm respectively it is found of the order of 3.75. At such a high I_m and HI the effect of ϕ on form factor becomes insignificant in case of Al–Mg alloy. It possibly happens because a larger arc heat significantly affects the fluidity of weld pool and largely suppresses the effect of ϕ. The variation in form factor as a function of ϕ, mean current, and heat input generally happens due to a change in fluidity of weld deposit primarily because of change in heat (Fig. 4.1) transfer to the weld pool that results an increased bead width.

5.1.7 Base Metal Fusion

The fraction of base metal fusion with unit mass of weld deposition can be considered as a measure of efficiency of a welding process to undergo in any application with optimum use of energy input. It becomes more useful especially in case of the process like P-GMAW, where it involves more number of parameters responsible for base metal fusion than those of conventional GMAW. In view of the influence of different parameters affecting the geometry of bead on plate deposit the fraction of base metal fusion as function of unit mass of weld deposition (F_b) can be estimated as

$$F_b = [(A_f \times \rho_b)/(A_d \times \rho_d)] \tag{5.1}$$

where, A_f is area of base plate fusion, A_d is area of weld deposit, ρ_b is density of base metal and ρ_d is density of weld deposit. The increase of ϕ from 0.05 to 0.1 comparatively enhances the F_b followed by a reduction in it with a further increase of ϕ up to 0.3 or 0.4. It is observed in both the cases of weld deposition of aluminium and Al–Mg alloy at a given energy input and mean current as shown in Fig. 5.12a, b respectively. At a given ϕ and heat input the increase of mean current and at a given ϕ and mean current the increase of heat input enhances the F due to increase in rate of heat transfer per unit time (Fig. 4.1) and per unit length of base plate causing its larger melting. At a given energy input the F_b reduces with increase of ϕ at a faster rate at higher mean current. It possibly happens due to increase in tendency of finger type penetration [3, 5, 7, 8], at higher mean current. However, the variation in F_b with increase of ϕ follows a similar trend to that observed in case of variation in heat transfer to weld pool (Fig. 4.1) with ϕ. The fraction of base metal fusion with unit mass of bead deposition is significantly larger in case of Al–Mg alloy than that of aluminium especially at higher I_m.

5.1.8 Dilution

Variation of weld geometry with respect to amount of weld deposition and fusion of base metal affects the weld bead dilution defined as a ratio of the (amount of fused base metal/ total amount of molten weld metal). They are generally estimated by measuring the area of fused region of base plate and total area of weld as revealed in transverse section of bead on plate weld deposit. The dilution of weld especially in case of using dissimilar filler wire significantly affects its chemical and metallurgical properties. Thus in case of variation in weld geometry it is imperative to study the dilution of weld metal. In the line of general understanding on weld geometry affecting the dilution, here also the increase of ϕ from 0.05 to 0.1 relatively enhances the dilution followed by a considerable reduction in it with a further increase of ϕ up to 0.3 or 0.4 at a given conventional energy input and mean

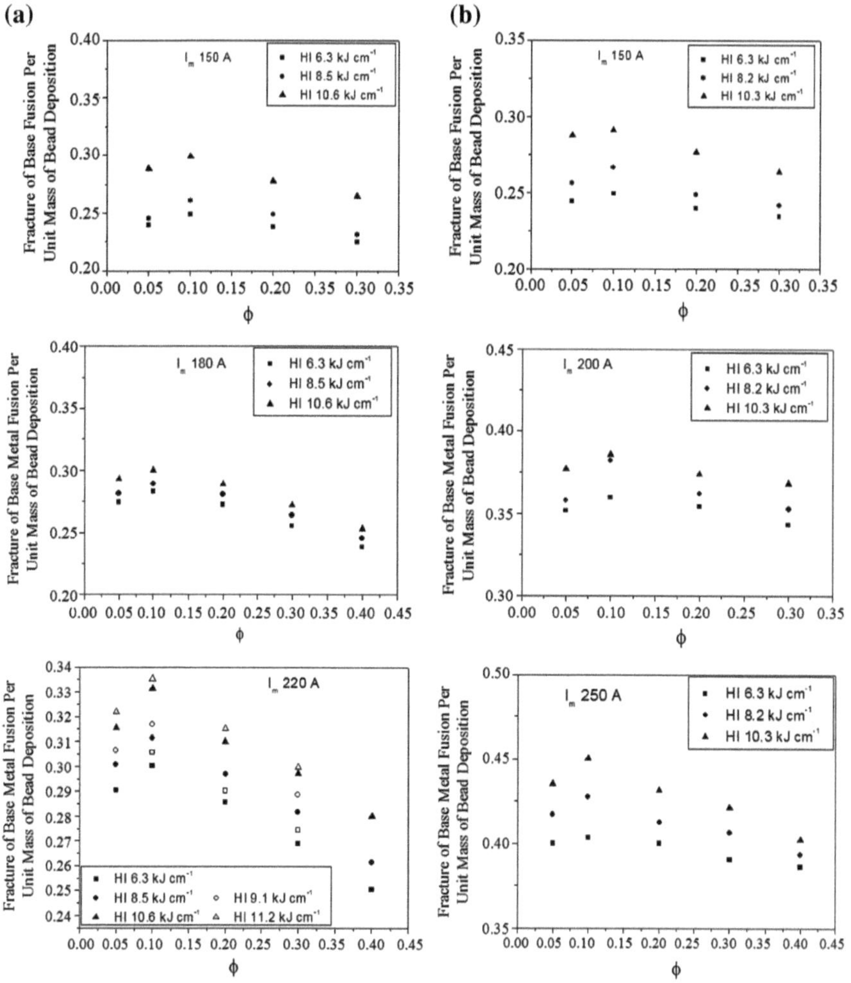

Fig. 5.12 Effect of ϕ on fraction of base metal fusion per unit mass of weld bead on plate deposition of **a** aluminium and **b** Al–Mg alloy with varying energy input at different mean current

current. It is observed in both the cases of aluminium and Al–Mg alloy as shown in Fig. 5.13a, b respectively. At a given ϕ and mean current the increase of energy input and at a given ϕ and energy input the increase of mean current enhances the dilution of weld bead primarily due to increased heat and heat build-up in the weld pool causing larger melting of base metal. Further due to comparatively larger melting of base plate, the Al–Mg alloy (Fig. 5.12) undergoes through relatively larger dilution of weld deposit than aluminium especially at higher mean current. But, the dilution being a function of several interactive aspects of weld geometry its

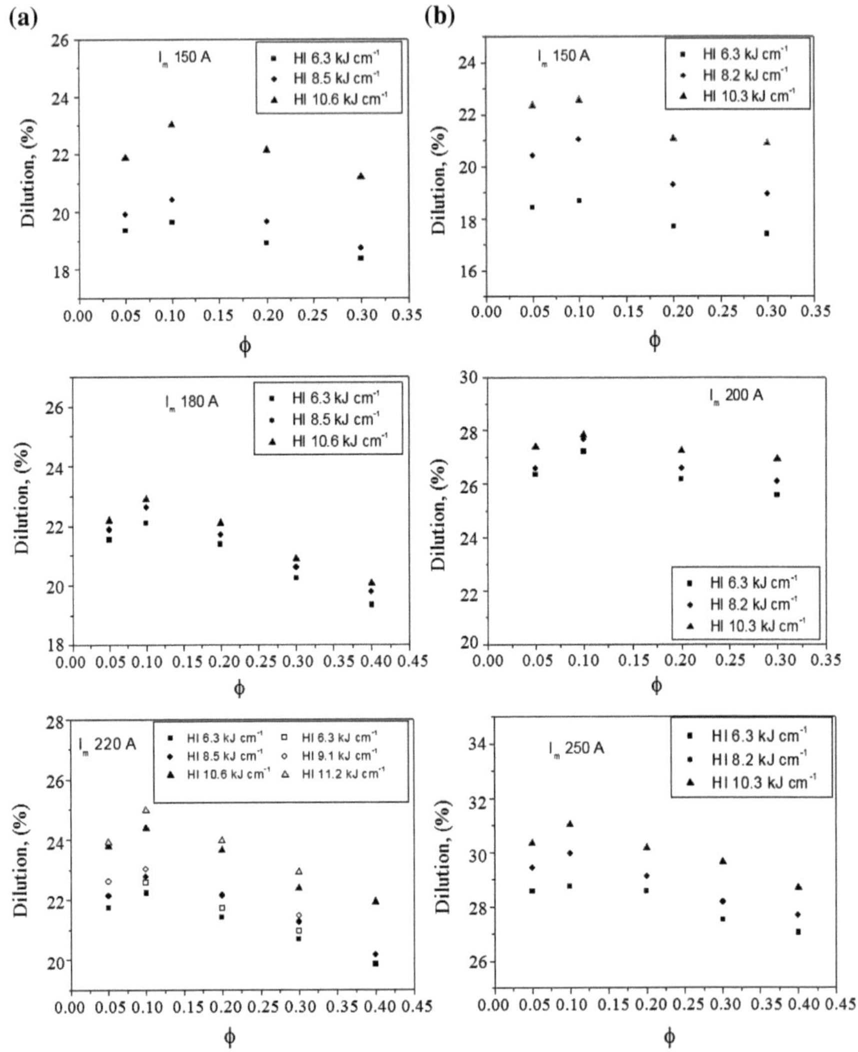

Fig. 5.13 Effect of ϕ on dilution of bead deposition of **a** aluminium and **b** Al–Mg alloy with varying energy input at different mean current

correlation with any individual pulse parameter except mean current is not very conclusive due to significant scattering as it is reported [2] in case of bead on plate deposition on Al–Zn–Mg alloy. However, while considering the summarized effect of pulse parameters, increase of ϕ clearly shows linear reduction of dilution (D_n) and penetration (P_n) but at a comparatively slower rate at lower I_m as shown in Figs. 5.14 and 5.15 respectively following the empirical correlations as follows.

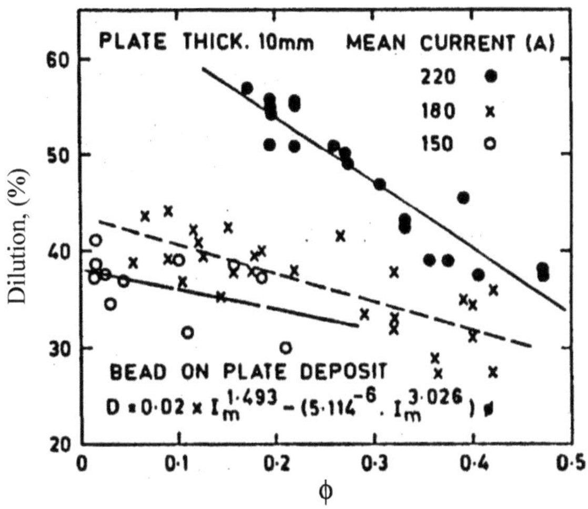

Fig. 5.14 Influence of the factor ϕ and I_m on dilution of bead on plate deposit

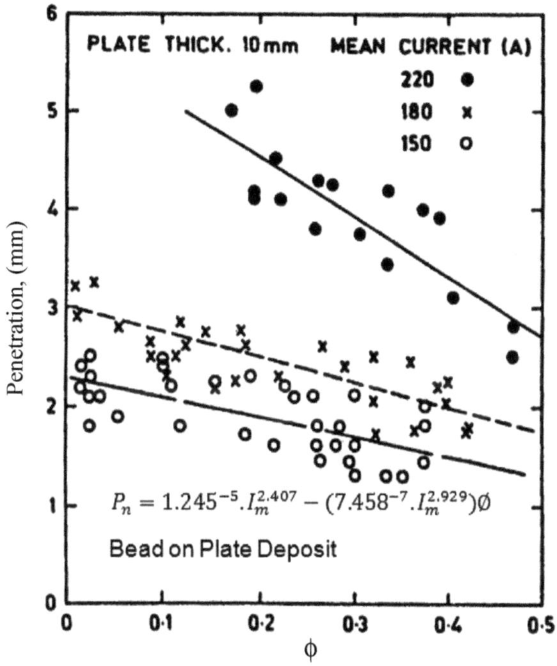

Fig. 5.15 Influence of the factor ϕ and I_m on penetration in bead on plate deposit

$$D_n = \left[0.02 \cdot I_m^{1.493} - (5.114^{-6} \cdot I_m^{3.026})\phi\right] \tag{5.2}$$

$$P_n = \left[1.245^{-5} \cdot I_m^{2.407} - (7.458^{-7} \cdot I_m^{2.929})\phi\right] \tag{5.3}$$

The expressions estimate the D and P_n as a function of ϕ and I_m with a difference lying in the range of 4–8%. From the Eqs. 5.2 and 5.3 it is realized that the characteristics of pulsed current weld deposit are well correlated to the basic pulse parameters such as the I_b, I_p, f and t_b which governs the energy available and its distribution in P-GMAW process [9].

The P-GMAW produces comparatively higher bead height, bead width and penetration (Fig. 5.1) than that observed in case of using GMAW (Fig. 5.16) by keeping the welding current similar to the mean current of the earlier process, especially when it is lying beyond the transition level. In consideration of the appreciable influence of conventional energy input (E_c), mean current (I_m) and ϕ on the bead height (H_b), bead width (W_b), toe angle (θ), area of bead deposits (A_d), form factor (F_f) of bead deposition, fraction of base metal fusion with unit mass of bead deposition (F_b) and dilution of weld bead (D_{nb}) of aluminium and Al–Mg alloy the empirical correlations of these geometrical aspects as a function of ϕ, I_m and E_c are found as follows. A comparison of the theoretically evaluated H_b, W_b, θ, A_d, F_f, F_b and D_{nb} on the basis of their empirical correlations at different welding parameters with their corresponding measured values at the same parameters in case of both the aluminium and Al–Mg alloy shows a good agreement to each other with a difference within a range of ±5–10%.

Fig. 5.16 The influence of welding current on bead geometry of Al–Mg alloy on 10 mm thick Al–Zn–Mg alloy plate in conventional GMAW

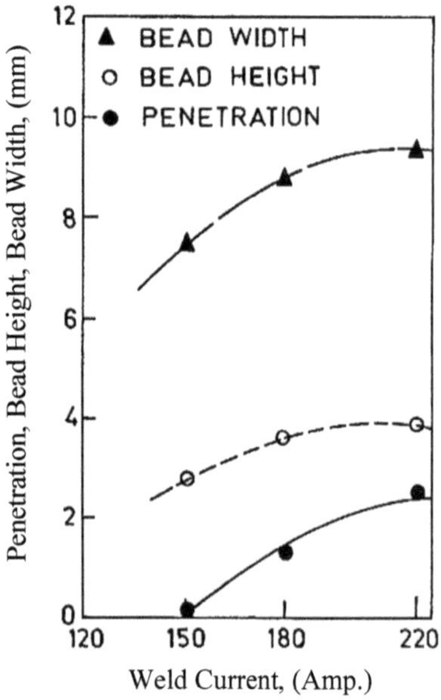

For aluminium

$$W_{b(Al)} = 0.0744\phi^{-0.0565} I_m^{0.6783} E_c^{0.6741} \quad (5.4)$$

$$H_{b(Al)} = 0.1018\phi^{-0.0319} I_m^{0.5208} E_c^{0.4737} \quad (5.5)$$

$$A_{d(Al)} = 0.4782\phi^{-0.0542} I_m^{0.7685} E_c^{0.6817} \quad (5.6)$$

$$\theta_{(Al)} = 63.4203\phi^{-0.0365} I_m^{0.1003} E_c^{0.119} \quad (5.7)$$

$$F_{f(Al)} = 0.7461\phi^{-0.026} I_m^{0.1581} E_c^{0.2034} \quad (5.8)$$

$$F_{b(Al)} = 0.0219\phi^{-0.0542} I_m^{0.386} E_c^{0.1972} \quad (5.9)$$

$$D_{nb(Al)} = 3.9547\phi^{-0.0386} I_m^{0.2383} E_c^{0.1738} \quad (5.10)$$

For Al–Mg alloy

$$W_{b(Al-Mg)} = 0.1703\phi^{-0.0357} I_m^{0.566} E_c^{0.6067} \quad (5.11)$$

$$H_{b(Al-Mg)} = 1.0781\phi^{-0.0212} I_m^{0.1325} E_c^{0.2794} \quad (5.12)$$

$$A_{d(Al-Mg)} = 1.0363\phi^{-0.0408} I_m^{0.5332} E_c^{0.921} \quad (5.13)$$

$$\theta_{(Al-Mg)} = 42.923\phi^{-0.0221} I_m^{0.1603} E_c^{0.1265} \quad (5.14)$$

$$F_{f(Al-Mg)} = 0.1934\phi^{-0.0204} I_m^{0.3969} E_c^{0.3105} \quad (5.15)$$

$$F_{b(Al-Mg)} = 0.00216\phi^{-0.029} I_m^{0.8776} E_c^{0.176} \quad (5.16)$$

$$D_{nb(Al-Mg)} = 0.3963\phi^{-0.0305} I_m^{0.7087} E_c^{0.1623} \quad (5.17)$$

5.2 Non Conventional Bead on Plate Weld Deposition

The possibility of employing P-GMAW process in various applications including the cases other than the conventional practices of down hand welding of different metals using conventional solid filler wire as discussed above has been explored. Number of non-conventional works has been reported analyzing the geometrical aspects of weld deposits as a function of various parameters. They may be broadly classified with respect to variation in type of consumable, welding procedure and welding position used at different welding parameters.

5.2.1 Weld Deposition of Flux Cored Wire

It is well known that the use of flux cored wire (FCW) in place of solid filler wire is beneficial primarily due to its advantage of high deposition rate that favours production economy, improved weld quality and better performance in positional welding [10–12]. In spite of certain questions about undesirable fume formation, the growing scope of GMAW using flux cored wire has raised curiosity about its performance under pulsed current welding [9, 10, 13, 14], due to paucity of comprehensive knowledge to identify the role of various pulse parameters to control the weld characteristics. Studies on geometry of weld bead deposited on 12 mm thick C-Mn steel plate by P-GMAW process using 1.2 mm diameter basic flux cored mild steel filler wire of specification Fluxofil-31:AWS E 70T-5 with argon-20% carbon dioxide shielding has been reported as a function of pulse parameters, arc voltage and welding speed. Increase of pulse frequency and pulse duration enhances the width but reduces height and penetration of weld bead as shown in Figs. 5.17, 5.18, 5.19 and 5.20 respectively when rest of the parameters out of the pulse duration, pulse frequency, mean current, arc voltage and welding speed are kept constant. At a given other parameters the increase of welding speed reduces the width, height and penetration of weld bead (Figs. 5.21 and 5.22), whereas the increase of arc voltage although relatively reduces the bead height but enhances the bead height and penetration (Figs. 5.23 and 5.24). However, at any parameters of welding the increase of mean current always enhances all these aspects of weld geometry with a comparatively stronger effect during initial enhancement from 100–150 A followed by a moderate influence at further increase to 200 A. It is realized that the rate of influence of these parameters on various aspects of geometry of pulsed current gas metal arc weld is also very much different. The variation in geometry of weld bead as usual affects its dilution. At a

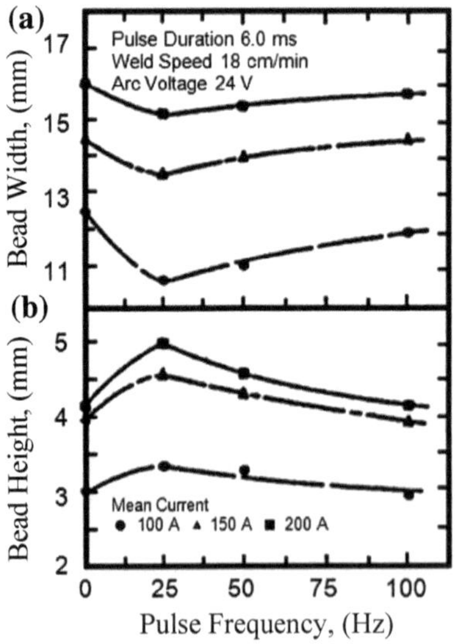

Fig. 5.17 Effect of pulse frequency on bead width and bead height in FCW deposition

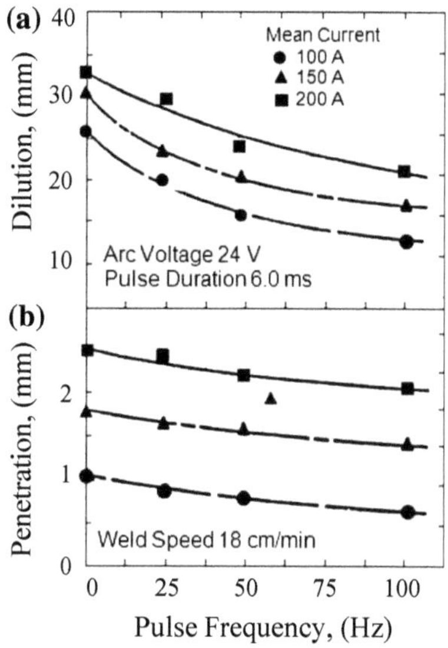

Fig. 5.18 Effect of pulse frequency on dilution and penetration of weld in FCW deposition

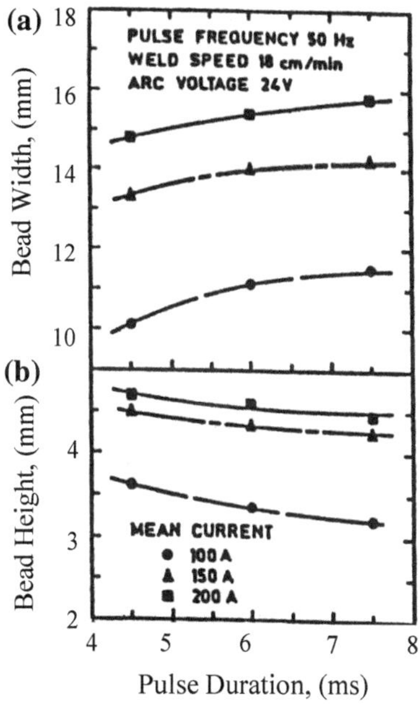

Fig. 5.19 Effect of pulse duration on bead width and bead height in FCW deposition

given pulse parameter increase of welding speed, mean current and arc voltage enhances dilution (Figs. 5.22 and 5.24) but at given other conditions of welding the increase of pulse frequency or pulse duration significantly reduces the dilution (Figs. 5.18 and 5.20).

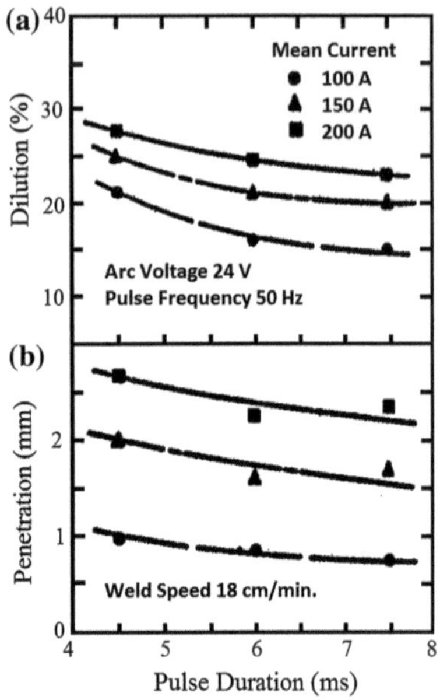

Fig. 5.20 Effect of pulse duration on dilution and penetration of weld bead in FCW deposition

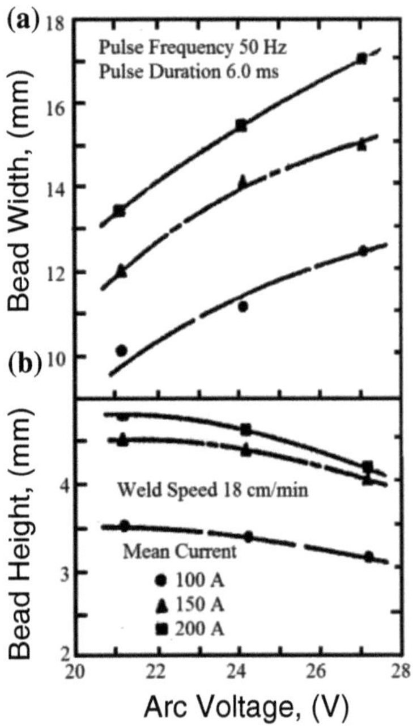

Fig. 5.21 Effect of welding speed on bead width and bead height in FCW deposition

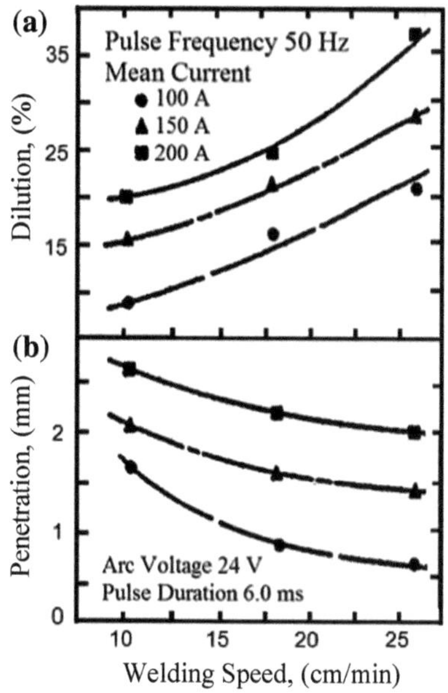

Fig. 5.22 Effect of welding speed on dilution and penetration of weld bead in FCW deposition

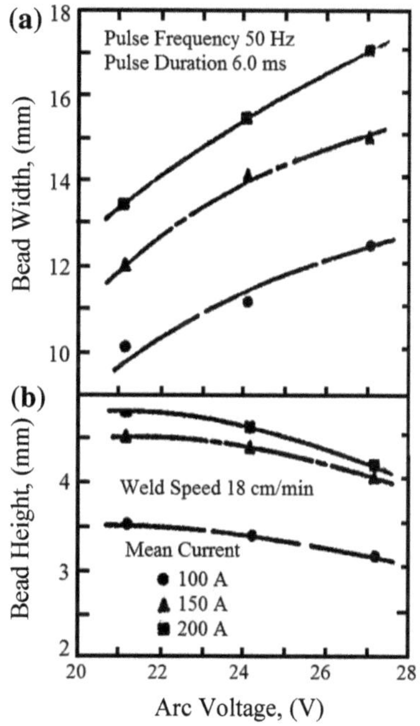

Fig. 5.23 Effect of arc voltage on bead width and bead height in FCW deposition

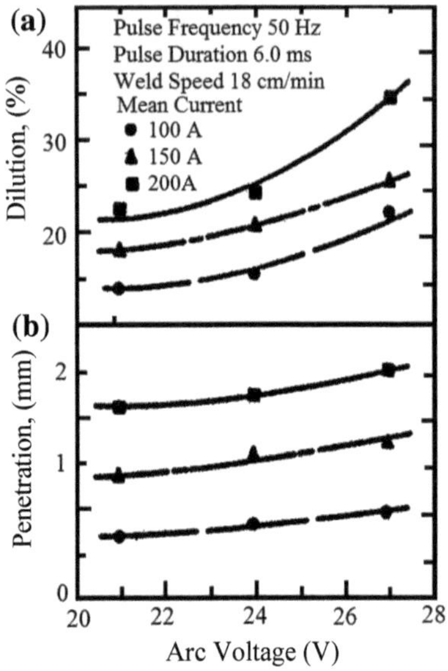

Fig. 5.24 Effect of arc voltage on dilution and penetration of weld bead in FCW deposition

The increase of pulse frequency and pulse duration shows (Figs. 5.17, 5.18, 5.19 and 5.20) similar trend of variation of width, height, dilution and penetration of weld bead. But, they affect considerably the I_p in opposite fashion along with an identical decreasing mode of influence on I_b as shown in Table 5.1. This has happened at a given mean current, arc voltage and welding speed that keeps the deposition rate, spreading of weld deposition and deposition per unit length constant. Thus, it is assumed that the change in bead geometry and dilution is primarily caused by the variation in I_p and I_b due to their influence on super heating of weld pool, arc characteristic and behaviour of metal transfer. Increase of them (especially I_p) may enhance the fluidity of weld pool, which reduces the bead height along with an enhancement of bead width and increases the penetration as well as dilution depending upon deposition per unit length. The variation in f or t_p also affects the process of interruption in metal deposition and consequently the heat buildup in weld pool that influences the bead geometry and dilution. Hence, it is realized that the nature of influence of pulse parameters, like f and t_p, on the bead geometry and dilution can not be similar because the variation of f or t_p do not affect the I_p and I_b in same fashion and proportion (I_b/I_p) as shown in Table 5.1. Under these circumstances the influence of pulse parameters on weld geometry and dilution could be to a great extent realized by considering their summarized effect on the bead characteristics with the help of the factor ϕ as given in Eq. 3.8.

At a given arc voltage and welding speed increase of ϕ at a mean current reduces bead width and enhances bead height, penetration and dilution (Figs. 5.25 and 5.26). Whereas, at a given f and t_p the increase of ϕ at a given range of arc voltage of 21–27 V enhances the bead width, penetration and dilution but reduces the bead height significantly at any mean current (Figs. 5.27 and 5.28). At constant f and t_p

Table 5.1 Typical P-GMA welding parameters using 1.2 mm diameter basic flux cored filler wire

Mean current (A)	Pulse frequency (Hz)	Pulse duration (ms)	Base current (A)	Peak current (A)	Arc voltage (V)	Welding speed (mm/s)	Factor ∅
100	0	–	–	–	24	3	–
100	25	6.0	90	271	24	3	0.282
100	50	6.0	70	250	24	3	0.196
100	100	6.0	30	212	24	3	0.057
100	25	4.5	90	260	24	3	0.268
100	50	7.5	60	272	24	3	0.138
100	100	6.0	70	280	21	3	0.175
100	25	6.0	90	233	27	3	0.270
100	50	6.0	70	255	24	1.67	0.192
100	100	6.0	75	243	24	4.33	0.216
150	0	–	–	–	24	3	–
150	25	6.0	120	367	24	3	0.278
150	50	6.0	110	332	24	3	0.232
150	100	6.0	70	300	24	3	0.093
150	25	4.5	120	294	24	3	0.316
150	50	7.5	100	362	24	3	0.173
150	100	6.0	100	356	21	3	0.197
150	25	6.0	120	320	27	3	0.263
150	50	6.0	100	340	24	1.67	0.206
150	100	6.0	110	332	24	4.33	0.232
200	0	–	–	–	24	3	–
200	25	6.0	170	380	24	3	0.380
200	50	6.0	120	360	24	3	0.233
200	100	6.0	70	327	24	3	0.086
200	25	4.5	160	347	24	3	0.357
200	50	7.5	110	372	24	3	0.185
200	100	6.0	120	404	21	3	0.208
200	25	6.0	140	327	27	3	0.300
200	50	6.0	130	370	24	1.67	0.246
200	100	6.0	130	362	24	4.33	0.252

the variation in ϕ with a change of arc voltage takes place due to consequent change in I_p and I_b as shown in Table 5.1. The variation of arc voltage significantly affects the ϕ. But in this case the observed ϕ along with the values of ϕ evaluated during the variations of f and t_p cannot be correlated to the bead characteristics. This is because the ϕ as a function of f, t_b, I_p and I_b does not consider any physical term accounting for the arc length. However, in all the cases the geometry of weld bead varies with ϕ almost linearly and the bead geometry including its dilution is comparatively more sensitive to ϕ resulting from a change in arc voltage than to ϕ

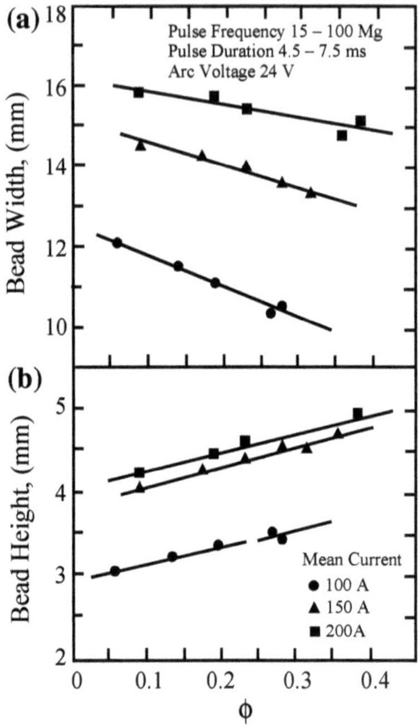

Fig. 5.25 Effect of ϕ on bead width and bead height during variation in pulse frequency and duration in FCW deposition

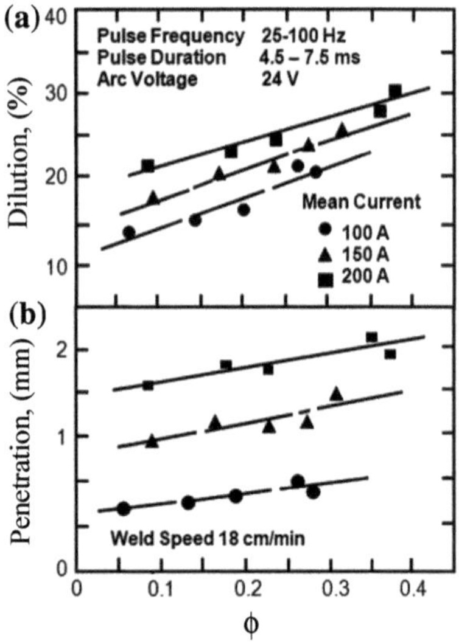

Fig. 5.26 Effect of ϕ on dilution and penetration of weld bead during variation in pulse frequency and duration in FCW deposition

arising from a change in f or t_p. The correlations of various characteristics of weld bead with the factor ϕ confirm the simultaneous influence of all the pulse parameters in governing weld characteristics and provide a basis to understand their summarized influence on the geometry and dilution of weld bead deposited by

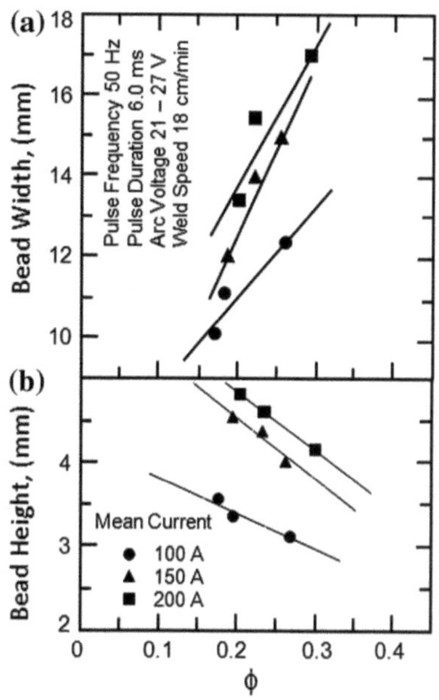

Fig. 5.27 Effect of ϕ on bead width and bead height during variation in arc voltage in FCW deposition

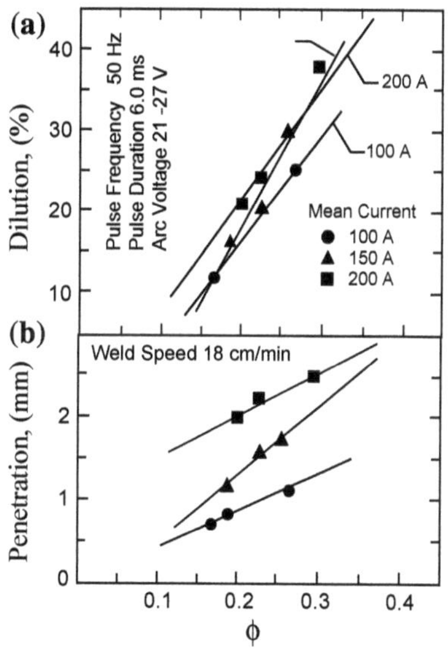

Fig. 5.28 Effect of ϕ on dilution and penetration of weld bead during variation in arc voltage in FCW deposition

pulsed current GMAW process. This knowledge can be used in solving the criticality for selection of pulse parameters to obtain a desired weld.

Effect of pulse parameters on geometry of weld bead of flux cored wire is somehow different from that observed in case of solid filler wire possibly due to their different nature of thermal and metal transfer. In case of using flux cored wire the metal drop transfers within an envelope of molten flux [15, 16] which significantly affects its nature of heat loss in flight prior to deposition in weld pool and subsequently the bead geometry. The pulse current GMAW generally results in a comparatively lower bead width, penetration and dilution but higher bead height of weld bead than those observed during conventional GMAW (0 Hz) at a given mean/welding current. However, the bead width, bead height and penetration of pulsed current deposit are comparable to those of the conventional GMA weld deposit produced at higher pulse frequency and pulse duration.

5.2.2 *Stainless Steel Cladding*

The knowledge of weld bead geometry as a function of welding parameters is of paramount importance in preparation of stainless steel cladding by bead on plate weld overlay in order to control its quality especially with respect to its chemistry and microstructure. Some observations in this regard have been made during stainless steel cladding on 12 mm thick structural steel plate by oscillatory movement of bead on plate weld deposition using P-GMAW with 1.2 mm diameter stainless steel filler wire of specification GRINOX S-SKOLA DIN 8556 (AWSA 5.4–81) under argon gas shielding at a flow rate of 18 l/min. Geometry of weld overlay as its thickness (T), average depth of fusion (P) and dilution defined as the ratio of P to (P+T) varies significantly with a change of pulse parameters as given in Table 5.2. In conventional GMAW the increase of welding current considerably enhances the thickness of cladding due to higher rate of deposition, along with a relative increase in Depth of fusionand dilution of cladding. At high welding current a minor increase of dilution occurs due to large increase of deposition of cladding material as compared to small amount of melting of base metal in relatively low depth of fusion. In P-GMAW the increase of mean current behaves in similar manner when the pulse frequency and duration remains constant. But, at a given mean current and pulse duration increase of pulse frequency from 0 to 100 Hz significantly enhances the thickness of cladding (Fig. 5.29). It shows that the P-GMAW is relatively more efficient for deposition of clad layer than the conventional GMAW process. However, the advantage of deposition of larger thickness by P-GMAW can be more effectively explored by the increase of pulse frequency than pulse duration.

At a given pulse frequency and duration increase of mean current significantly enhances depth of penetration of weld overlay by enhancing the plate fusion due to larger energy input. But, at a given mean current increase of pulse frequency reduces depth of fusion, whereas increase of pulse duration enhances the same (Fig. 5.30). It possibly happens due to reduction and enhancement of peak current with the increase of pulse frequency and pulse duration respectively (Table 5.2). Out of the three parameters as mean current, pulse frequency and pulse duration, increase of mean current moderately enhances but pulse frequency considerably reduces the dilution when the pulse duration influences it insignificantly at a given

Table 5.2 Typical parameters for P-GMA weld cladding using 1.2 mm diameter stainless steel filler wire

Ws (m/min)	I_m (A)	I_p (A)	I_b (A)	f (H_z)	t_p (ms)	Arc voltage (V)
6.6	160	–	–	0	–	30
6.6	160	239	150	25	4.5	30
6.6	160	222	142	50	4.5	30
6.6	160	184	140	100	4.5	30
6.6	160	245	145	25	6.0	30
6.6	160	225	132	50	6.0	30
6.6	160	190	115	100	6.0	30
6.6	160	264	136	25	7.5	30
6.6	160	228	119	50	7.5	30
6.6	160	195	55	100	7.5	30
8.0	180	–	–	0	–	30
8.0	180	267	169	25	4.5	30
8.0	180	256	158	50	4.5	30
8.0	180	215	151	100	4.5	30
8.0	180	288	161	25	6.0	30
8.0	180	273	140	50	6.0	30
8.0	180	217	125	100	6.0	30
8.0	180	327	146	25	7.5	30
8.0	180	280	120	50	7.5	30
8.0	180	220	60	100	7.5	30
10.7	200	–	–	0	–	30
10.7	200	279	190	25	4.5	30
10.7	200	269	180	50	4.5	30
10.7	200	237	170	100	4.5	30
10.7	200	341	175	25	6.0	30
10.7	200	293	160	50	6.0	30
10.7	200	240	140	100	6.0	30
10.7	200	373	160	25	7.5	30
10.7	200	317	130	50	7.5	30
10.7	200	243	70	100	7.5	30

other parameters (Fig. 5.31). In general at a given energy input the P-GMAW produces comparatively lower depth of fusion and dilution than conventional GMAW, which is beneficial for maintaining desired chemical composition of clad layer. For example a single pass stainless steel weld overlay of thickness of 3–4.5 mm on 12 mm thick mild steel plate can be produced by P-GMAW with a penetration and dilution of the order of 0.52–0.53 mm and 10–13% respectively. Whereas, in case of the SAW strip cladding the penetration and dilution lies in the

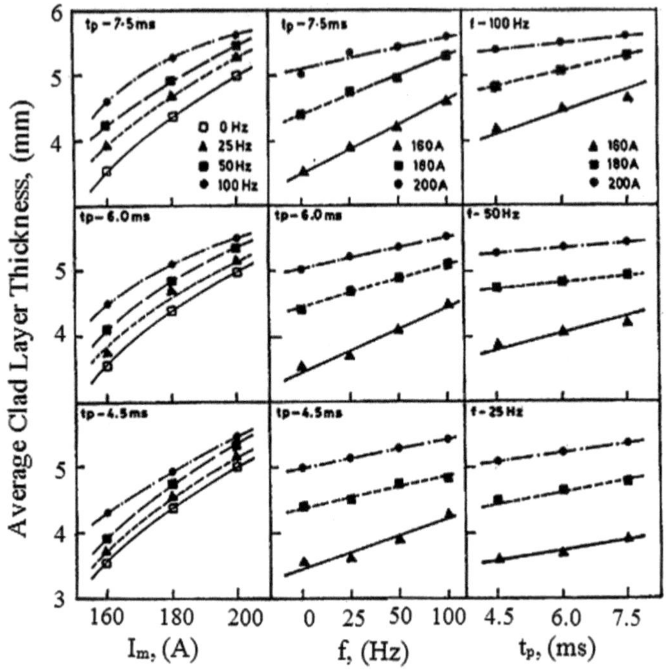

Fig. 5.29 Influence of pulse parameters on thickness of P-GMA weld cladding

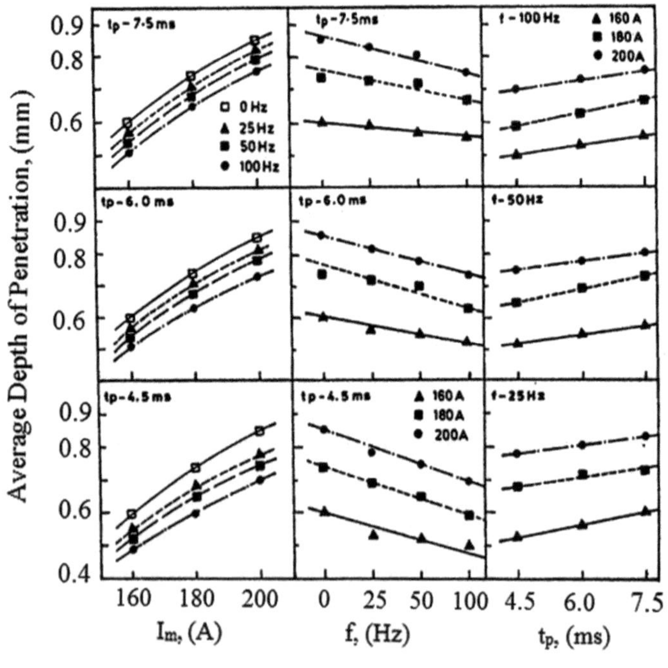

Fig. 5.30 Influence of pulse parameters on depth of fusion of P-GMA weld cladding

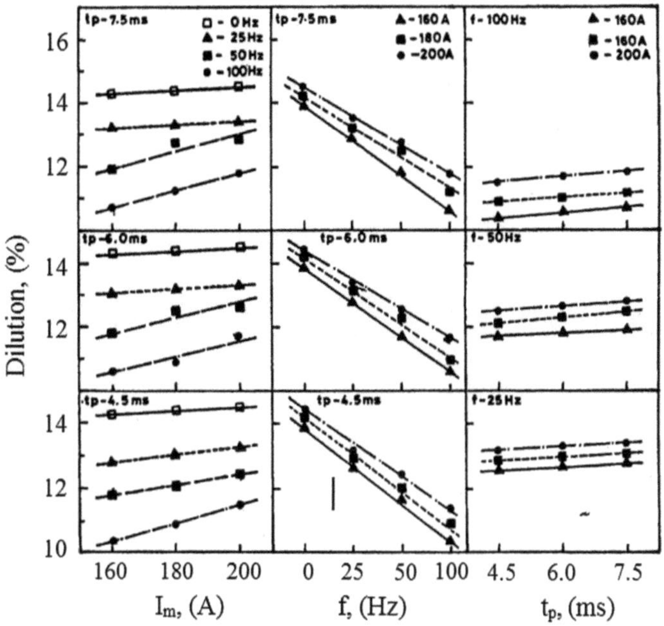

Fig. 5.31 Influence of pulse parameters on dilution of P-GMA weld clading

range of 0.74–1.69 mm and 19–27% respectively [15, 16]. A lower dilution of weld overlay is generally considered less detrimental to its corrosion resistance and other desired properties.

5.2.3 *Out of Position Weld Deposition*

The thermal behaviour of droplet, nature of metal transfer and rate of metal transfer significantly influences the characteristic of weld deposition in positional or out-of-position welding. The criticality of such characteristics in reference to pulse parameters has been well understood during analysis of bead on plate weld deposition of steel in vertical-up pulsed current GMA welding [17, 18] as a function of ϕ (Fig. 5.32). The variation of pulse parameters affecting rate of weld deposition including its temperature and heat content along with impact of droplet on weld pool considerably influences the surface appearance and geometrical characteristics of weld bead. Excessive thermal intensity enhances the fluidity of weld deposit that makes it difficult to hold in position. An analytical model study [18] indicates a solution to resolve the critical selection of pulse parameters for acceptable bead on plate weld deposit. The geometrical characteristics of weld bead involving the penetration, height and width of reinforcement, toe angle and dilution is largely

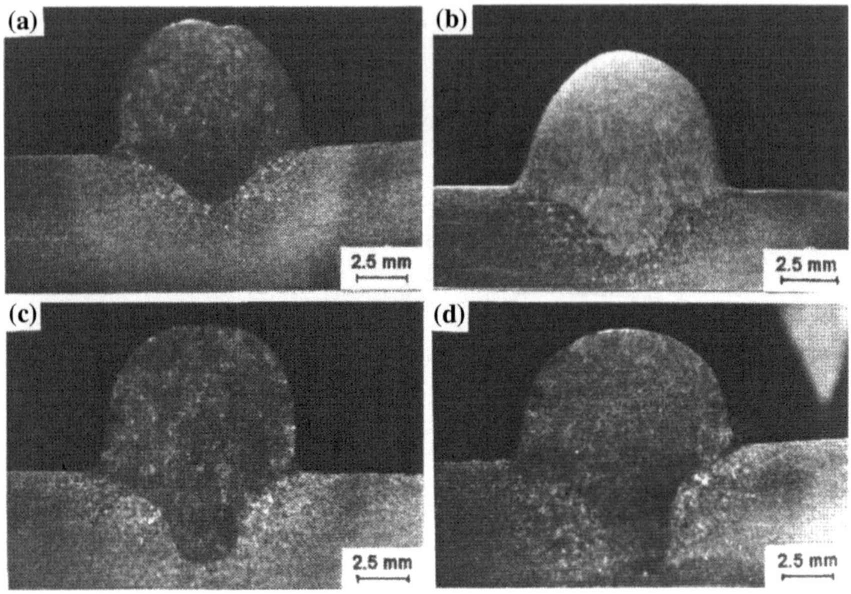

Fig. 5.32 At a given mean current of 100 A typical appearance of transverse section of pulsed current vertical-up bead on plate weld deposit of steel at different ϕ of **a** 0.206, **b** 0.179, **c** 0.085 and **d** 0.072

governed by $[(\phi k_p)/I_m]$ where $(k_p = t_p/t_{pul})$ is the pulse duty cycle. For an acceptable quality of bead on plate steel weld deposition using 1.2 mm diameter filler wire the value of $[(\phi k_p)/ I_m]$ lies in the range of 2.81 ± 0.14 and 1.77 ± 0.07 in case of welding with and without inductance respectively at a close control of arc voltage of about 21 ± 0.5 V.

5.3 Weld Joint

In view of the significant influence of pulse parameters on amount of fusion of base metal and the shape of depositing weld bead, governed by the nature of delivering (amount and location) heat by molten metal droplets to the weld pool, it is very much expected that they can also affect the shape of weld in a joint. But due to change in dynamics of the molten droplets with respect to the direction of gravitational force acting on it during its path of deposition in weld pool the nature of heat transfer may get significantly affected which may consequently change the weld geometry. Thus in case of preparation of weld joint by P-GMAW the weld geometry may depend not only on the welding parameters but also on the position of welding.

5.3.1 *Flat Butt Welding*

A desired weld quality could be maintained by controlling the value of an objective function defined as a function of weld geometry, which decreases with the increase of welding speed and complementary rise of adequate wire feed speed during pulse current MIG welding of aluminium sheet [19]. The P-GMAW is largely characterized by its comparatively high deposition rate at relatively low heat input [20, 21], than conventional GMAW. Welding at low heat input in case using the pulsed current primarily results from an interruption in metal deposition [22] that leads to low heat build-up in weld pool. At a given mean current, welding speed and arc voltage [23–26] a change in pulse frequency from 0–100 Hz significantly changes the weld geometry. This is typically observed in case of a single pass V-groove (Fig. 5.33) butt welding of about 7 mm thick section of Al–Zn–Mg alloy using 1.6 mm diameter Al–Mg filler wire under argon shielding and copper backing. Increase of pulse frequency from 0–50 Hz narrows down the weld root but it broadens with a further increase of pulse frequency to 100 Hz. It especially happens during the use of mean current higher than the transition current (210 A) of the filler fire (Fig. 5.34). Change of pulse frequency from 25–100 Hz at the mean current within 180–220 A significantly varies the weld geometry defined by a ratio (R_G) of the width of root to top of the weld in the range of about 0.08–0.69. At a given conventional energy input such a change in trend of variation of weld geometry at certain point of pulse frequency again indicates that there exists a criticality of the influence of pulse parameters which could be realized by considering their summarized effect defined by the factor ϕ as discussed above. The increase of pulse frequency up to 50 Hz possibly causes a lowering of net heat input [23, 25, 26] or heat buildup in weld pool and thereby reduces the melting of base metal which narrows down the weld root. It is reported that pulse off time fraction (t_b/t_{pul}), where ($t_{pul} = t_p + t_b$), plays a significant role in this regard due to its appreciable influence on heat build-up and heat concentration especially at higher peak current of higher pulse frequency. This behaviour is prominently visible (Fig. 5.35) in case of welding of 10 mm thick Al–Zn–Mg alloy plate at a given pulse frequency of 100 Hz with different pulse off time fraction. Here the effect of droplet transfer mechanism on the width and depth of fusion in base plate is more pronounced at high peak current of 100 Hz. This is because the spray transfer of the droplets leads to deep up to finger type penetration (Fig. 5.35iii) in P-GMAW process [8]. But, at

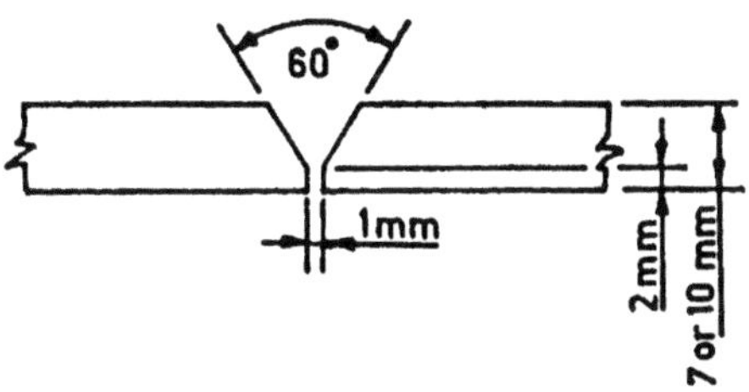

Fig. 5.33 Weld groove for single pass deposit

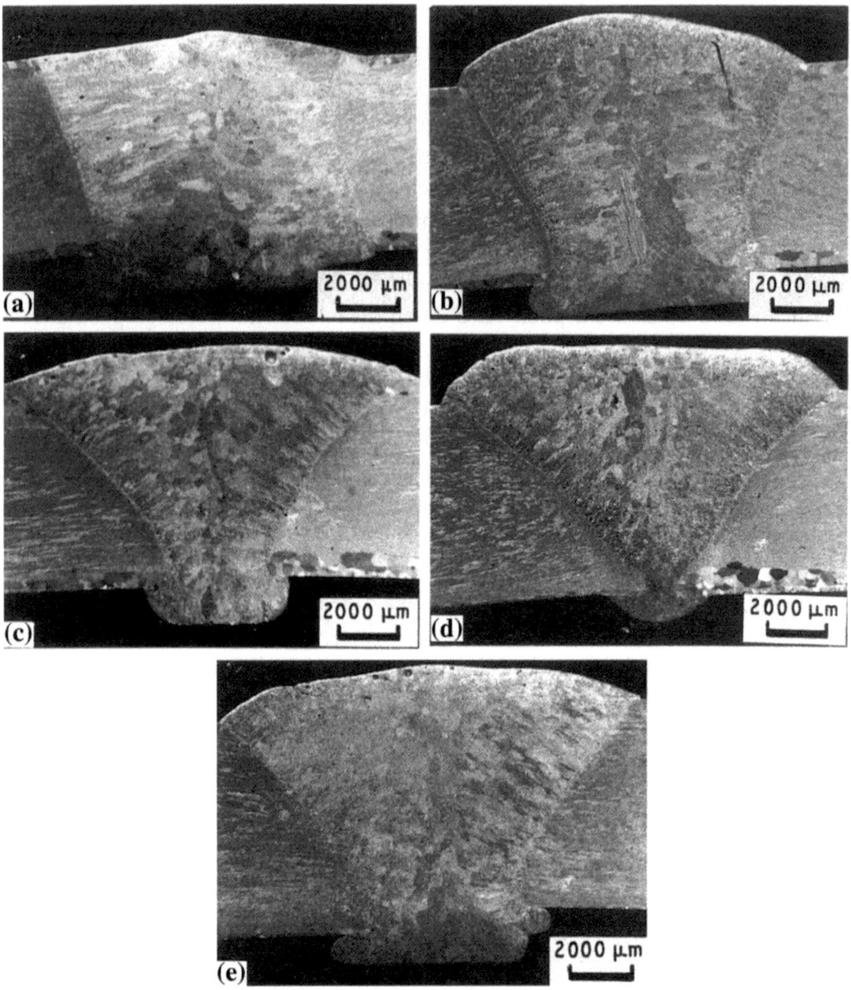

Fig. 5.34 At a given mean current of 210 A the influence of pulse frequency on the weld geometry of Al–Zn–Mg alloy; **a** 0, **b** 25, **c** 33, **d** 50 and **e** 100 Hz

excessively low pulse off time fraction of higher pulse frequency the influence of interruption in solidification becomes practically ineffective and the weld pool starts behaving as a conventional weld, and thus widens the weld root as it is revealed in Figs. 5.34e and 5.35i during welding at relatively higher pulse frequency of 100 Hz.

In case of multipass butt welding of a thick plate (25 mm) (Fig. 5.36a, b) of Al–Zn–Mg alloy using the similar welding parameters as stated above [27, 28], but with preheating and inter pass temperatures of 393 K, the dilution also reduces linearly with ϕ at a given I_m of 220 A (Fig. 5.37). The significantly low dilution of

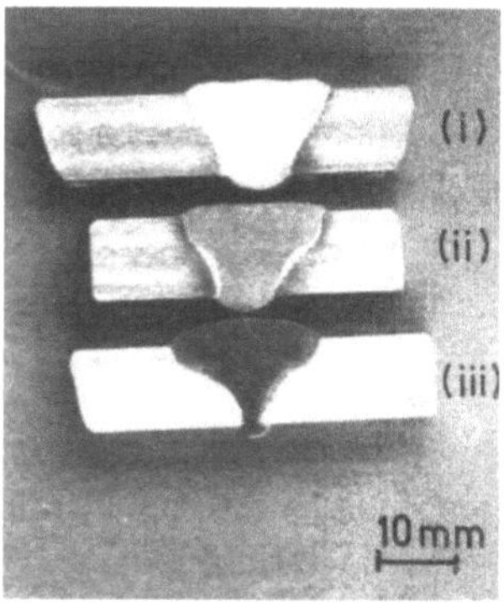

Fig. 5.35 At a given pulse frequency of 100 Hz effect of (tb/t) on typical variation of RG of the weld *i* 0.1, RG = 0.5, *ii* 0.3, RG = 0.3 and *iii* 0.5, RG = 0.1

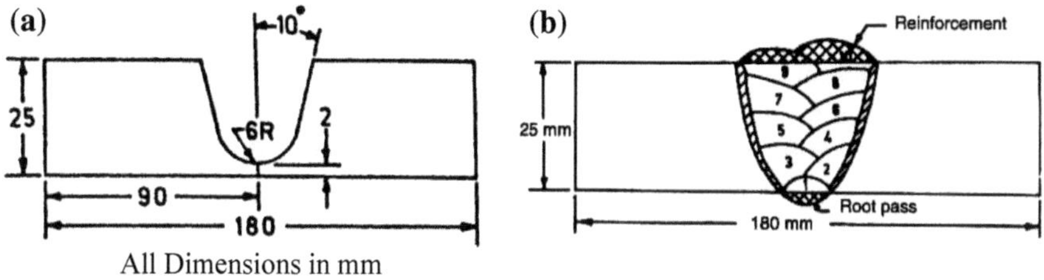

Fig. 5.36 **a** Weld groove and **b** multi-pass weld joint of thick section of Al–Zn–Mg alloy

Fig. 5.37 Influence of the factor ϕ on dilution of multi-pass weld of Al–Zn–Mg alloy

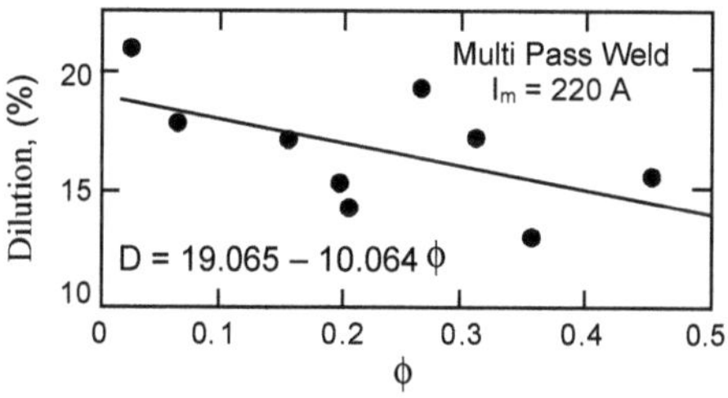

multi-pass weld than the single pass weld (Fig. 5.14) may be understood from a large amount of deposition in multi-filled weld groove, where the fusion of base plate primarily takes place only from the deposition adjacent to the groove surface.

During P-GMA welding an inappropriate selection of pulse parameters as the mean current, pulse duration and pulse frequency can produce undesirable penetration, bead width and bead height. The use of rotating arc P-GMAW technique gives rise to comparatively flat weld bead with less penetration [29, 30]. In consideration of the possibilities of favourable control of weld geometry by varying the pulse parameters, the use of P-GMAW process with multiple drop transfer per pulse can be effectively explored in welding of relatively thin sheet to thick plate. Use of comparatively lower ϕ, I_m and arc voltage can satisfactorily weld thin sheets [31], whereas for welding of thick sections [32, 33] an intermediate level of I_m and arc voltage with lower ϕ is found quite effective.

5.3.2 *Positional Welding*

Construction of an engineering structure often requires the use of positional or out-of-position welding such as welding in vertical, horizontal and overhead positions. Execution of this type of welding is a complicated phenomenon due to difficulties arises from (i) transportation of molten filler metal from electrode tip to weld pool and (ii) retaining the weld pool in required location. P-GMAW is often preferred for positional welding due to its better directional control over metal transfer at relatively low mean current than that of conventional GMAW [34, 35] which is generally employed with short circuit or spray mode of metal transfer. The short circuit welding always carries a threat of significant spattering, cold lapping, lack of penetration and low process economy [36] when the spray transfer uses relatively high current that often makes the process control difficult due to undesirably high fluidity of weld pool.

Positional welding by employing pulsed current is much more convenient and operator friendly in control than doing the same by using conventional spray transfer GMAW. This is because in P-GMAW process the spray transfer at extra high current of each pulse can be discontinued at a desired interval that gives an opportunity for execution of weld deposition with relatively low heat build-up facilitating a better control and manipulation of weld pool. Here it is interesting to note that in spite of low heat buildup the pulsed current process improves the groove wall and root fusion because it operates with larger energy of much higher average (mean) current than GMAW and extremely high current (peak) for a while during deposition. This is followed by a considerably low base current with or without weld deposition facilitating the energy dissipation or loss to continue low heat build-up in weld pool. It immensely helps in elimination of common difficulty of preparation of out-of-position weld joint as incomplete fusion along with enjoying other major benefits of arc characteristics as generation of low spatter and fume which are finally attributing to higher process economy and low operators fatigue. However, selection of right pulse parameters in positional welding becomes quite complex due to difficulties to control the bead profile in appropriate location, undercutting, humped bead formation and spill-over of molten weld metal. Thus, selection of right pulse parameters for desired weld characteristics is very much critical [10, 12, 29, 30] which requires knowledge of summarized influence of pulse parameters on it [10, 11].

The better performances of P-GMAW in out-of-position welding produces weld joint of superior mechanical properties to those of the weld joint prepared by commonly known short-circuiting or spray mode GMAW process. Use of short-circuiting arc often significantly reduces fatigue life of weld joint due to occurrence of undercut at the weld toe and incomplete side-wall fusion of base material. Basic factors like microstructure and weld geometry primarily govern mechanical properties of the P-GMA weld joint being influenced by the pulse parameters. Increase of factor ϕ favours refinement of microstructure that enhances the tensile strength, Cv toughness, and fatigue life of weld joint [37]. Further as a relatively low heat process it reduces distortion and minimise development of residual stresses in weld joint especially when the high travel speed is not needed or cannot be employed.

Critical studies on vertical up welding (Fig. 5.38a) convincingly show the suitability of pulsed current GMAW over the conventional GMAW in positional welding [17, 38–40]. The pulse parameters play integrated role to determine geometrical characteristics of positional weld as observed in square butt welding (Fig. 5.38a) of 6 mm thick structural steel plate carried out under Ar-5% CO_2 shielding. Conventional GMAW at an energy input of 2.3 kJ/mm results severe undercut in vertical up weld produced at the welding current of 100 and 130 A as shown in Fig. 5.39a, b respectively. But at mean currents of the same level the P-GMAW can produce sound weld with acceptable geometry. However, the P-GMAW can produce flawless weld with satisfactory root penetration and weld geometry only at certain pulse parameters. At the relatively low and high mean currents of 100 and 130 A variation in ϕ significantly affects the weld geometry with respect to its width, toe angle and the top and root reinforcements as shown in

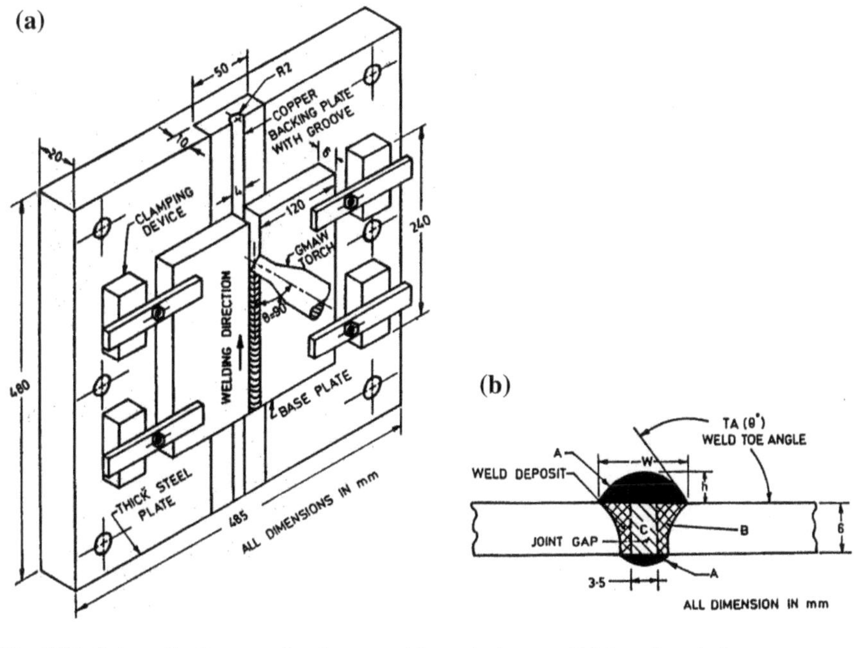

Fig. 5.38 Schematic diagram of a fixture and b vertical-up weld joint of steel plate

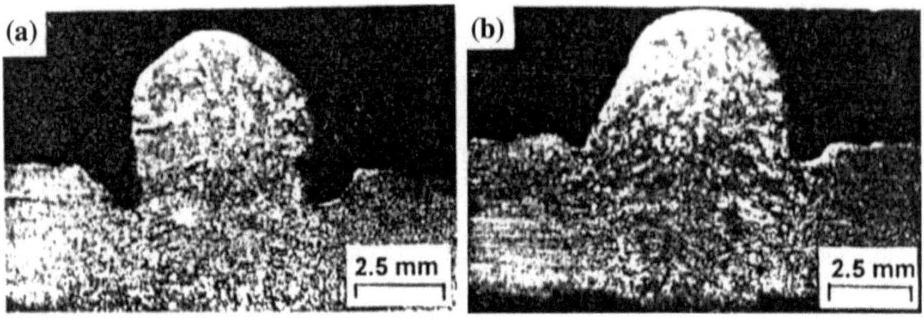

Fig. 5.39 Typical appearance of vertical-up GMA welds of steel at different welding currents of **a** 100 and **b** 130 A

Figs. 5.40 and 5.41 respectively. The root reinforcement reduces with decrease of ϕ and increase of mean current leading to lack of penetration as observed at ϕ and I_m of 0.071 and 130 A respectively. At a given energy input the observed irregularities in weld quality primarily happens due to variation in temperature of droplet along with a change in behaviour of metal transfer at different pulse parameters. Thus, to produce an acceptable weld by P-GMAW it may be very much useful to have a clear knowledge of the influence of pulse parameters in terms of ϕ on different geometrical characteristics of positional weld.

During vertical up welding of 6 mm thick steel plate carried out by using 1.2 mm diameter filler wire at the mean currents of 100 and 130 A under Ar-5% CO_2 shielding, the weld geometry maintains a good correlation with the

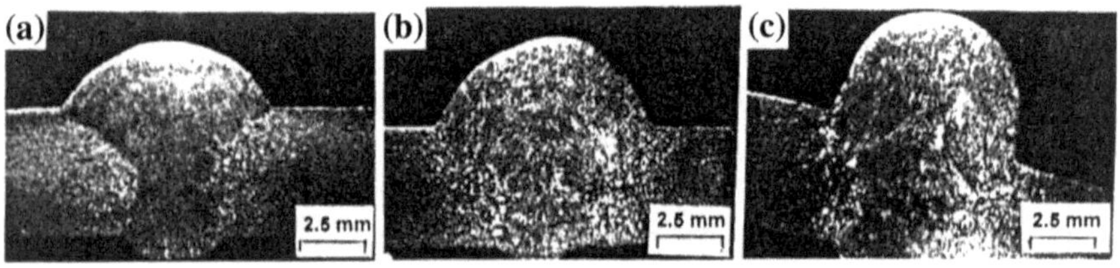

Fig. 5.40 Typical appearance of pulsed current vertical-up steel welds at different ϕ of **a** 0.171, **b** 0.097 and **c** 0.053 at 100 A mean current

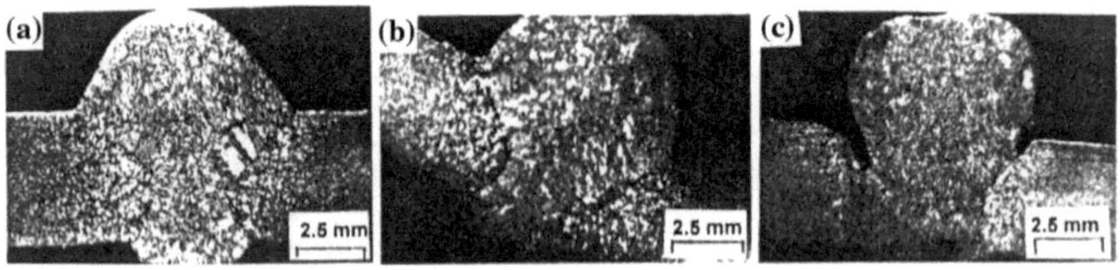

Fig. 5.41 Typical appearance of pulsed current vertical-up steel welds at different ϕ of **a** 0.226, **b** 0.125 and **c** 0.071 at 130 A mean current

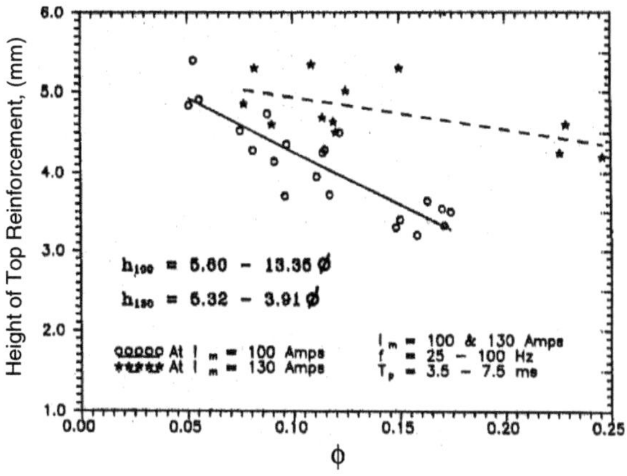

Fig. 5.42 Influence of ϕ on height of top reinforcement of vertical-up steel welds deposited at different mean currents

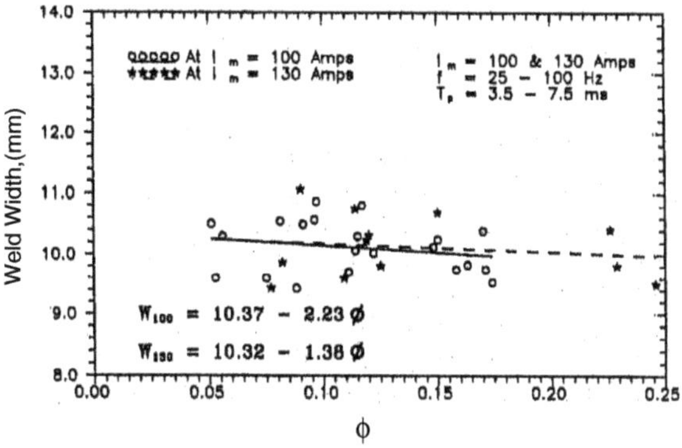

Fig. 5.43 Influence of ϕ on width of vertical-up steel welds deposited at different mean currents

summarized influence of pulse parameters of P-GMAW. Increase of ϕ significantly reduces height of weld reinforcement (Fig. 5.42), but marginally reduces weld width (Fig. 5.43) possibly due to predominant influence of rate of deposition ((D_r) defined as follows and the extent of root reinforcement.

$$D_r = \left[\left[(\pi d^2/4) \times V_w \times \rho\right]/v\right] \tag{5.18}$$

where d is wire diameter, V_w is wire feed rate, ρ is mass density of filler wire and v is the welding speed. However, at a given ϕ increase of I_m markedly enhances height of weld reinforcement (Fig. 5.42) especially at higher ϕ up to about 0.15–0.25. But it insignificantly affects the weld width (Fig. 5.43). It appears to be justified because at a given mean current an increase of ϕ reduces rate of deposition (Fig. 5.44), but enhances area of root reinforcement (Fig. 5.45) which keeps the weld width low by consuming more material at root of the weld. The enhancement of area of root reinforcement with increase of ϕ up to about 0.2–0.25 primarily happens due to increase in drop temperature with the increase of ϕ as depicted in Fig. 4.6. However, in spite of decrease in drop temperature (Figs. 4.5 and 4.6) an

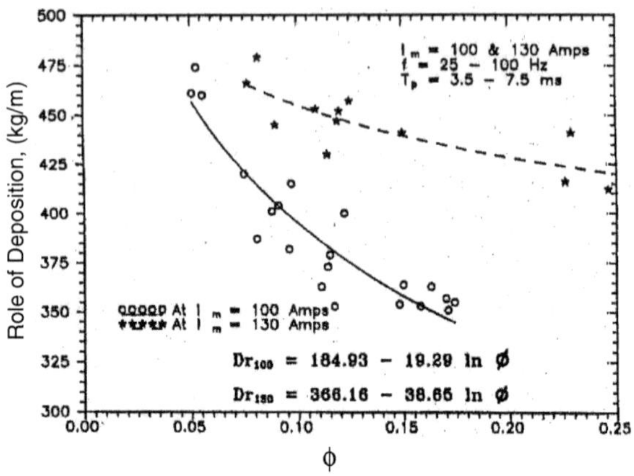

Fig. 5.44 Influence of ϕ on rate of deposition per unit length of vertical-up steel welds deposited at different mean currents

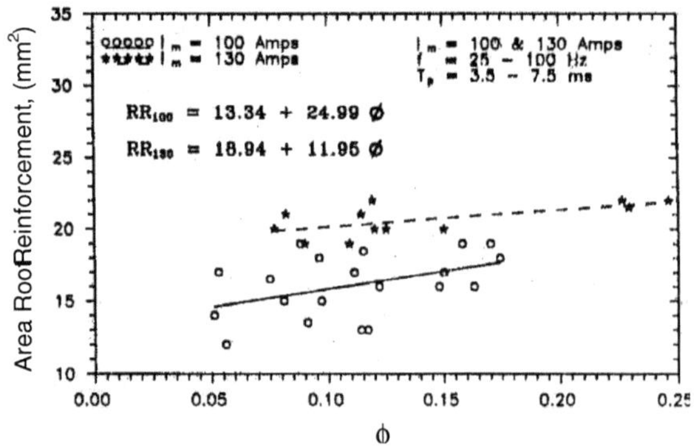

Fig. 5.45 Influence of ϕ on area of root reinforcement of vertical-up steel welds deposited at different mean currents

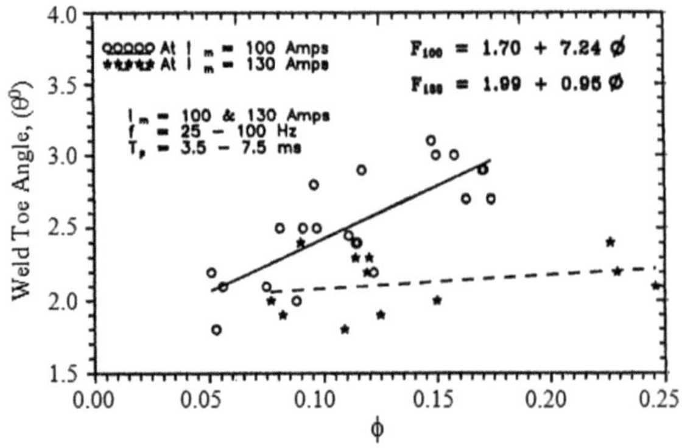

Fig. 5.46 Influence of ϕ on form factor of vertical-up steel welds deposited at different mean currents

increase in area of root reinforcement at a given ϕ of higher mean current indicates that the area of root reinforcement is not only governed by the drop temperature, but also by some other factors [41]. These factors may be considered as arc heating, transfer behaviour of droplets and flow characteristics of weld pool under the vertical up deposition. At a given ϕ an increase of mean current considerably enhances rate of deposition (Fig. 5.44) especially at higher ϕ, which is in concurrence to the observed enhancement of height of weld reinforcement with the increase of mean current. But increase in area of root reinforcement opposes in principle the phenomenon of increase of height and width of top reinforcement of weld with increase of mean current at a given ϕ. The increase of height of reinforcement without affecting the weld width at higher I_m possibly also happens due to predominant vertically down flow characteristics of the weld pool. Thus it is realized that height and width of top reinforcement of positional weld are primarily governed by a competitive influence of the rate of deposition and the area of root reinforcement. Here the former one plays predominant role by over compensating consumption of weld metal through increase of root reinforcement.

At a low mean current of 100 A an increase of ϕ enhances the form factor significantly (Fig. 5.46) by raising it up to a value of about 3.0 at ϕ in the range of about 0.145–0.160. Such a form factor is recommended [42], as an acceptable value for the practical purposes of application. But at higher mean current increase of ϕ marginally affects the weld form factor by maintaining its value within about 2 ± 0.2. However at a given mean current the toe angle (Fig. 5.38b) of top reinforcement of weld increases significantly with increase of ϕ, but at a given ϕ toe angle reduces with increase of mean current from 100 to 130 A (Figs. 5.47 and 5.48). Both the weld form factor and toe angle are important criteria due to their potential influence on notch effect when a weld joint is especially subjected to bend and/or dynamic loading conditions that introduces tensile load at the notch. Thus, increase of weld form factor and reduction in inner toe angle is beneficial to minimize the notch effect, which can be achieved by using higher ϕ at lower mean current and pulse frequency [43]. A low mean current of 100 A may give deficient fusion of base metal at weld toe in vertical-up GMA weld of mild steel plate as typically shown in (Fig. 5.49). The variation in weld geometry as usual influences the dilution of weld (D_{nw}) deposit defined as follows.

$$D_{nw} = [[(A+B+C) - (A+C)]/(A+B+C)] \tag{5.19}$$

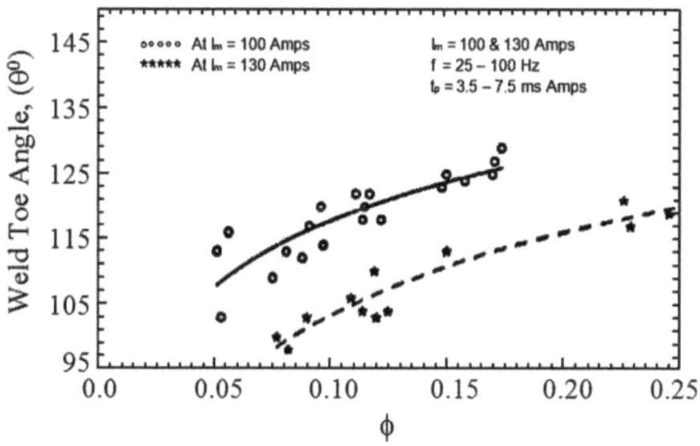

Fig. 5.47 Influence of ϕ on toe angle of vertical-up steel welds deposited at different mean currents

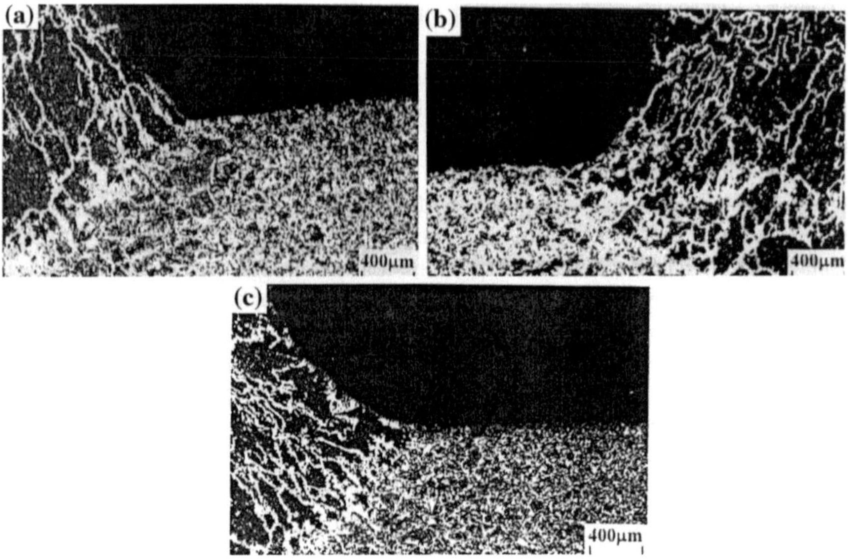

Fig. 5.48 At a given mean current of 100 A typical influence of ϕ on toe angle of vertical-up steel weld **a** 0.056, **b** 0.115, and **c** 0.175

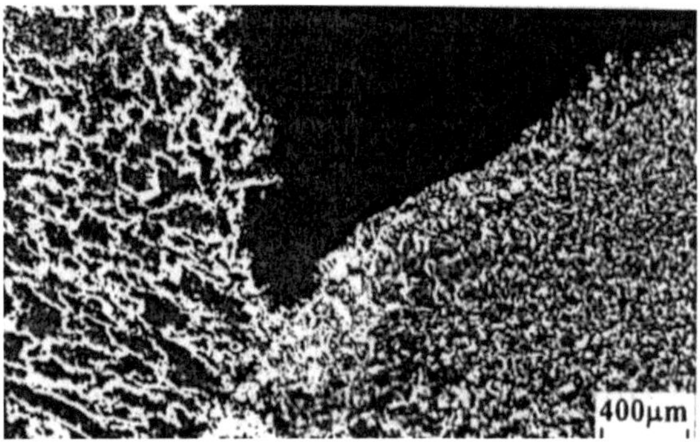

Fig. 5.49 Typical incomplete fusion of base metal at weld toe of low current (100 A) vertical-up GMA weld of mild steel plate

where A is area of top and bottom reinforcement, B is area of fused base metal, and C is the area of joint gap (Fig. 5.38b). At a given mean current dilution of weld enhances significantly with increase of ϕ (Fig. 5.50) but it relatively reduces with decrease of mean current from 130 to 100 A at a given ϕ. The increase in dilution with the increase of ϕ possibly arises due to increase of drop temperature (Fig. 4.6). But the anomaly observed in correlation of dilution and drop temperature at different mean currents further indicates that beside drop temperature the other factors as discussed above also play significant role that govern dilution of weld deposit in out of position welding.

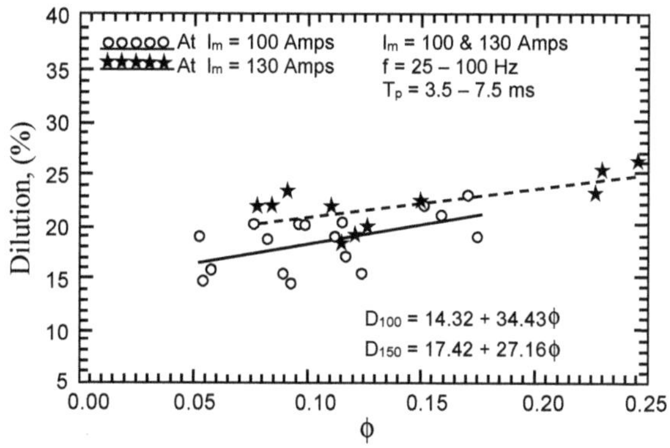

Fig. 5.50 Influence of ϕ on dilution of the welds deposited at different mean currents

The positional welding with P-GMAW process can be successfully used in practically all kinds of ferrous and non ferrous filler electrodes. A synergic programmable pulse power supply can be conveniently used with customized pulse waveform for commonly known welding consumables. But, for any advanced challenging application of the process by negotiating non conventional design and properties of weld joint of special material the consultation of the factor ϕ may be helpful in proper selection of pulse parameters.

References

1. Cho, M.H., Lim, Y.C., Farson, D.F.: Simulation of weld pool dynamics in the stationary pulsed gas metal arc welding process and final weld shape. Weld. J. **85**(12), 271s–283s (2006)
2. Ghosh, P.K., Gupta, P.C., Somani, R.: Influence of pulse parameters on bead geometry and HAZ during bead on plate deposition by MIG welding process. Z. Metallkde **82**(10), 756–762 (1991)
3. Goyal, V.K., Ghosh, P.K., Saini, J.S.: Analytical studies on thermal behaviour and geometry of weld pool in pulsed current gas metal arc welding. J. of Mater. Processing Tech. **209**, 1318–1336 (2009)
4. Dillenbeck, V.R., Castagno, L.: The Effects of Various Shielding Gases and Associated Mixtures in GMA Welding of Mild Steel. Weld. J. **66**(9), 45–49 (1987)
5. Goyal, V.K., Effect of thermal and solidification behaviour on characteristics of pulsed current GMA weld, PhD thesis, Indian Institute of Technology Roorkee, India (2007)
6. Goyal, V.K., Ghosh, P.K., Saini, J.S.: Met. Mater. Trans. A **38A**(8), 1794–1805 (2007)
7. Shackleton, D.N., Lucas, W.: Shielding gas mixtures for high quality mechanized GMA welding of Q&T steels. Weld. J. **53**(12), 537s–547s (1974)
8. Ghosh, P.K., Dorn, L., Issler, L.: Fatigue crack growth behaviour of pulsed current MIG weld of Al-Zn-Mg alloy. Int. J. Joining Mater. **6**(4), 163–168 (1994)
9. Needham, J.C.: Pulse controlled consumable electrode welding arcs. Br. Weld. J. **12**(4), 191–197 (1965)
10. Ghosh, P.K., Rai, B.K.: Characteristics of pulsed current bead on plate deposit in flux cored GMAW process. ISIJ Int. **36**(8), 1036–1045 (1996)

11. Ghosh, P.K.: An analysis of weld characteristics as a function of pulse current MIG welding parameters. Int. J. Joining Mater. **8**(4), 157–161 (1996)
12. Amin, M.: Pulse current parameters for arc stability and controlled metal transfer in arc welding. Metal Constr. **15**(5), 272–287 (1983)
13. Ghosh, P.K., Rai, B.K.: Correlations of pulse parameters and bead characteristics in pulsed current flux cored GMAW process. Ind. Weld. J. **31**(4), 30–39 (1998)
14. French, I.E., Bosworth, M.R.: A comparison of pulsed and conventional welding with basic flux cored and metal cored welding wires. Weld. J. **74**(6), 197s–205s (1995)
15. Widgery, D.J.: Flux cored wire: An update welding and met. Fabrication, April, 188–124 (1988)
16. Liu, H.Y., Li, Z.X., Li, H., Shi, Y.W.: Study on metal transfer modes and welding spatter characteristics of self-shielded fluxcored wire. Sci. Technol. Weld. Joining **13**(8), 777–780 (2008)
17. Ghosh, P.K., Gupta, S.R., Randhawa, H.S.: Characteristics and criticality of pulsed current vertical up GMA weld in steel. Int. J. Join. Mater. **11**(4), 99–110 (1999)
18. Ghosh, P.K., Gupta, S.R., Randhawa, H.S.: Analytical studies on characteristics of vertical up bead on plate weld deposition using pulsed current GMAW. Int. J. for the Joining of materials **12**(3), 76–85 (2000)
19. Ghosh, P.K.: The influence of dilution and heat input on the characteristics of SAW stainless steel cladding on mild steel. Int. J. Joining Mater. **4**(3), 90–99 (1992)
20. Ghosh, P.K., Gupta, P.C., Breazu, M., Gupta, R.K.: The influence of some welding parameters on the properties of stain-less steel strip cladding deposited by a submerged arc process. Trans. JIM **28**(7), 579–584 (1987)
21. Waszink, J.H., Piena, M.J.: Experimental investigation of drop detachment and drop velocity in GMAW. Welding Journal **65**(11), 289s–298s (1986)
22. Park, H.J., Kim, D.C., Kang, M.J., Rhee, S.: Optimization of the wire feed rate during pulse MIG welding of Al sheets. J. Achievements Mater. Manufact. Eng. 27, 1 March 83–86 (2008)
23. Gray, T.G.F., Spence, J.: Rational welding design, Butterworths, London, Revised Ed., (1983)
24. Ghosh, P.K., Gupta, P.C., Jain, N.K.: Effect of pulse frequency on the weld seam properties in pulsed arc MIG welding of Al–Zn–Mg alloy. Aluminium **64**(9), 933–935 (1988)
25. Ghosh, P.K., Gupta, S.R., Gupta, P.C., Rathi, R.: Influence of pulsed current welding on the microstructure and porosity content of Al–Zn–Mg alloy weldment. Pract. Met. **27**, 613–626 (1990)
26. Ghosh, P.K., Gupta, S.R., Gupta, P.C., Rathi, R.: Pulsed MIG welding of Al–Zn–Mg alloy. Mater. Trans. JIM **31**(8), 723–729 (1990)
27. Hussain, H.M., Ghosh, P.K., Gupta, P.C., Potluri, N.B.: Weld characteristics of multipass pulse current MIG welded Al–Zn–Mg alloy. Int. J. Join. Mater. **9**(2), 74–79 (1997)
28. Hussain, H.M., Ghosh, P.K., Gupta, P.C., Potluri, N.B.: Properties of pulsed current multipass GMA welded Al–Zn–Mg alloy. Weld. J. **75**(7), 210s–215s (1996)
29. Bagryanskii, K.V., Royanov, V.A.: Svar. Proiz. **10**, 23–24 (1968)
30. Smati, Z.: Automatic pulsed MIG welding. Met. Constr. **18**(1), 38R–44R (1980)
31. Ghosh, P.K., Kulkarni, S.G., Kumar, M., Dhiman, H.K.: Pulsed current GMAW for superior weld quality of austenitic stainless steel sheet. ISIJ Int. **47**(1), 138–145 (2007)
32. Ghosh, P.K., Devakumaran, K., Goyal, V.K., Kulkarni, S.G., Ghosh, Aritra K.: Procd. 14th ISME Int. Conf. Mechanical Engineering in knowledge and age, New Delhi, India, December, Delhi college of engineering, Paper 338 (2005)
33. Ghosh, P.K., Hussain, H.M.: Morphology and porosity content of multipass pulsed current GMA weld of Al–Zn–Mg alloy. Int. J. Join. Mater. **14**(1/2), 16–27 (2002)
34. Matthews, J.R., Lassaline, E.E., Porter, J.F., Leewis, K.G.: Evaluating frequency modulated GMA welding for HY 80 steel. Weld. J. **71**(9), 49–53 (1992)
35. Alum, C.J.: MIG welding—Time for a reassessment, Metal Constr. **15**(6), 347–353 (1983). [2.12128.]
36. Craig, E.: A unique mode of GMAW transfer. Weld. J. **66**(9), 51–55 (1987)

37. Ghosh, P.K., Randhawa, H.S., Gupta, S.R.: Met. and Mater. Trans. A **31A**(9), 2247–2259 (2000)
38. Ghosh, P.K., Gupta, S.R., Randhawa, H.S.: Analytical studies on characteristics of vertical up bead on plate weld deposition using pulsed current GMAW. Int. J. Joining Mater. **12**(3), 76–85 (2000)
39. Randhawa, H.S., Ghosh, P.K., Gupta, S.R.: Geometric characteristics of pulsed current positional GMA weld. ISIJ Int. **38**(4), 276–284 (1998)
40. Randhawa, H.S., Ghosh, P.K., Gupta, S.R.: Some basic aspects of geometrical characteristics of pulsed current vertical up GMA weld. ISIJ Int. **40**(1), 71–76 (2000)
41. Allum, C.J., Quintino, L.: Control of fusion characteristics in pulsed current MIG welding, Part-II, Simple model of fusion characteristics. Metal Constr. **17**(5), 314R–317R (1985)
42. Cornu, Jean: Advanced welding system, IFS Publication Ltd. UK, **2**, 127–165 (1988)
43. Maruo, H., Hirata, Y.B.: Study of pulsed MIG welding, IIW Doc. SG 212-585-84, Welding Department, Osaka University, Japan, July (1984)

Chapter 6
Characteristics of P-GMA Weld Joint

Abstract Metallurgical and mechanical characteristics of weld joints of aluminium, mild steel and stainless steel governed by the parameters of P-GMA welding have been discussed for control as a function of the summarised influence of pulse parameters defined by the factor ϕ. The metallurgical characteristics are systematically addressed in terms of weld chemistry, weld porosity and solidification behaviour of weld pool. The mechanism of the variation of microstructure in aluminium, mild steel and stainless steel welds as a function of pulse parameters is thoroughly discussed. The mechanical characteristics of weld joint are addressed by the control of various stresses generated in it with a change in pulse parameters during P-GMA welding.

Keywords Weld joints · Weld chemistry · Solidification · Porosity · Microstructure · Aluminium · Mild steel · Stainless steel · Distortion · Shrinkage stresses · Residual stresses

6.1 Metallurgical Characteristics

In structural application of materials the consideration of metallurgical characteristics of weld joint is of utmost importance. The P-GMAW process has become widely popular in welding of nonferrus and ferrous materials due to its more versatility in application to produce quality weld with better economy. However, as it is discussed in earlier chapters, the benefits of this process depends upon better control of metal deposition and weld thermal behaviour giving rise to more favourable influence on various metallurgical characteristics of weld and heat affected zone (HAZ) and mechanical characteristics of the joint. Appropriate use of this process can improve the metallurgical characteristics of weld joint in comparison to those of conventional GMA weld mainly with respect to the chemical composition, solidification and porosity content of weld deposit and microstructures of weld and HAZ.

P.K. Ghosh, *Pulse Current Gas Metal Arc Welding*, Materials Forming, Machining and Tribology, DOI 10.1007/978-981-10-3557-9_6

6.1.1 Chemistry of Weld

At a given combination of base metal and welding consumables, the chemical composition of weld is primarily determined by the dilution, loss of element by oxidation and solid state diffusion depending upon the type of material as well as welding process, procedure and parameters. As the P-GMAW is a comparatively low intensity thermal process than GMAW, it generally affects the weld chemistry relatively less intensively by lower dilution. But the thermal behaviour of P-GMAW varies significantly with pulse parameters as discussed in Chap. 4. By inert shielding of arc environment the GMAW in general protects the liquid weld pool from the oxidation loss of elements, but the situation becomes somehow different in case of P-GMAW due to fluctuation of arc pressure under pulsed current affecting dynamics of shielding gas flow. Under adverse conditions of pulse parameters disturbing the protective shielding gas jacket up to an extent to form a vortex in it may cause air aspiration in the arc environment. Thus, it can encourage the loss of element from the molten metal especially those are having high affinity to oxidation. The control of such conditions largely depends upon maintaining the stability of arc and shielding jacket as well as using appropriate thermal and transfer behaviours of molten metal at proper pulse parameters and arc voltage as discussed earlier in Chap. 3.

The above mentioned phenomena can be interestingly realized by considering the case of welding of Al–Zn–Mg alloy containing vaporizable Zn by using non Zn content Al–Mg filler wire. During P-GMA welding of Al–Zn–Mg (7005) alloy containing about 4.5 wt% Zn the weld deposition from 1.6 mm diameter Al–Mg (5183) filler wire under argon (99.98%) shielding produces weld containing appreciable amount of zinc pick-up in it through dilution of base metal. The amount of zinc pick-up in weld very much depends upon pulse parameters as observed in single and multipass welds of 10 and 25 mm thick sections respectively (Fig. 6.1a, b). At a given mean current the increase of pulse frequency and pulse duration enhances zinc pick-up in weld, but the amount of zinc content of the weld can be regulated to high or low level with respect to that of the conventional GMA (0 Hz) weld by a control of the pulse parameters [62]. The flexibility in control of the zinc pick-up in weld may be considered as an advantage of the P-GMAW because the presence of sufficient zinc can enhance strength of the weld [63] by introducing precipitation strengthening phase ($Mg_3Zn_3Al_2$) in the matrix whereas low zinc content can protect the corrosion resistance of Al–Mg alloy weld deposit. Zinc pick-up of the weld maintaining a Mg/Zn ratio lying in the range of 3.7–6.0 may give rise to formation of $Mg_3Zn_3Al_2$ precipitate in the matrix [64]. However, X-ray diffraction analysis of central part of a multipass P-GMA weld deposit of Al–Mg alloy joining the Al–Zn–Mg alloy plates shows the presence of $(Al,Zn)_{49}Mg_{12}$ phase after more than 30 days of natural ageing even at a low zinc content (0.53–0.78 wt%). It arises out of relatively low dilution (D) caused by lesser contact area of multipass deposit (Fig. 6.2) with the groove wall than single pass weld. The $(Al,Zn)_{49}Mg_{32}$ phase is in close approximation by nature [1] to the $Mg_3Zn_3Al_2$.

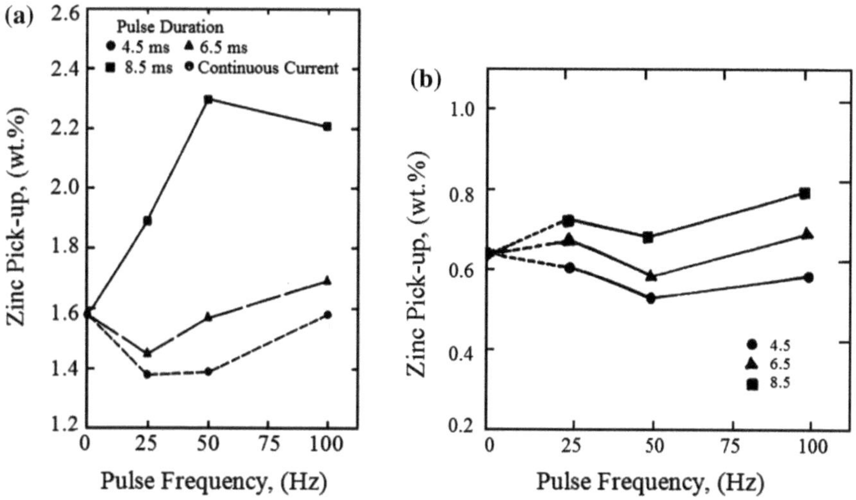

Fig. 6.1 At a given mean current influence of pulse frequency and pulse duration on zinc pick-up in Al–Mg deposit in welding of Al–Zn–Mg alloy; **a** Single-pass at Im = 235 A and **b** Multi-pass at Im = 220 A

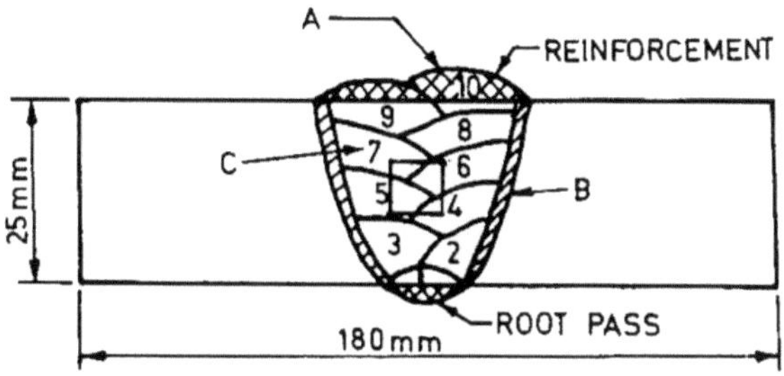

Fig. 6.2 Schematic diagram of multipass weld deposit

$$\text{Dilution (\%)} = \frac{(A+B+C)-(A+C)}{A+B+C} \times 100$$

The stoichiometric amount of the precipitate enhances with the increase of zinc pick-up at more dilution as shown in Table 6.1. The measured zinc pick-up (Z_p) in multipass weld deposit of Al-4.5 Mg alloy as a function of dilution (D) is generally found [65] comparatively lower than the theoretically estimated zinc pick-up with respect to the zinc content (4.5 wt%) of Al–Zn–Mg base metal as shown in Fig. 6.3. It happens primarily due to the loss of zinc by vaporization and oxidation [63] during welding. At this point it should be remembered that during GMA welding of alloys often the filler metal is selected giving due importance not only to its strength but also to some other properties. For example the zinc free (Al–Mg) filler metal is generally recommended for welding of Al–Zn–Mg alloy due to its good mechanical properties and corrosion resistance, where the excessive

Table 6.1 At different pulse parameters zinc pick-up and precipitate content of 5183 alloy weld deposit in multipass weld joint of 7075Al-alloy

Pulse frequency (Hz)	Pulse duration (ms)	Dilution (%)	Zinc pickup (wt%)	Amount of $(Al,Zn)_{49}Mg_{12}$ (wt%)
0	–	17.5	0.64	1.0539
25	4.5	15.5	0.60	0.9880
50	4.5	13.0	0.53	0.8563
100	4.5	14.5	0.56	0.9180
25	6.5	17.5	0.67	1.1033
50	6.5	15.5	0.58	0.9551
100	6.5	18.0	0.69	1.1362
25	8.5	19.5	0.72	1.1856
50	8.5	17.5	0.68	1.1198
100	8.5	21.0	0.78	1.2845

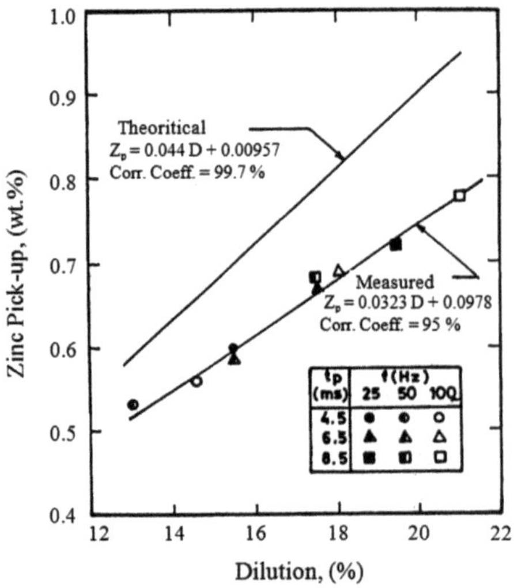

Fig. 6.3 Influence of dilution on zinc pick-up at multipass Al–Mg deposit in welding of Al–Zn–Mg alloy

precipitation strengthening may enhance its susceptibility to corrosion. Thus, the optimum control of zinc level in weld by appropriate selection of pulse parameters is necessary through understanding of their influence on zinc pick-up. In this regard the knowledge of influence of pulse parameters, as a function of energy (I_b/I_p) and its distribution in the system through pulse off-time fraction ($f.t_b$), on specific aspects of weld chemistry may be of significant importance. Due to clear correlation of ϕ with dilution of weld (Chap. 5), its basic components (Eq. 3.1) also keep meaningful correlation with chemistry of weld deposit as shown in Fig. 6.4 for the zinc pick-up in Al–Mg weld deposit as a function of the pulse off time fraction $[(t_b/(t_p + t_b)]$ in multipass welding of Al–Zn–Mg alloy [65]. The increase of $f.t_b$ at a given pulse frequency significantly reduces the zinc pick-up in weld. A similar

Fig. 6.4 At a given pulse frequency effect of pulse off-time fraction on zinc pick-up by Al–Mg deposit in welding of Al–Zn–Mg alloy

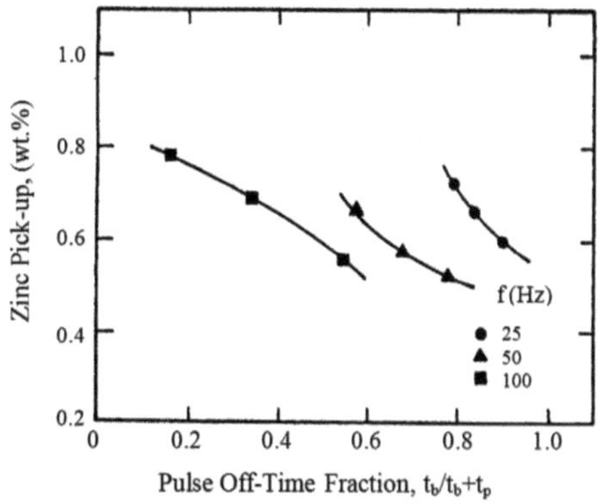

understanding about correlation of pulse parameters with dilution and chemistry of weld deposit especially in case of other dissimilar metal deposition by P-GMAW process is necessary to produce weld joint of desired quality.

In case stainless steel cladding by weld overlay on C-Mn structural steel the control of chemical composition is of immense importance to protect it from the loss of corrosion resistance due to disturbance in compositional balance resulting from dilution of base metal. Although due to its comparatively milder thermal behaviour, the P-GMAW gives rise to relatively lower dilution with respect to other commonly used welding processes in this area as discussed in Chap. 5, but its control is necessary to maintain the desired composition of weld overlay. The dilution primarily enhances C and reduces Si, Mn, Cr and Ni content of overlay (Fig. 6.5) with respect to chemical composition of the stainless steel filler wire and base metal [2] (Table 6.2) respectively. As per Delong diagram of transformation, the change in chemical composition of weld overlay affects its chromium and nickel equivalents. The increase in dilution significantly reduces the chromium equivalent but moderately enhances nickel equivalent of the cladding (Fig. 6.6), where the nitrogen content generally lies within a range of 0.05–0.1 wt% [3–5]. The variations of chromium and nickel equivalents with change of dilution and vis a vis the welding parameters strongly affects the presence of magnetic micro constituents as δ-ferrite and martensite in the matrix of weld overlay. According to ferrite meter analysis of stainless steel overlay by magnetic induction principle, the use of P-GMAW gives rise to formation of comparatively larger amount of combined magnetic constituents in matrix [3, 66] with respect to that observed in cladding produced by conventional (0 Hz) GMAW process (Fig. 6.7). However, increase of mean current or pulse duration reduces amount of magnetic constituent, whereas at a given mean current and pulse duration increase of pulse frequency enhances the same in agreement to their effects on dilution of the matrix. The decrease of dilution reduces nickel equivalent at a relatively slower rate than the rate of enhancement of chromium equivalent and thus, as per Delong diagram, it makes the weld overlay

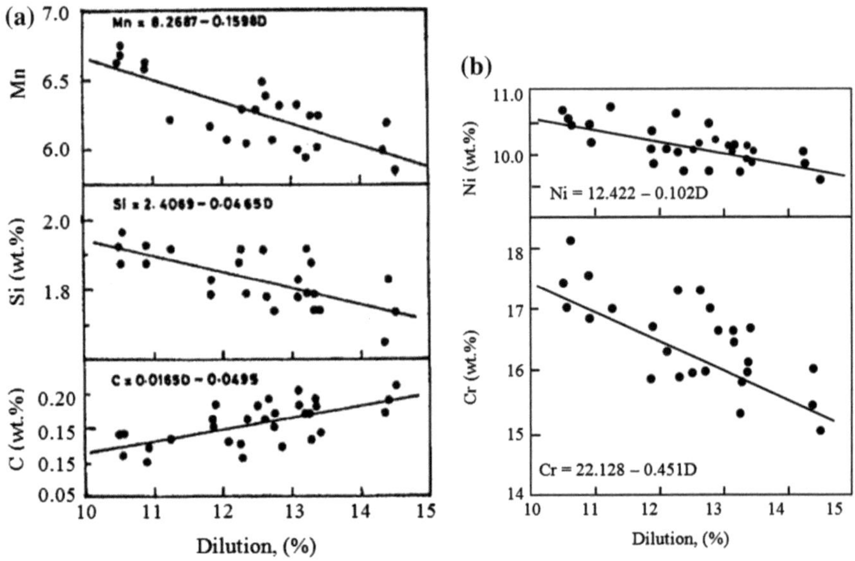

Fig. 6.5 Effect of dilution on different chemical constituents **a** for (C,Si and Mn) and **b** for (Cr and Ni) of stainless steel weld overlay on C–Mn Steel

Table 6.2 Chemical compositions of C–Mn base metal and stainless steel filler metal

Material	Chemical composition (wt%)				
	C	Mn	Si	Cr	Ni
Base metal	0.23	0.94	0.45	–	–
Filler metal	0.02	7.0	2.3	21.0	11.0

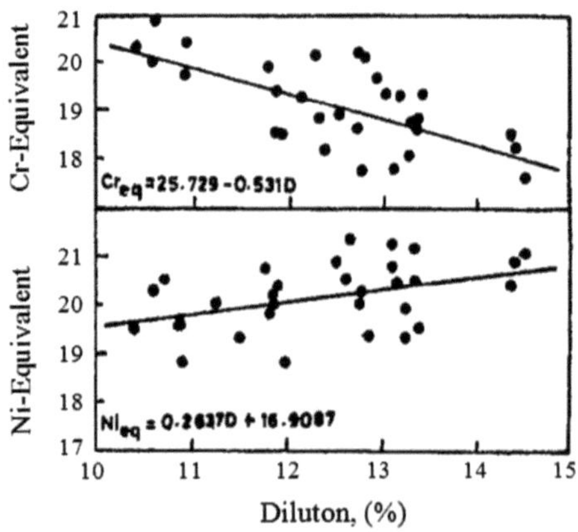

Fig. 6.6 Effect of dilution on Cr and Ni equivalents of stainless steel weld overlay on C–Mn steel

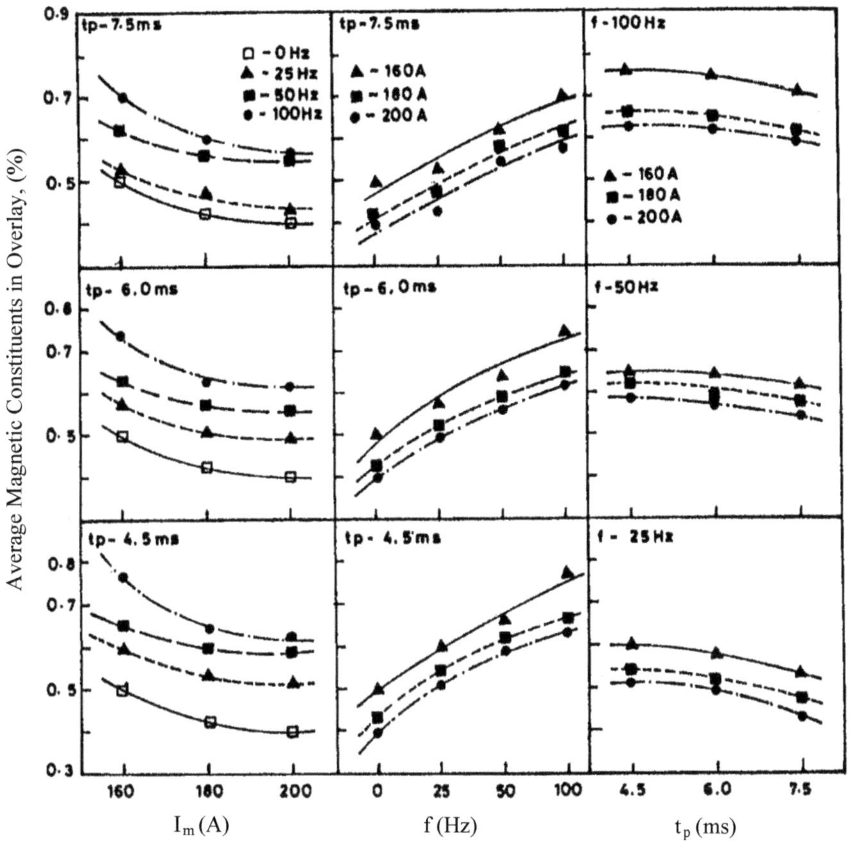

Fig. 6.7 Influence of pulse parameters on amount of magnetic constituent in stainless steel weld overlay on C–Mn steel

more prone to transformation of magnetic constituents in it. At the dilution of the order of 10–14% the weld overlay as per Delong diagram should not have any δ-ferrite or martensite in the matrix. But, variation of heterogeneity of arc welding, transportation of alloying elements by diffusion and heterogeneity of weld solidification dictated by the welding parameters may result in local inhomogeneity of chemical composition of the weld overlay that causes formation of some amount of magnetic phases in it.

However, for a more precise understanding of the control of chemistry of alloy weld an intensive study on the characteristics of fluid dynamics of shielding environment of GMAW process containing metallic and non metallic plasma is very much necessary. This may be especially required in case of the P-GMAW process operating under different pulse parameters influencing the arc physics significantly.

6.1.2 Solidification of Weld Pool

Based on the chemistry and thermal cycle, the solidification behaviour of weld varies significantly influencing its microstructure, porosity content and mechanical properties. Being governed by welding parameters, presence of coarse dendrites in weld metal weakens it significantly [6]. Thus, to achieve desired properties of weld a control over the coarsening of microstructure of weld deposit often becomes a matter of considerable interest. Various techniques including vibration of solidifying weld pool [67], alloying of weld deposit by grain refiner [7, 8, 68], lowering of current and heat input [63] to maintain low heat build-up in weld pool are largely used to minimize coarsening of microstructure of weld deposit. However, lately the refining of cast structure only by interruption in solidification of weld pool through interruption in metal deposition by introducing pulse current [1, 69, 71] during GMAW has been found quite effective in this regard. The interruption in metal deposition significantly affects refinement of microstructure largely depending upon pulse parameters influencing the heat input and heat build-up in weld pool. In general the effective heat that finally builds up in weld pool for carrying out welding by conventional GMAW process primarily depends upon heat input varies as a function of energy input. Whereas the heat build-up in weld pool of P-GMAW process largely depends also upon heat consumed in re-melting of solid fraction grown in it during interruption in energy input at the pulse off period. Thus, a critical control of pulse parameters of P-GMAW process may be useful to produce finer microstructure of weld deposit containing grain refiner or not.

The pulse parameters dictate certain physical and mechanical aspects of solidification of molten weld pool causing refinement of its cast microstructure. The physical aspects are the matters associated with the control of thermal behaviour of weld metal deposited at each pulse governing growth of dendritic arms of primary solids in absence or partial metal transfer under pulse off period. In case of partial metal transfer during pulse of period the solidification behaviour of weld deposit is further modified by the different amount and thermal nature of the deposits under high pulse current and low base current. The mechanical aspects are considered from the view point of disintegration of primary dendrite arms of solidifying weld metal by thermal shock of interruption of weld deposit under pulsed current and dispersion of the fragments in molten matrix providing large number of nucleation sites for refinement of cast structure. Thus, it is understood that for refining of microstructure of weld deposit control of simultaneous influence of all the primary pulse parameters as application of pulse current and their duration and frequency is of absolute necessity. Here also in this regard the summarized influence of pulse parameters defined by the factor ϕ (Eq. 3.1) re-conciliates the control of P-GMAW process to use appropriately for refinement of microstructure of weld deposit. However, the overall thermal characteristics of pulsed current weld deposit should be primarily considered as a function of mean current (I_m) affecting the conventional energy input (HI) per unit length (Eq. 4.56) at a given arc voltage and welding speed.

Variation in solidification behaviour of weld pool as a function of pulse parameters significantly affects coaxial growth of dendrite in it. The morphology of primary dendrite is defined by its arms length (l_d), aspect ratio (A_r) and arms spacing (D_s). The increase in aspect ratio along with decrease in arm spacing and length of primary dendrites shows increase in refinement of its equiaxed microstructure. The change in average characteristic features of dendrites maintains meaningful correlations with the conventional heat input (HI), I_m and the factor ϕ. The increase of E_c and I_m increases the length and arm spacing but reduces the aspect ratio of dendrite as typically observed in aluminium and Al–Mg (5086) alloy as shown in Figs. 6.8, 6.9 and 6.10. At a given I_m and HI the effect of ϕ on all the characteristics of dendrite maintains a criticality in the range of about 0.1–0.2 where the trend of its influence changes to opposite direction showing a reduction of length and arm spacing but enhancement in the aspect ratio with an increase of it. It happens as a result of their influence on thermal behaviour of weld, when the total heat of metal transfer increases with the increase of ϕ up to about 0.2 followed by a decrease in it with a further increase of ϕ and with increase of I_m at given ϕ (Fig. 4.1) as discussed earlier. The increase of heat transfer to the weld pool either by change of I_m and ϕ or by increase of HI results a coarsening of microstructure. The empirical correlations of the HI, I_m and the factor ϕ with l_d, A_r and D_s of dendrite in P-GMA weld of aluminium and Al–Mg are found as follows (Fig. 6.11).

For aluminium

$$l_d = 4.62 \times 10^{-4} (I_m)^{1.9197} (HI)^{0.8441} (\phi)^{-0.1009} \tag{6.1}$$

$$A_r = 2901.32 (I_m)^{-1.3942} (HI)^{-0.6875} (\phi)^{0.0841} \tag{6.2}$$

$$D_S = 1.44\mathrm{x}10^{-3} (I_m)^{1.6343} (HI)^{0.5648} (\phi)^{-0.0642} \tag{6.3}$$

(a) (b)

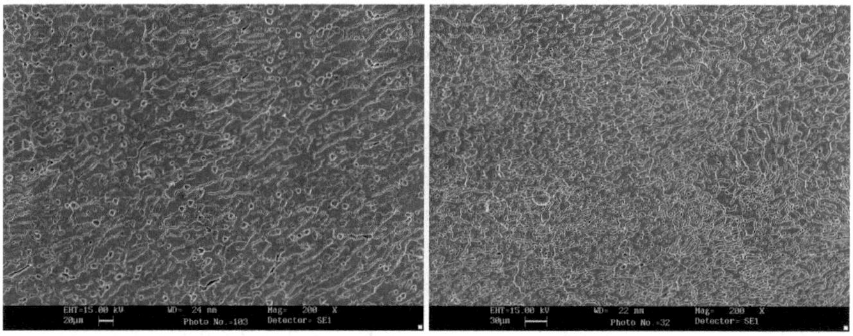

Fig. 6.8 Typical appearance of cast morphology of comparatively **a** coarser and **b** finer dendrite of aluminium weld deposit under SEM

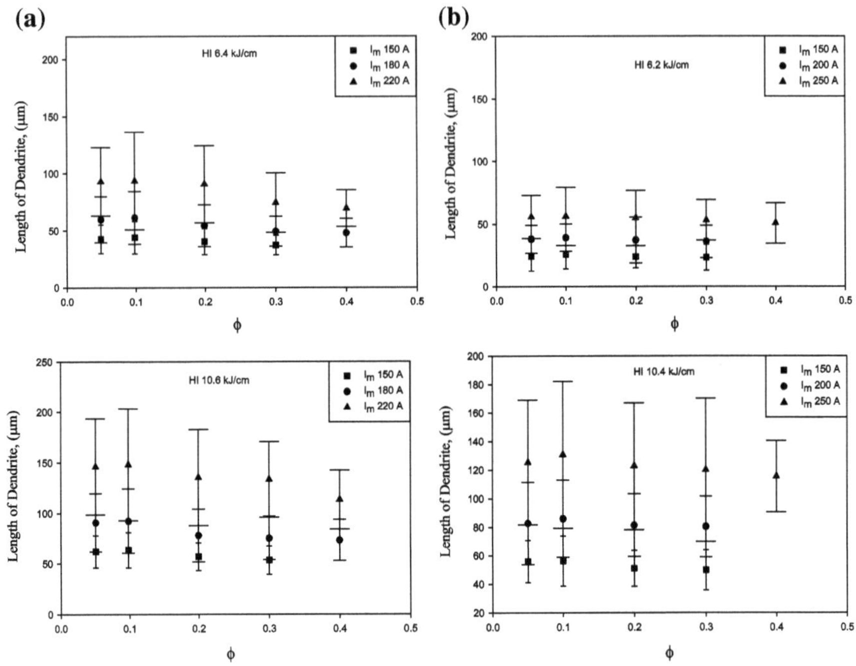

Fig. 6.9 Effect of ϕ on length of dendrite in weld deposits of **a** aluminium and **b** Al–Mg alloy at different energy inputs and mean currents

For Al–Mg alloy

$$l_d = 3.68x10^{-4}\,(I_m)^{1.567}(HI)^{1.6975}(\phi)^{-0.0454} \tag{6.4}$$

$$A_r = 386.8754(I_m)^{-1.0167}(HI)^{-0.5947}(\phi)^{0.0579} \tag{6.5}$$

$$D_S = 0.0124(I_m)^{1.11}(HI)^{0.7507}(\phi)^{-0.0695} \tag{6.6}$$

At the similar range of welding parameters the measured values of l_d, A_r and D_S lie well in agreement to their corresponding predicted values (Eqs. 6.1–6.6) within a difference of ±10% [9, 72].

The change of primary arm spacing of dendrites is suppose to result from variation of cooling rate of weld deposit with change of HI, I_m and ϕ. Cooling rate of weld metal enhances with increase of ϕ but, it reduces with the increase of HI or I_m at a given other two respectively relevant parameters as it is shown in Fig. 6.12a, b for aluminium and Al–Mg alloy respectively. Thus, it confirms that increase of cooling rate (C_R) of weld deposit reduces its primary dendrite arm spacing (D_s) as shown in Fig. 6.13 following a correlation as given in Eqs. 6.7 and 6.8 for aluminium and

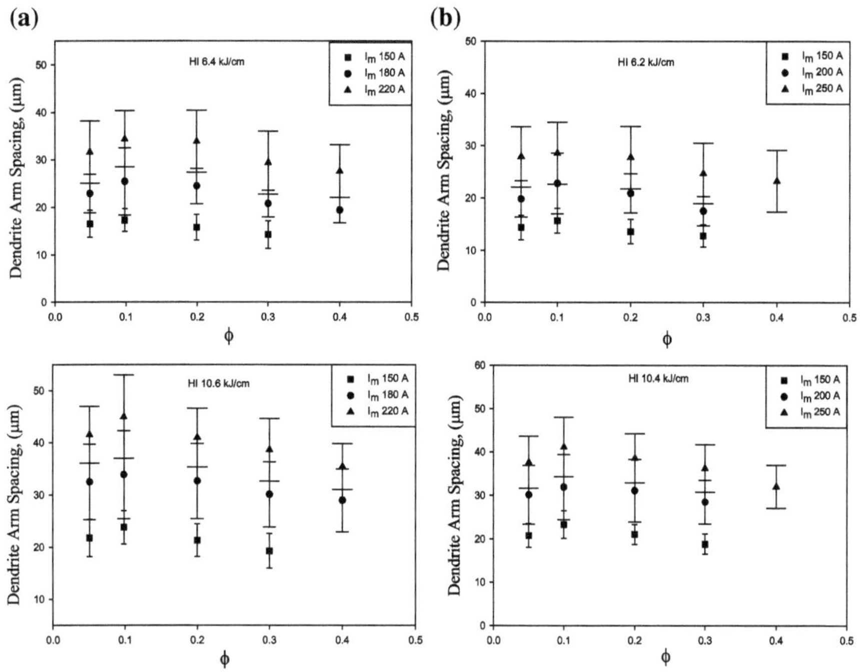

Fig. 6.10 Effect of ϕ on dendrite arm spacing in weld deposits of **a** aluminium and **b** Al–Mg alloy at different energy inputs and mean currents

Al–Mg alloy respectively [9, 10, 72, 73]. However, at a given cooling rate of weld pool dendrite arm spacing of aluminium becomes comparatively higher than Al–Mg alloy which may have primarily happened by influence of comparatively more effective constitutional super cooling of Al–Mg alloy. It is in confirmation to the well known [10] fact that at a given cooling rate D_s becomes relatively more in case of pure aluminium than its alloys. These basic observations provide ample scope to systematic designing of pulse parameters that can control microstructure of P-GMA weld deposit for its desired properties.

For aluminium

$$D_{s(Al.)} = 2354.7823\ C_R^{0.8937} \tag{6.7}$$

For Al–Mg alloy

$$D_{s(Al-Mg.)} = 1132.133\ C_R^{0.8459} \tag{6.8}$$

A certain degree of irregularities in behaviour of metal transfer, as discussed in Chaps. 3 and 4, affects the degree of thermal shock produced in solidifying weld pool and heterogeneity in solidification process. It possibly happens due to

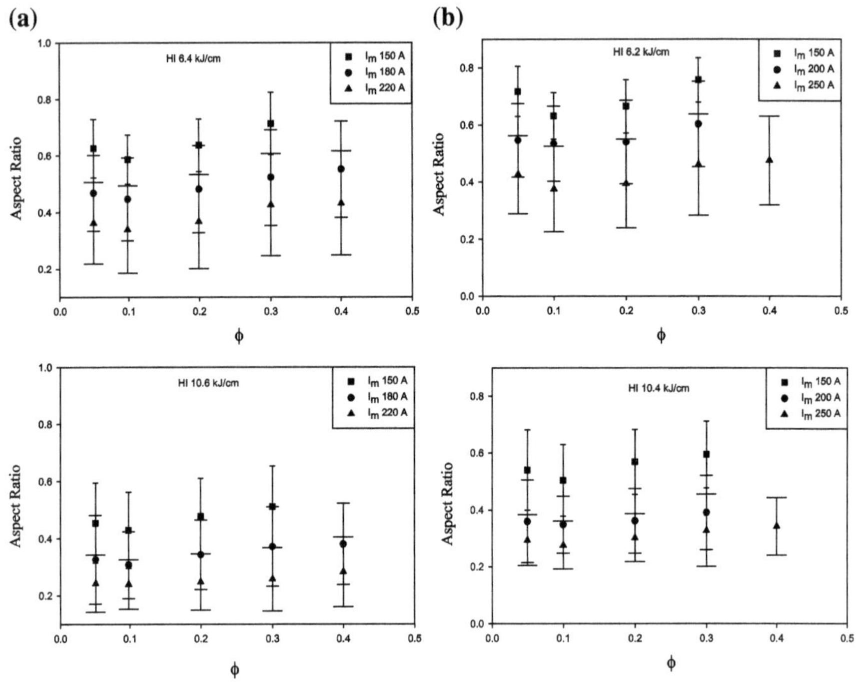

Fig. 6.11 Effect of ϕ on aspect ratio of dendrites in weld deposits of **a** aluminium and **b** Al–Mg alloy at different energy inputs and mean currents

irregularities in convective path of metal flow through different locations of weld pool during its consecutive heating and cooling under the pulse current causing an heterogeneity in morphological characteristics of weld deposit especially in case of the coarser cast structure. Hence, it varies the amount and distribution of significantly coarse and finer coaxial dendritic regions in the matrix as a function of welding parameters as typically shown in Fig. 6.14. An increase of HI or I_m considerably reduces the area fraction of finer dendrite but at a given heat input and I_m increase of ϕ enhances it appreciably and thus refines microstructure of weld deposit as shown in Fig. 6.15a, b in case of bead on plate weld deposits of aluminium and Al–Mg alloy respectively.

Solidification of weld pool in P-GMAW process primarily depends upon heat input and heat build-up in weld pool as a function of interruption in deposition of super heated weld metal under pulsed current. Here solidification mechanism of weld during P-GMAW basically differs from that of GMAW by an interruption in heat flow resulting from pulse off period after peak current duration. The extent of solidification of a droplet after deposition depends primarily on the superheating, mass of the droplet and the extent of pulse off duration [11, 74, 75]. When the subsequent pulsation sets in, the deposition of later droplet on the earlier one causes a localised melting and produces a thermal shock to the region adjacent to it.

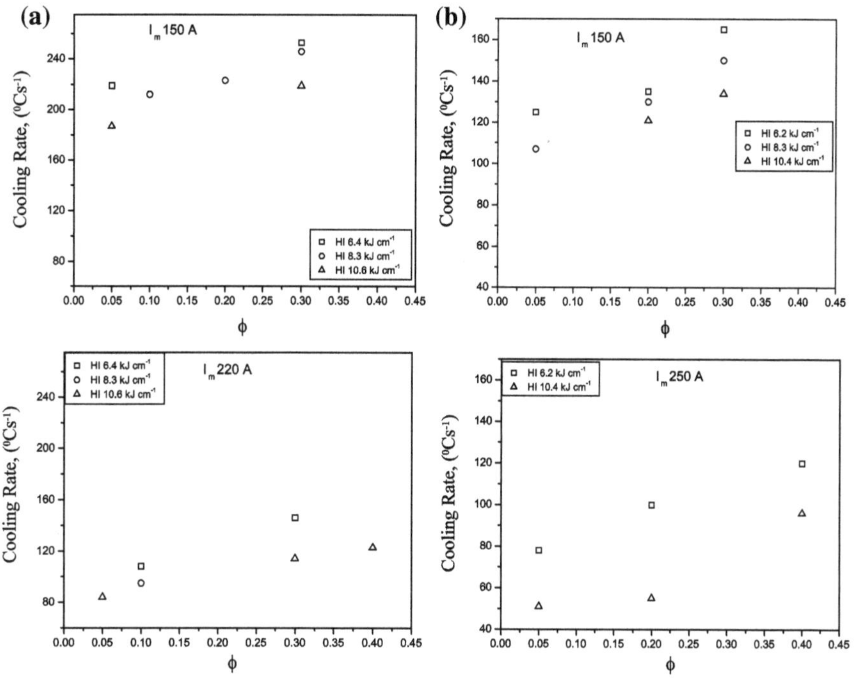

Fig. 6.12 Effect of ϕ on cooling rate in weld deposits of **a** aluminium and **b** Al–Mg alloy at different energy inputs and mean currents

The local melting may cause necking and pinching of the dendrite arms and the crystallites may be dispersed in the liquid metal caused by its convection driven by thermal gradient, magneto-hydrodynamic and/or arc forces. These crystallites provide more nucleation sites for solidification and growth to new randomly oriented crystals resulting in fine-grained cast structure of the weld by a grain multiplication process [12, 13]. However, during solidification of weld deposit such process of microstructure refinement by breaking of co-axially growing dendrite under thermal shock of interruption of metal deposition largely depends upon strength of dendrite arm in solidifying weld pool. The intensity of thermal shock is basically governed by thermal nature of droplet and its interval of interrupted deposition under pulsed current process. Intensity of thermal shock may be significantly dictated by heat input and heat build-up in weld pool. Increase of them reduces temperature gradient between depositing droplet and weld pool during interruption in metal deposition that causes reduction of thermal shock and consequently lowers refinement of dendrite. In this context it is also to be realized that at a given arc voltage the heat input affecting solidification mechanism of weld deposit can also be appreciably varied by changing welding speed and mean current of P-GMAW process. Nature of influence of heat input on weld solidification may vary with change of pattern of heat distribution in weld happens during its control

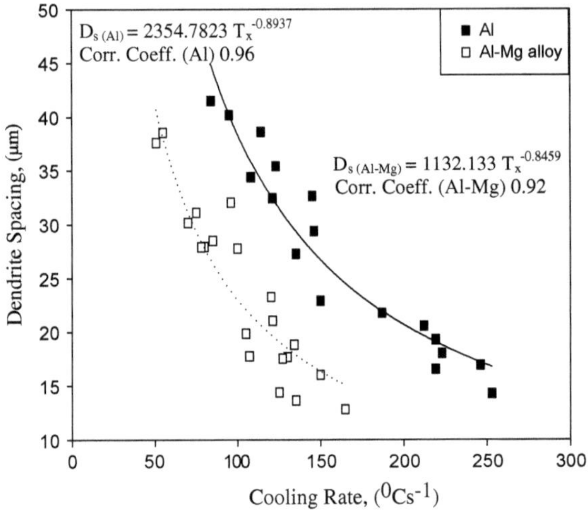

Fig. 6.13 Effect of cooling rate on dendrite arm spacing of aluminium and Al–Mg alloy weld metal

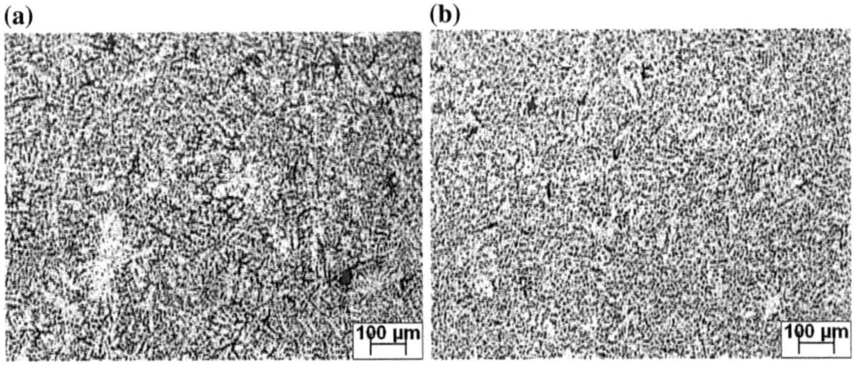

Fig. 6.14 Varied distribution of comparatively coarse and fine coaxial dendrites in Al–Mg weld deposit at different pulse parameters; **a** I_m = 150 A and ϕ = 0.2 and **b** I_m = 250 A and ϕ = 0.4 at a given heat input of 10.4 kJ/cm

by welding speed or mean current. Effect of higher heat input on rate of solidification at slower welding speed is generally understood by its adverse influence on interruption in solidification of weld pool that reduces refinement of dendrite while mean current is kept constant. But, at a given welding speed interruption in

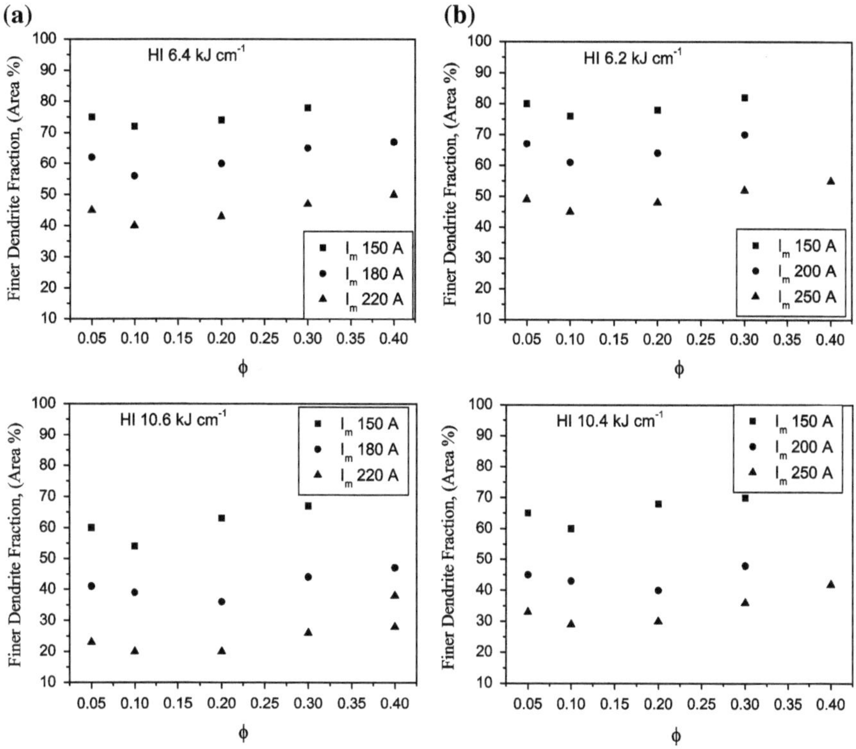

Fig. 6.15 Effect of ϕ on finer dendrite fraction in weld deposits of **a** aluminium and **b** Al–Mg alloy at different energy inputs and mean currents

solidification of weld pool due to change of heat input caused by variation of mean current may be further affected by simultaneous change in peak and base current because of their significant influence on heat build-up in it. However, degree of interruption in solidification affecting microstructure of weld deposit largely depends upon pulse off time fraction, ft_b, [65] and nature of metal transfer that primarily takes place during peak current. Though the metal transfer mostly takes place within duration of peak current but, an interruption of metal deposition with possibility of metal transfer during base current period may also significantly affect the microstructure of weld deposit [71] by influencing its nature of solidification.

6.1.3 Porosity in Weld

During welding of ferrous and nonferrous materials the control of porosity formation in weld deposit is often considered as a primary concern due to its significant influence on mechanical properties of weld joint, especially under cyclic loading. In comparison to other arc welding processes, the GMA welding produces a relatively cleaner weld with respect to its porosity and inclusion content primarily due to shielding of arc environment from outer atmosphere. But the inherited nature of heterogeneity of arc welding process due to fluctuation of arc characteristics causes certain degree of turbulence in flow of shielding gas and contaminates the arc environment and consequently the weld. The contamination of weld deposit, which also occurs under adverse condition of thermal and transfer behaviours of weld metal, may introduce dissolved gases in molten weld pool resulting into formation of blow hole and shrinkage porosity in the matrix (Fig. 6.16) depending upon the amount of gas and solidification mechanism of the weld. However, the porosity formation in weld is finally determined by many mechanisms of absorption of gases and their escape exists in the process under the influence of welding parameters. In case of GMAW the porosity content of weld generally reduces with increase of welding current as shown in Fig. 6.17. However, the vital role of such mechanisms becomes further critical [76] during the use of pulse current in GMAW. This can make it more favourable or worse than the conventional GMAW by creation and elimination of pores of weld deposit depending upon welding parameters (Fig. 6.18). In case of an appropriate use of pulse parameters the pulsed

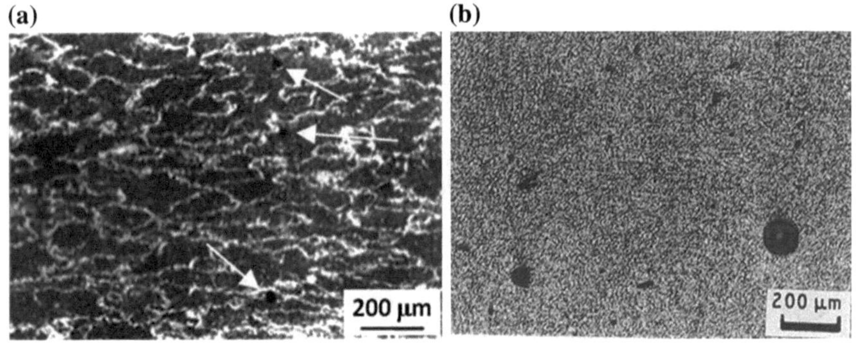

Fig. 6.16 Typical presence of porosity in GMA weld deposit of **a** mild steel and **b** Al–Mg alloy

Fig. 6.17 Influence of welding current on porosity content in Al–Mg weld deposit of conventional GMAW

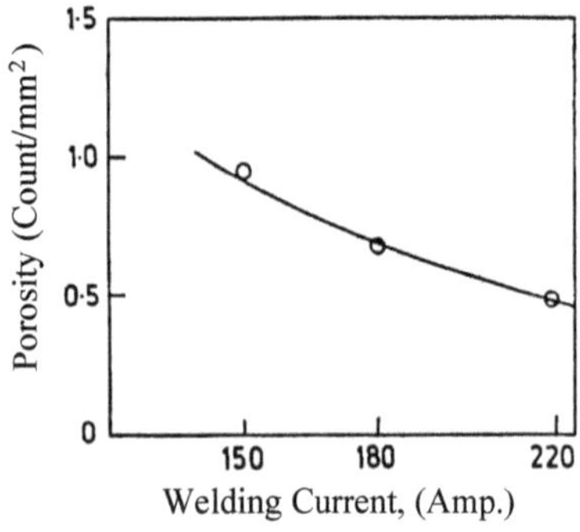

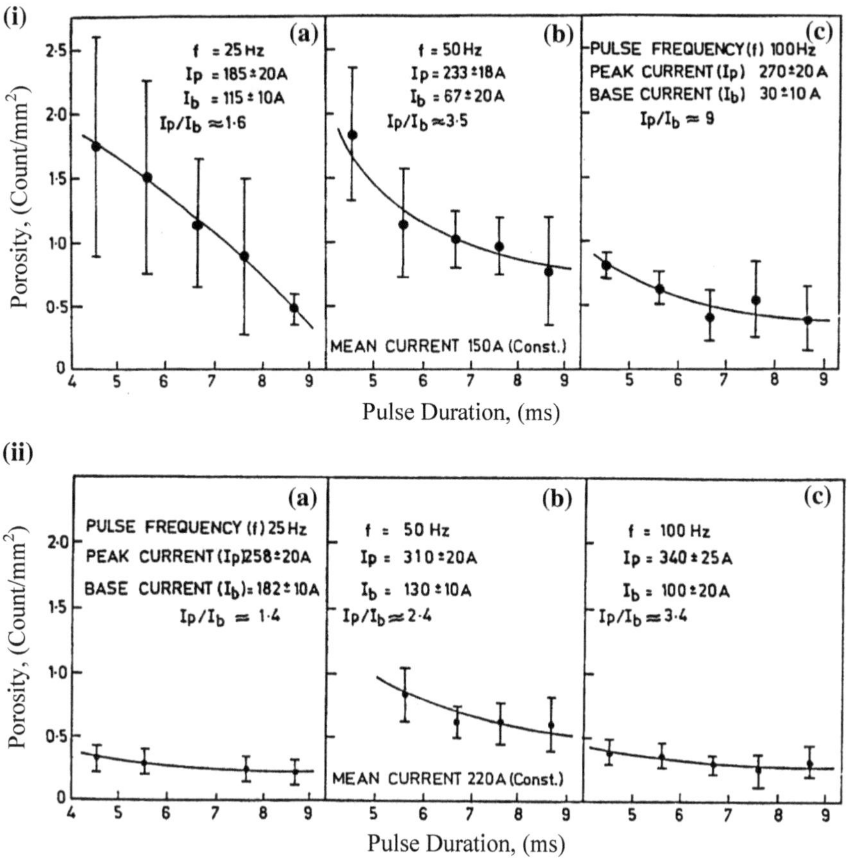

Fig. 6.18 Effect of pulse parameters on porosity content in Al–Mg weld bead on Al plate deposited at different mean currents of **i** 150 A and **ii** 220 A

current weld deposition at a given mean current similar to the welding current of conventional GMA welding gives comparatively lower porosity content in weld metal than that of the GMA weld. The use of higher mean current especially beyond transition current of globular to spray mode of metal transfer from filler wire is beneficial for reduction in porosity content of PGMA weld. In welding of Al–Zn–Mg alloy plates by single-pass or multi-pass techniques using Al–Mg filler wire, increase of pulse frequency initially increases the porosity formation followed by a decrease in it, whereas the increase of pulse duration generally reduces the porosity formation in weld deposit [14, 15, 62] as typically shown in Fig. 6.19. It is understood that relatively more porosity forms in Al–Mg weld deposit in contact to Al–Zn–Mg alloy base material than it happens in Al–Mg weld deposit on aluminium plate (Fig. 6.18). It is generally understood as besides the absorption of hydrogen from arc environment the vaporization of low melting element as Zn

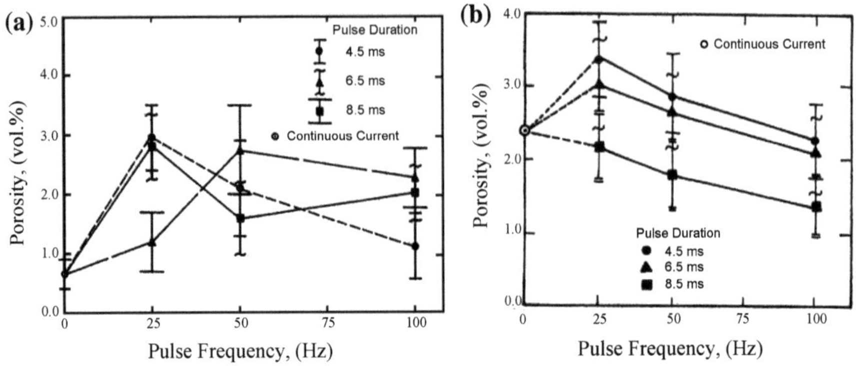

Fig. 6.19 Influence of pulse frequency and pulse duration on porosity content in Al–Mg weld deposits of Al–Zn–Mg weld joint; **a** single-pass at I_m = 235 A and **b** multipass at I_m = 220 A

under the arc heat and/or super heated weld pool largely influences the porosity formation in weld [13, 16–18]. However, an EPMA analysis of the pores in PGMA weld deposit of Al–Mg on Al–Zn–Mg alloy reveals that both the matrix and pores contains practically similar zinc concentration [77] which does not support the mechanism of pore formation primarily due to zinc vapor in weld deposit. Thus, it is argued that the variation of porosity formation in pulse current weld of Al–Zn–Mg alloy is basically caused by the dissolution of hydrogen in weld deposit from the air aspiration in shielding gas dictated by the pulse parameters. In this context the role of the presence of zinc vapour in arc shielding with respect to its further influence on stability of inert jacket under the pulse current affecting the hydrogen pick-up in it may not be ignored to realize enhanced porosity formation in the weld. But, the vaporisation of low melting constituents like zinc in Al–Zn–Mg alloy under intense excessive heating may result significant pores in the weld adjacent to the fusion line.

A similar trend of the influence of pulse frequency and pulse duration at different mean currents on porosity formation in steel weld deposit using flux cored wire (FCW) during welding of steel has been observed as shown in Fig. 6.20. But the flux cored arc weld deposit behaves in opposite manner to that discussed above where the increase of mean current increases the porosity content possibly due to more absorption of gases at higher superheating of molten metal but less possibility of releasing the gases from weld pool at the time of its solidification under flux cover [78]. Variation of arc voltage insignificantly influences porosity content of weld, whereas, the increase of welding speed reduces the porosity content of weld deposit at high mean current beyond its transition level [78]. Thus, it is understood that the control of porosity in P-GMA weld by controlling any pulse parameter is not straight forward in nature, where an appropriate use of pulse parameters can reduce the porosity content of weld below that of the conventional (0 Hz) GMA weld. As such in many occasions of earlier days of research on pulse current GMA welding some qualitative statements of highly speculative and non assertive nature

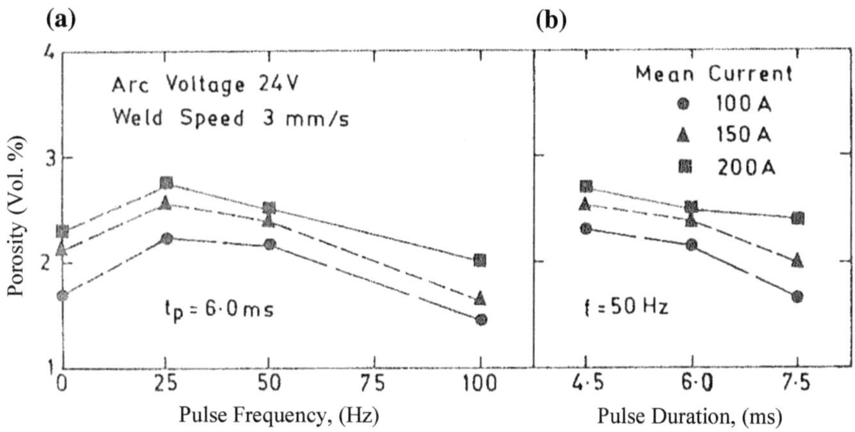

Fig. 6.20 At different mean currents effect of **a** pulse frequency and **b** pulse duration on porosity content of steel weld deposit from flux cored filler wire

are reported regarding the control of porosity content of weld deposits [14–16]. However, under a proper use of pulse parameters the porosity content of P-GMA weld of aluminium and its alloy can lie within an acceptable range, as per BS8118, part 2, satisfying many applications criteria [15]. Hence, in this regard to understand the influence of interactive role of pulse parameters on porosity formation in pulse current GMA weld is very much necessary.

In case of P-GMAW a severity in fluctuation of arc pressure driven by ratio of peak current and base current (I_p/I_b) along with their duration may lead to formation of vortex in inert jacket as discussed in Chap. 3. It may allow air aspiration in arc environment [79] and porosity formation in weld metal [80]. But on the other hand pulse current may also eliminate the pores by allowing easy escape of the absorbed gas from the molten weld pool [81] by remelting and providing sufficient thermal shock to it through appropriate interruption in metal deposition at certain pulse parameters. The occurrence of all such possibilities is clearly realized in Fig. 6.21, which indicates the necessity of proper control of interactive pulse parameters when the (I_p/I_b) ratio varies as a function of pulse frequency and pulse duration as shown in Fig. 6.22. It appears that inverse of the ratio of (I_p/I_b) varying in the range of about 0.15–0.3 depending upon I_m holds the criticality of controlling porosity formation in aluminium alloy weld at a given heat input by regulating the arc stability. Similarly the pulse off time fraction [$(t_b/(t_p + t_b)$] also significantly influences the porosity content of weld with different nature of response at different pulse frequency as it is typically shown in Fig. 6.23 for multipass Al–Mg alloy weld that was deposited during welding of Al–Zn–Mg alloy at a given mean current of 220 A. Thus, it is well understood that the porosity formation in P-GMA weld can be controlled by critical application of certain combination of simultaneously

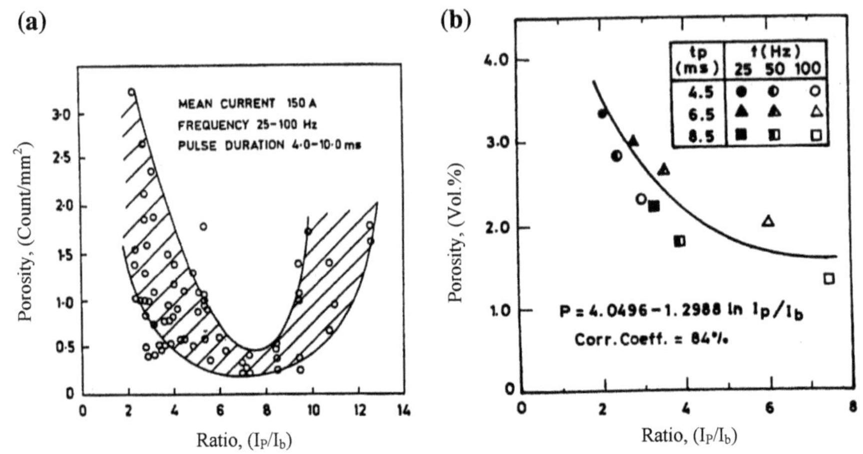

Fig. 6.21 Effect of (I_p/I_b) on porosity content in Al–Mg weld metal deposited at different conditions of welding **a** single-pass, I_m = 150 A and HI = 10.12 kJ/cm, **b** multipass, I_m = 220 A, and HI = 9.1 kJ/cm at arc voltage 23–24 V

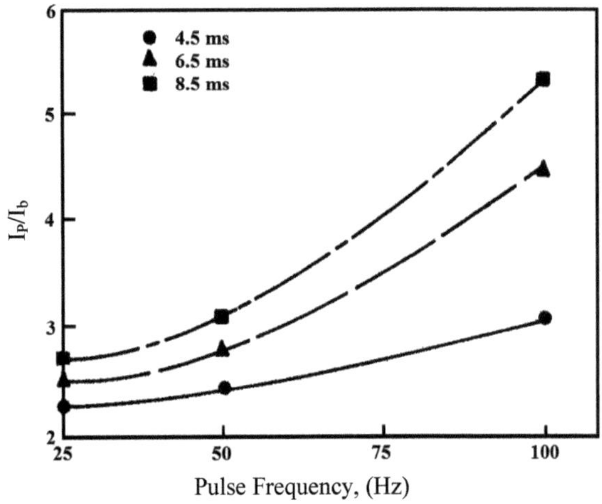

Fig. 6.22 Influence of pulse frequency and duration on I_p/I_b ratio during welding with 1.6 mm diameter Al–Mg filler wire at I_m = 235 A

interactive pulse parameters giving rise to a practically smoother arc and a well regulated interruption in metal deposition at a given heat input.

Amongst different pulse parameters the basic energy parameters I_p and I_b influence the arc characteristics, nature of metal transfer and thermal behaviour of weld deposit by functioning through a process of distribution of energy within the pulse system based on the pulse off time fraction (ft_b). In consideration of all these criterion it is difficult to find out a meaningful correlation of susceptibility to porosity formation in P-GMA weld with any individual pulse parameter but with the (ft_b) and ratio of I_p and I_b. The ratio of I_p and I_b simultaneously varies with a

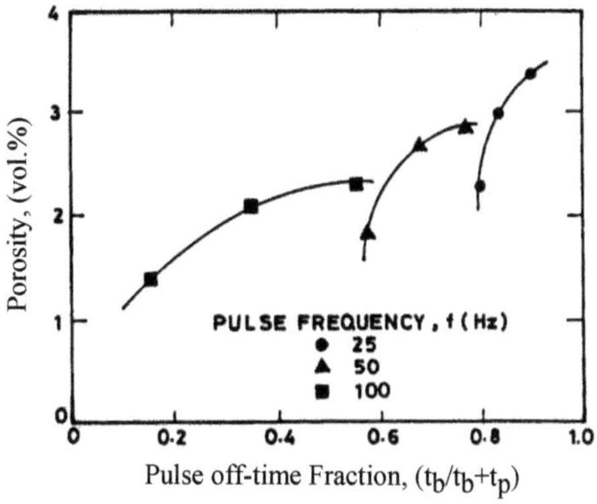

Fig. 6.23 At a given pulse frequency effect of pulse off time fraction on porosity content of multipass Al–Mg alloy weld deposit at I_m = 220 A

change in other pulse parameters (Fig. 6.22) influencing the (ft_b), thus their integrated effect on weld characteristics is finally better understood through the factor ϕ as summarized influence of pulse parameters. In number of studies on P-GMAW of aluminium and steel the porosity content of weld has been found to be linearly correlated to ϕ irrespective of considerable variation in welding parameters in different cases of welding. In general the increase of ϕ enhances the amount of weld porosity as studied in singlepass and multipass weld deposits of Al–Mg alloy under largely varied welding parameters as shown in Fig. 6.24a, b respectively. Weld deposit of FCW of steel also shows a similar linear relationship of ϕ with the porosity content of weld deposit (Fig. 6.25). But, in pulse current vertical-up welding increase of ϕ at a given mean current linearly reduces (Fig. 6.26) porosity content of weld deposit as it is observed [82–84] during singlepass welding of structural steel plate. The porosity formation in pulsed current weld under various

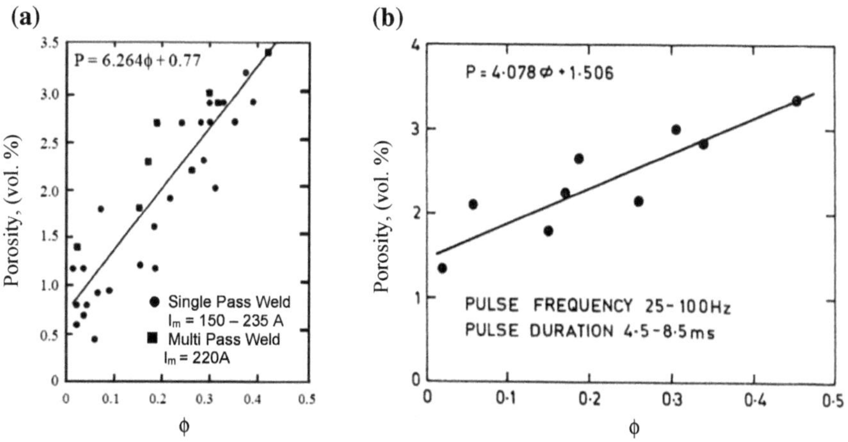

Fig. 6.24 Influence of ϕ on porosity content of **a** single and multi pass and **b** multi pass Al–Mg weld deposits

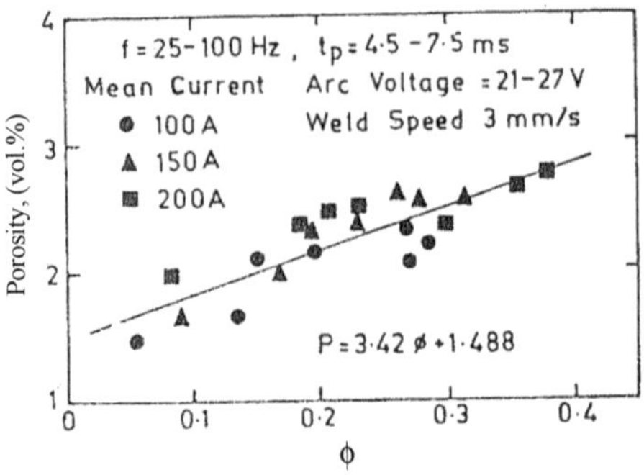

Fig. 6.25 Influence of ϕ on porosity content of steel weld deposit from FCW

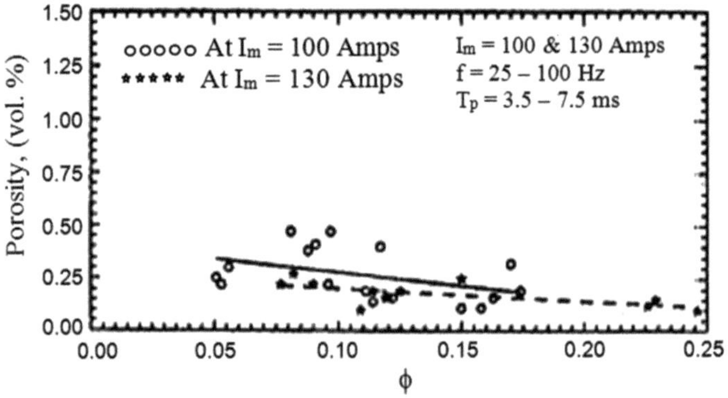

Fig. 6.26 At different mean currents influence of ϕ on porosity content of vertical-up steel weld deposit

kinds of interactive influences of pulse parameters can be effectively understood as a function of ϕ, when the nature of proportionality of ϕ with the porosity formation may depend upon other conditions of welding. They may largely comprise of efficiency and stability of welding arc. In order to understand the basic functions affecting the desired stability and efficiency of welding arc in such an interactive environment, it is necessary to understand advanced arc physics under fluid dynamics of different plasma shielding in pulse current gas metal arc welding. It should be understood in view of physical and chemical characteristics of different ferrous and non ferrous filler metal during their deposition under the plasmatic state of shielding environment of P-GMAW.

A low addition of active gas as nitrogen or oxygen to pure argon gas shielding of the order of 0.01% significantly increases the weld width and penetration along with reduction of microporosity compared to that observed during using pure argon during P-GMA welding of 2.5 mm thick AlMg5Mn sheets using AlMg5 filler wire. While the use of nitrogen gives insignificant change in spatter and the presence of oxygen appreciably enhances the same. However, the low addition of active hydrogen in argon shielding increases the porosity but remarkably reduces the spatter formation [19].

6.1.4 Microstructure of Weld and HAZ

It is well understood that solidification behaviour of weld pool largely controls the microstructure of weld metal and its homogeneity in the matrix. During solidification of weld metal the growth of co-axial dendrite varies considerably at different locations of weld pool depending upon its chemical homogeneity largely governed by the fluid flow based on temperature distribution in it. The P-GMA welding has the ability to regulate the microstructure of weld metal [18–23, 85] and HAZ through its influence on characteristics of metal transfer and thermal behaviour of weld metal at the time of deposition. It can refine microstructure of weld metal by controlling the nucleation and growth of co-axial dendrites into small equiaxed primary solids. Under considerable superheating of weld pool, it occurs in presence of stable nuclei of suitable size for equiaxed solidification after duration of each pulse current. In this regard the high frequency pulsing is more effective under controlled effect of thermal inertia [75] by regulation of other pulse parameters.

6.1.4.1 Aluminium

At high mean current of 235 A increase of pulse frequency refines dendritic microstructure of weld up to certain extent but its efficiency of refining microstructure weakens [62] at higher pulse duration (Fig. 6.27) as it is observed in aluminium and aluminium alloy weld deposition. The refinement of microstructure with change in pulse parameters is governed by an effective interruption in solidification primarily dictated by superheating and size of droplets and pulse-off time. At a given I_m and t_p the increase of pulse frequency reduces the peak and base currents and thus reduces superheating and enhances the size of the droplet which makes the interruption in solidification more effective by establishing a comparatively faster solidification of the droplets. Whereas, the increase in t_p reduces the pulse off-time and decreases the effect of interruption in solidification by allowing less time for the droplets to solidify before deposition of next droplets.

Microstructure of weld metal changes appreciably with a change of energy input, mean current and ϕ as it is typically shown in Figs. 6.28a–c and 6.29a, b for bead on plate weld deposition of aluminium and Al–Mg alloy respectively using 1.6 mm diameter filler wire. The increase of energy input due to change of welding speed at a given mean current or vice versa relatively coarsens the microstructure of weld metal. But, it appears that at a given mean current and energy input increase of ϕ up to about 0.1–0.2 initially coarsens microstructure of weld deposit followed by a relative refinement in it with further increase of ϕ. At a given mean current and heat

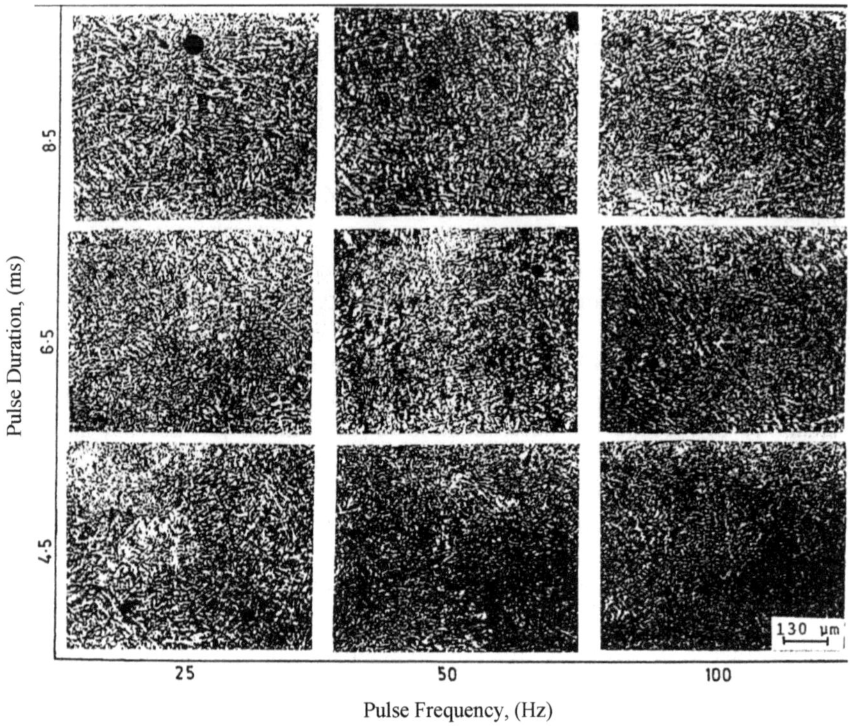

Fig. 6.27 Influence of pulse frequency and duration on microstructure of Al–Mg weld deposit at I_m = 235 A

input influence of ϕ on temperature of droplet at the time of deposition (Fig. 4.5) and heat transfer to the weld pool (Fig. 4.1) affecting its solidification mechanism suitably changes microstructure of weld deposit [11, 12, 74, 86]. From the earlier discussions it may be realized that the coarsening or refining of cast microstructure of weld metal largely depends upon extent of continuous or interrupted mechanism of solidification process of the weld pool. The varied nature of interruption of metal deposition under different pulse parameters can interrupt the solidification mechanism repeatedly in a practically similar manner and accordingly may form repeated zones of identically refined microstructure in the matrix. Thus depending upon size, superheating, and interruption in deposition of droplets they dictate the refinement of microstructure of weld deposit. Microstructure of weld can be coarsened by long co-axial growth of primary dendrite or it may be refined by the repeated pattern of interrupted solidification depending upon pulse parameters and energy input. This may be clearly realized in the light of the understanding on mass and thermal behavior of depositing metal droplets as function of various pulse parameters as discussed in Chaps. 3 and 4.

The repetitive presence of the zones of same pattern of microstructure, conforms the repetition of a given mechanism of solidification under the interruption of metal

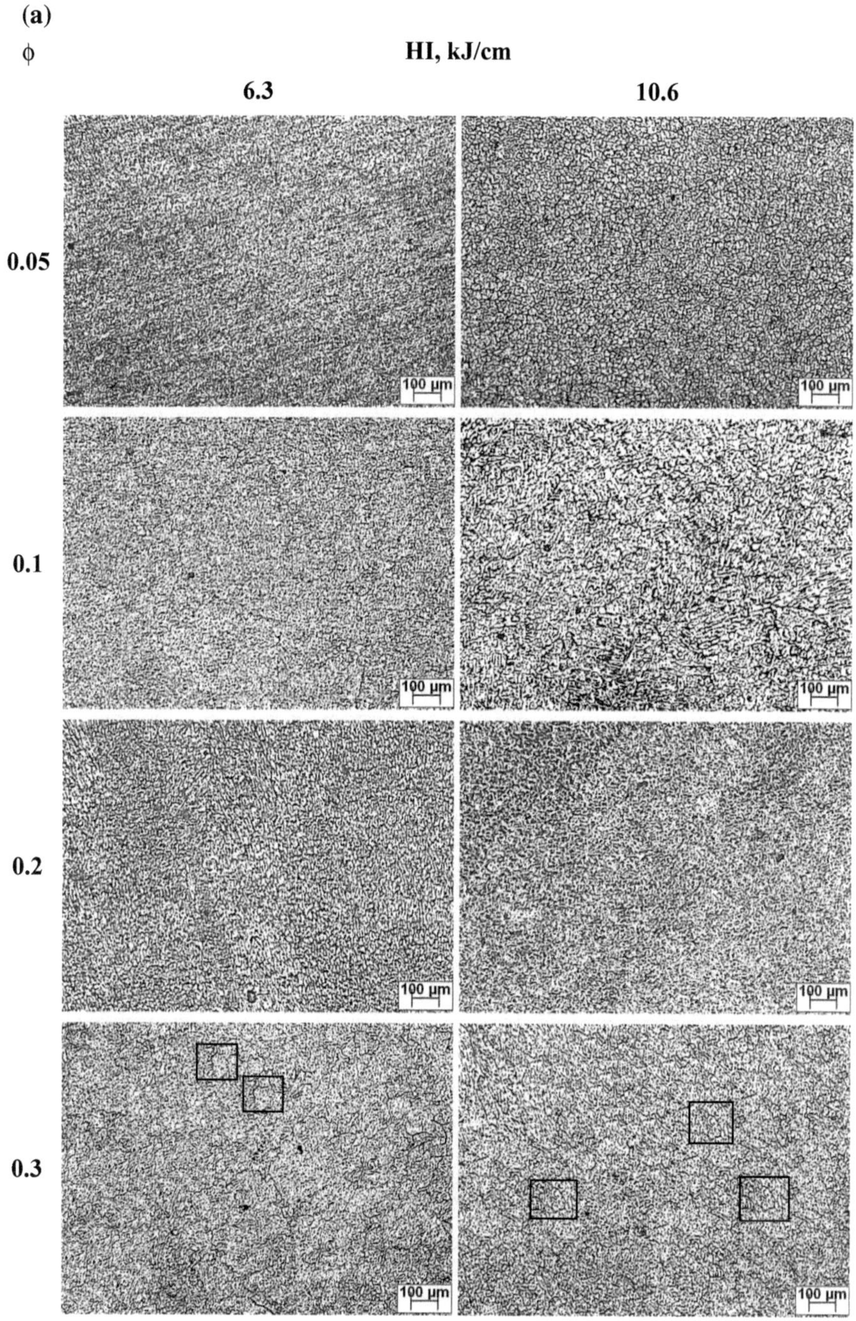

Fig. 6.28 **a** Effect of ϕ and energy input on microstructure of bead on plate weld deposit of aluminium at I_m = 150 A. **b** Effect of ϕ and energy input on microstructure of bead on plate weld deposit of aluminiumat I_m = 180 A. **c** Effect of ϕ and energy input on microstructure of bead on plate weld deposit of aluminium at I_m = 220 A

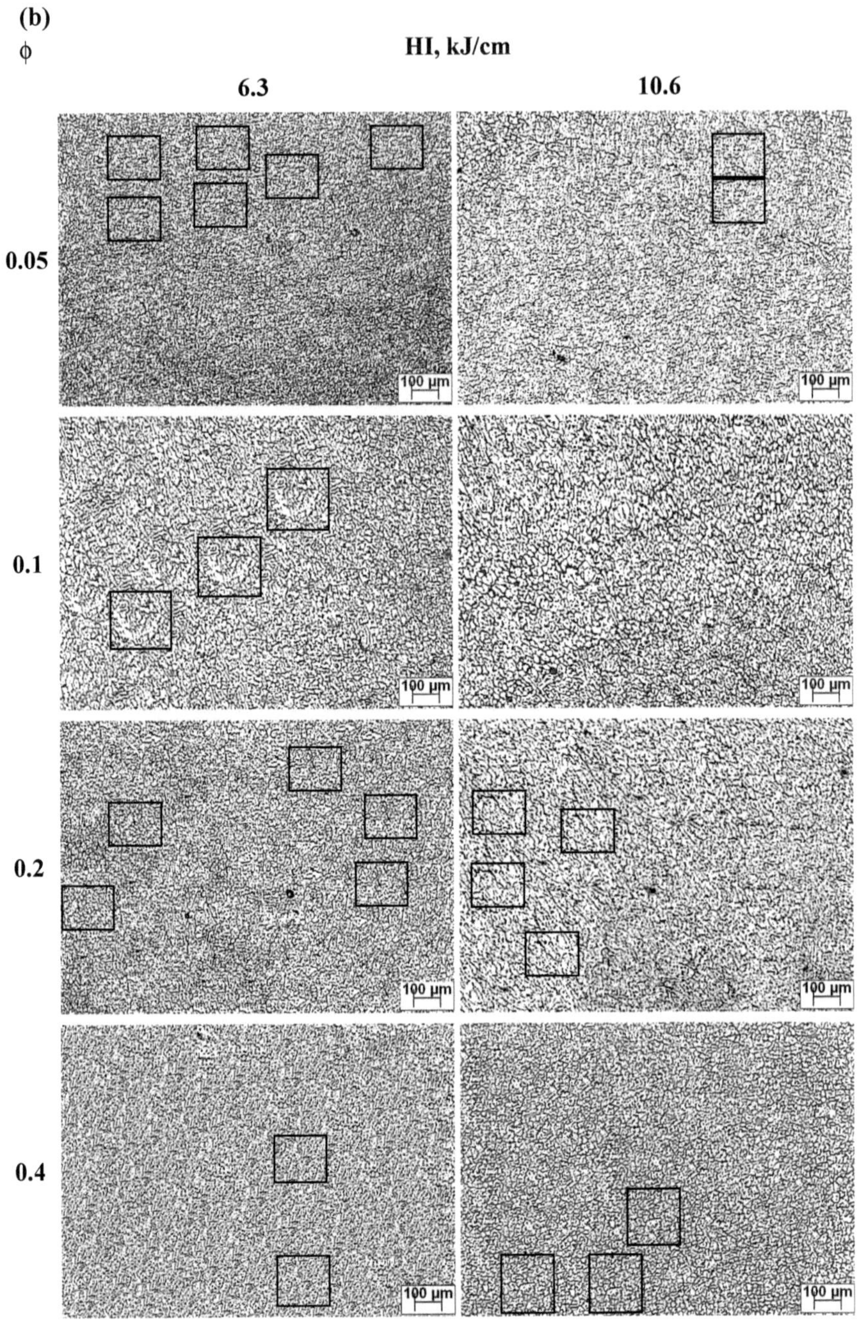

Fig. 6.28 (continued)

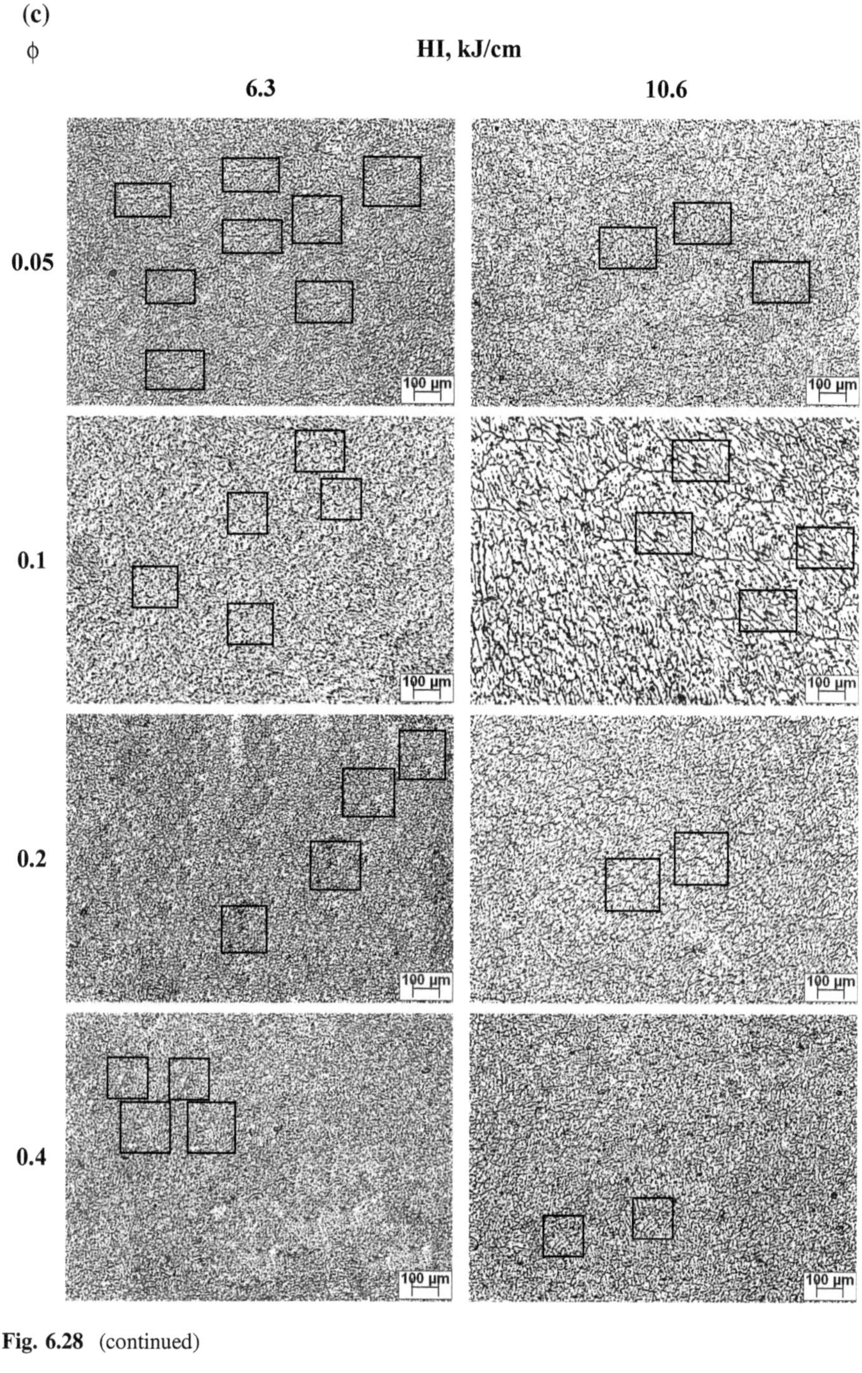

Fig. 6.28 (continued)

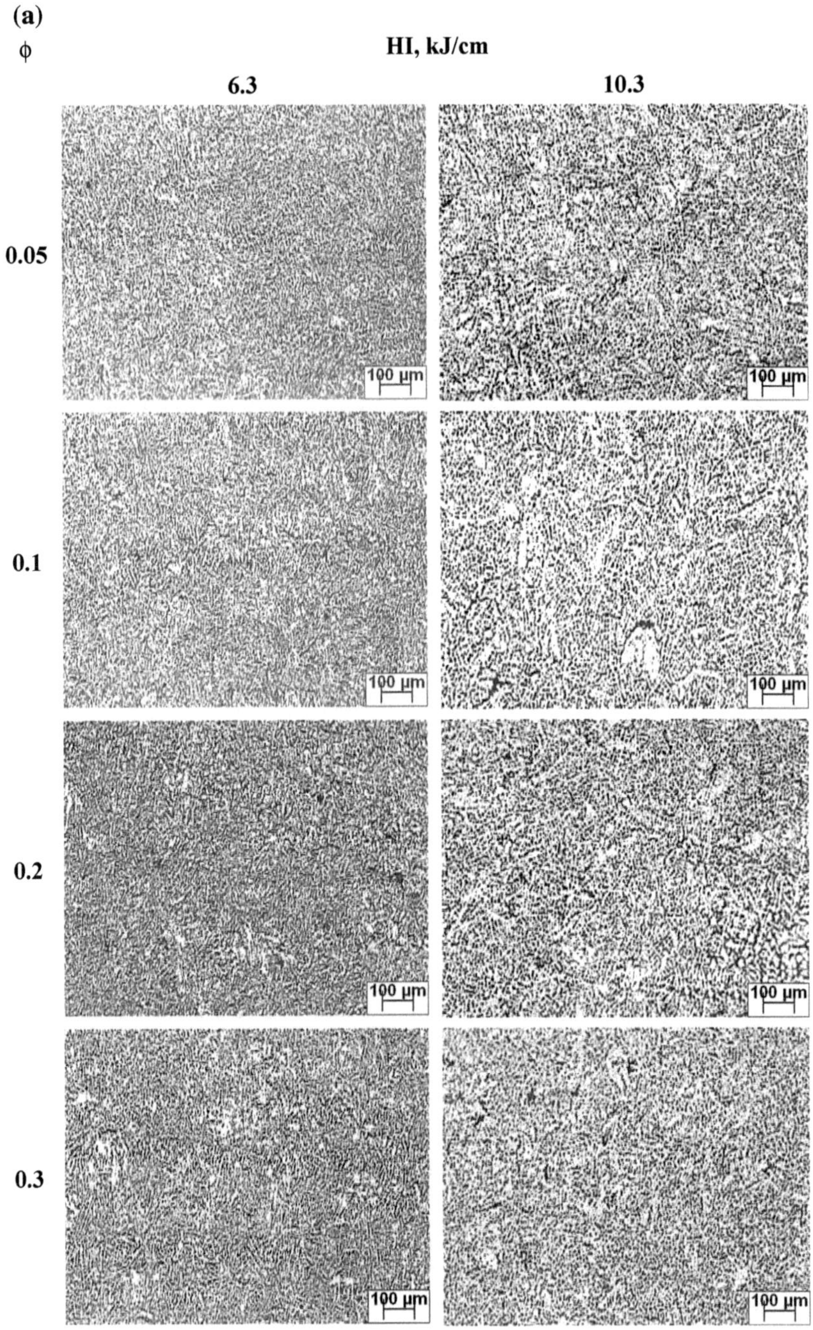

Fig. 6.29 **a** Effect of ϕ and energy input on microstructure of bead on plate weld deposit of Al–Mg alloy at I_m = 150 A. **b** Effect of ϕ and energy input on microstructure of bead on plate weld deposit of Al–Mg alloy at I_m = 200 A

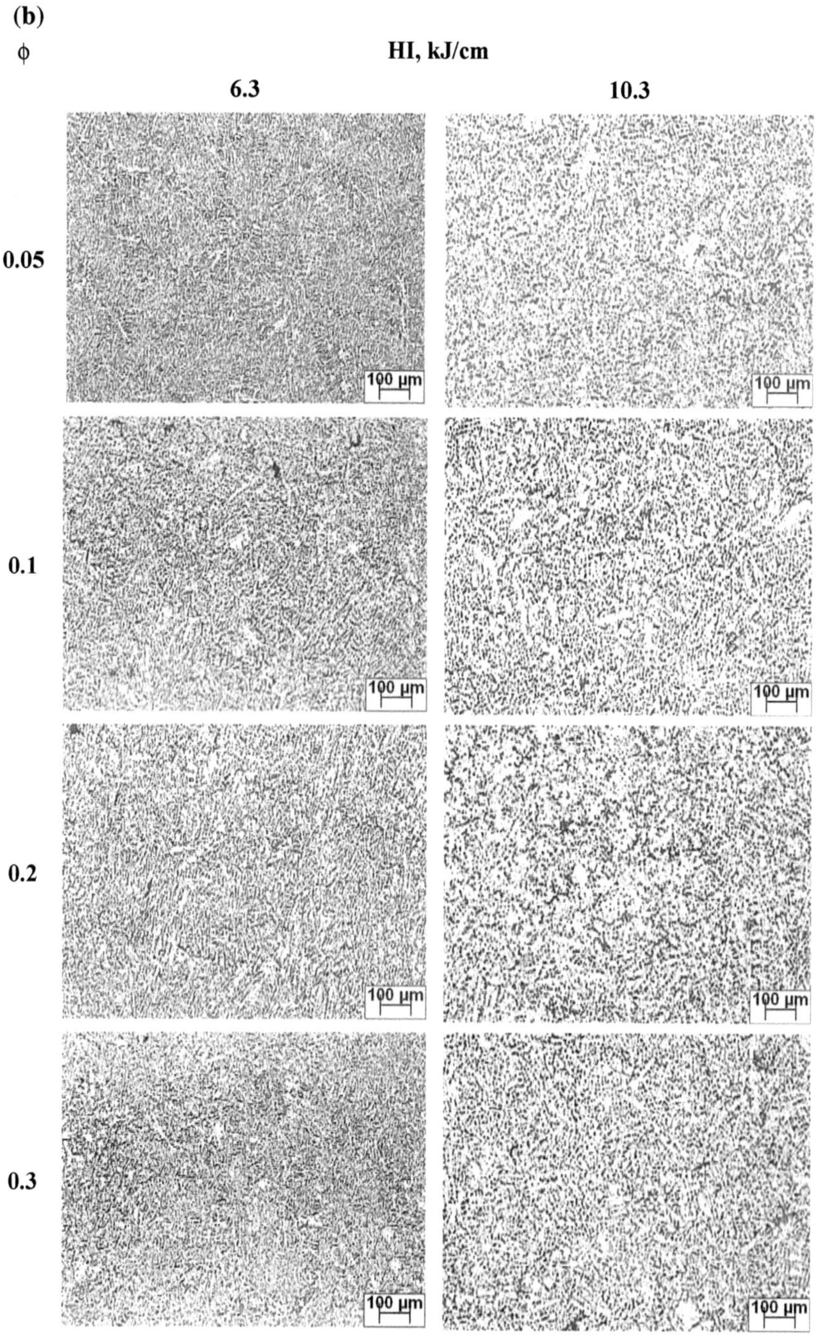

Fig. 6.29 (continued)

deposition from a given pulse parameter. However, the recurring pattern of microstructure [marked] and its distribution in the matrix, arising out of interruption of metal deposition changes with the pulse parameters and they are often more clearly visible in the matrix of weld metal deposited at high mean current and energy input as typically marked in the micrographs presented in Fig. 6.28b, c. As a result of such an interrupted nature of solidification the microstructure of weld metal may consist of a mixture of appreciably coarse and relatively finer dendritic regions in the matrix when their morphology and amount varies with a change in welding parameters. Increase of mean current and energy input significantly reduces finer dendrite fraction whereas at a given I_m and energy input increase of ϕ relatively enhances it in the matrix of both the aluminium and Al–Mg alloy as shown in Fig. 6.30a, b respectively. Nevertheless role of chemical composition of weld metal, influencing its melting point and constitutional super cooling, cannot be ignored in understanding the refinement of microstructure. Thus, it is more clearly evident in the matrix of pulsed current gas metal arc weld of aluminium than in Al–Mg alloy. It possibly happens due to heterogeneous micro segregation of Mg during solidification of Al–Mg alloy which prevents the repetitive formation of the zones of same pattern of microstructure in it during pulsed current gas metal arc

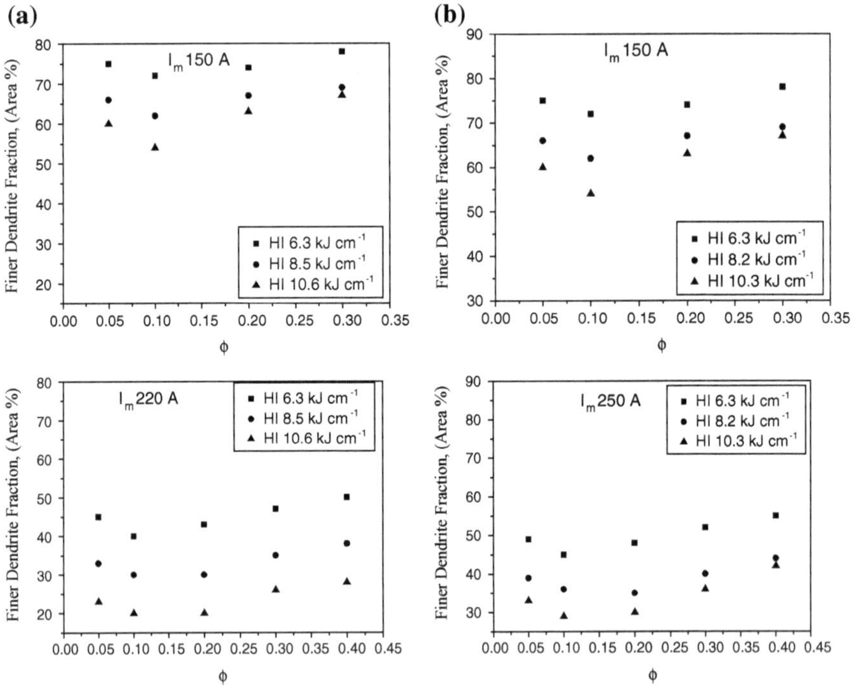

Fig. 6.30 Effect of ϕ on finer dendrite fraction in bead on plate weld deposits of **a** aluminium and **b** Al-Mg alloy at different mean currents and energy inputs

welding. This is observed even when it is showing comparatively more homogeneous and wide scale refinement of microstructure in the matrix.

Microstructure of multipass weld consists of a mixture of the regions of comparatively coarser (C) coaxial dendrite and finer (F) reheat-refined dendrite or grain structure (Fig. 6.31). The morphologies of the coarse coaxial and finer dendrites are largely defined by their different interdendritic spacing and aspect ratios of the primary dendrite arms. In Al–Mg alloy weld deposit the dendrite arm spacing of coarse coaxial dendrite is of the order of 4.78 ± 1.27 μm with an aspect ratio 0.22 ± 0.024, where the dendrite arm spacing of relatively finer dendrite is of the order of 1.54 ± 0.69 μm with an aspect ratio of 0.144 ± 0.020 [87]. However, the amount and distribution of the comparatively coarser and finer dendritic regions in the matrix of a multipass weld depends upon welding parameters. The variation in both the pulse frequency and duration affects the coarse and fine dendrite content (Fig. 6.32) as well as their distribution in the matrix. The use of pulse current welding results comparatively larger amount finer dendrite fraction in the weld than that observed in case of multipass conventional GMA weld. It primarily happens due to a change in shape of weld beads and temperature of depositing weld metal

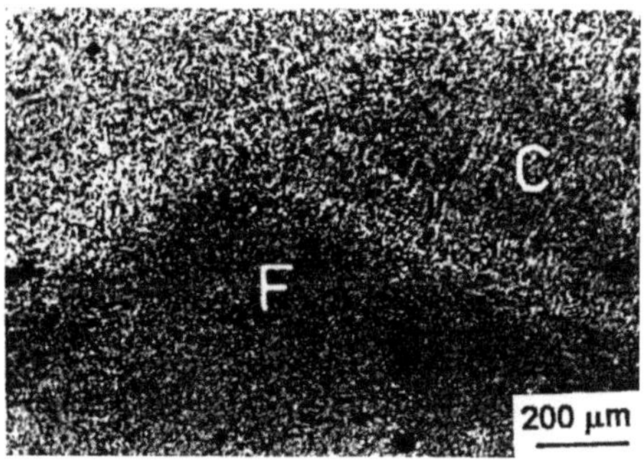

Fig. 6.31 Comparatively coarser (*C*) and finer (*F*) dendritic regions in the matrix of multipass conventional GMA weld of Al-Mg alloy

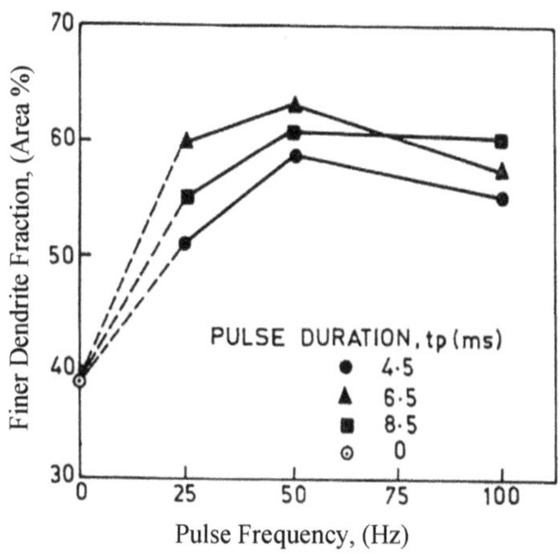

Fig. 6.32 Effect of pulse parameters on nature of variation of finer dendrite fraction in multipass Al–Mg weld deposit

[23, 88] depending upon welding parameters. A variation in the size and shape of weld bead vary with change in welding process from the conventional GMAW (Fig. 6.33) to P-GMAW using different pulse parameters (Fig. 6.34a–c). In case of multipass P-GMA welding both the reheat refinement of dendrite and interruption in solidification of weld deposit play significant role in controlling the area fraction of fine microstructure other than coaxial dendritic microstructure in the matrix. The finer dendrite content of the weld is quite sensitive to the pulse off time fraction (ft_b) and its response become relatively sharper with a decrease in pulse frequency (Fig. 6.35). Thus, the microstructure of multipass P-GMA weld is also significantly well correlated to summarized influence of pulse parameters estimated as the factor

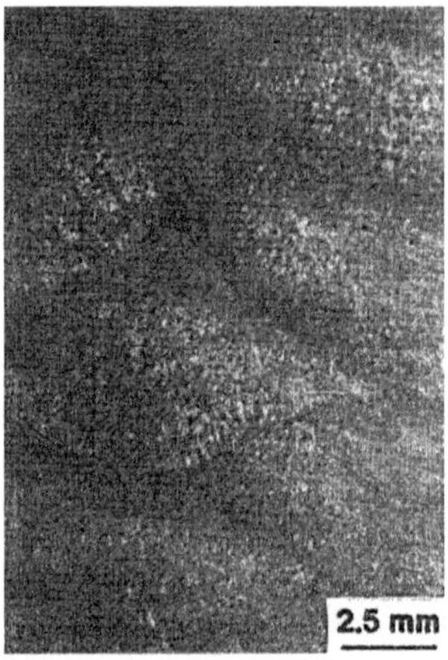

Fig. 6.33 Multipass deposit of Al–Mg in conventional GMAW at welding current of 220 A

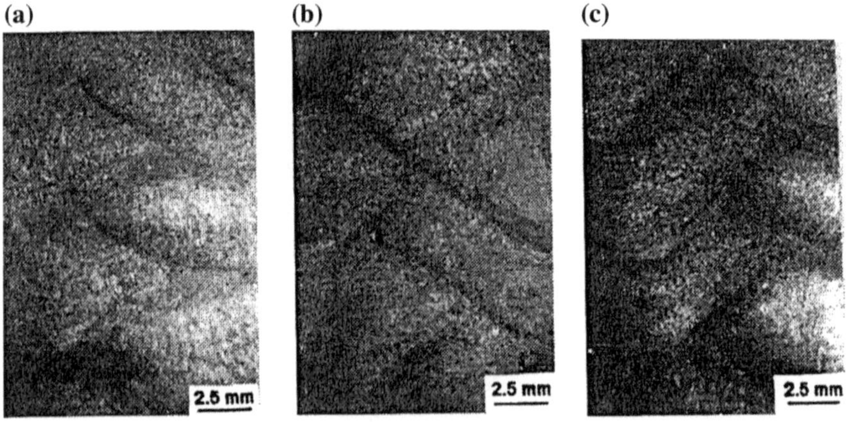

Fig. 6.34 Multipass deposit of Al–Mg in P-GMAW (I_m = 220 A) at different pulse parameters, **a** t_p = 4.5 ms and f = 100 Hz, **b** t_p = 6.5 ms and f = 50 Hz and **c** t_p = 8.5 ms and f = 50 Hz

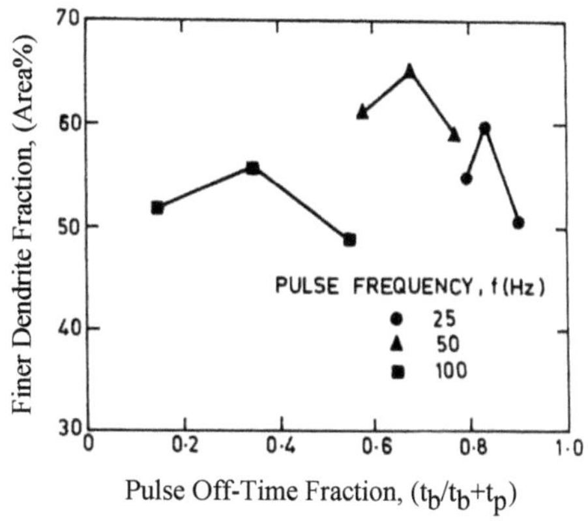

Fig. 6.35 At a given pulse frequency the influence of pulse off time fraction on finer dendrite fraction of the weld

ϕ. The increase of ϕ up to about 0.2 enhances the finer dendrite content up to about 62% followed by decrease in it with further increase of ϕ up to 0.45 (Fig. 6.36) as it is observed in case of the multipass Al–Mg weld deposit. It happens under a competitive influence of super heating and size of droplet, amount of metal deposition per unit length and shape of weld bead affecting reheat refinement of dendrite in a multipass weld deposit. Increase of ϕ reduces super heating but enhances size of droplet along with a decrease in amount of metal deposition per unit length of each weld bead due to reduction in number of droplet transfer per pulse [89–92]. Thus they significantly influence the super heating of molten weld metal, size and shape of weld bead and amount of metal deposition per unit length of weld bead affecting thermal distribution in weld. In multipass welding variation in these aspects of weld deposit considerably affects the partial re-melting and reheat refinement of dendritic microstructure of an earlier deposited weld bead by a later one [23, 24, 93], which governs the amount and distribution of the coarse and finer dendritic regions in the matrix.

The thermal cycle of weld deposition also re-crystallizes the microstructure of base metal close to the fusion line as typically shown in Fig. 6.37. The width of

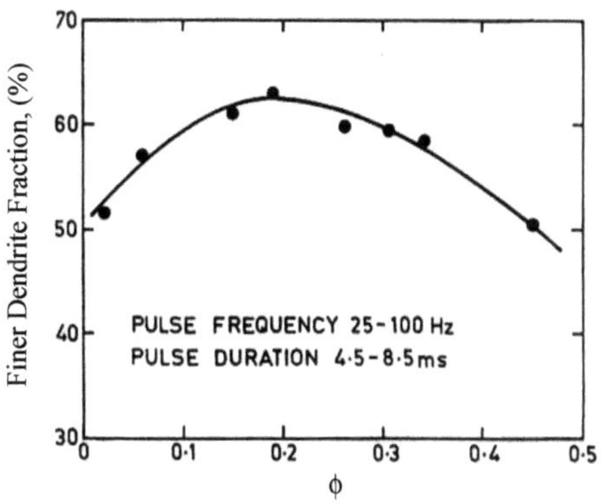

Fig. 6.36 The influence of ϕ on finer dendrite fraction of the pulsed current multipass Al–Mg weld deposit

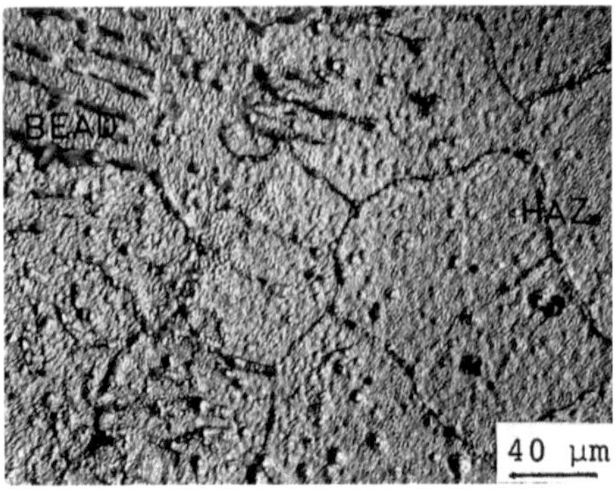

Fig. 6.37 Recrystallization of HAZ adjacent to fusion line of weld bead on Al–Zn–Mg alloy

recrystallized region and the extent of grain growth depend upon weld isotherm that varies with a change in welding parameters. However, the pulse current welding significantly lowers the extent of re-crystallization of HAZ compared to that of the conventional GMA welding of ferrous and nonferrous metals carried out at a given welding current [80, 81]. It happens primarily due to intermittent heat flow leading to lowering of peak temperature of HAZ. This is considered as one of the most beneficial effect of P-GMAW over the conventional GMAW. Extent of re-crystallization and its width in HAZ adjacent to fusion line varies significantly with change of pulse parameters as typically shown in Fig. 6.38a, b for a multipass weld of Al–Zn–Mg alloy. Increase of both the pulse frequency and pulse duration

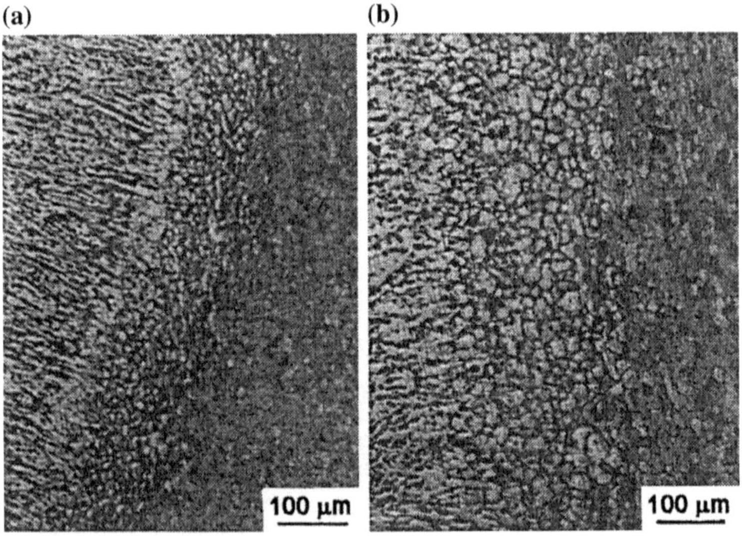

Fig. 6.38 Typical variation of morphology and width of recrystallization in HAZ adjacent to fusion line with a change in pulse parameters, **a** f = 100 Hz and t_p = 8.5 ms and **b** f = 25 Hz and t_p = 4.5 ms

enhances re-crystallization of HAZ but, pulse frequency is having considerably less influence than the pulse duration. Thus, it may be realized that an appropriate adjustment of pulse frequency and pulse duration at desired arc stability can effectively control the recrystallization of HAZ. The variation in recrystallization of HAZ with change of pulse frequency and duration may be attributed to variation of effective heat buildup in the weld bead deposited adjacent to the groove wall of the base metal.

6.1.4.2 Steel

Thermal behaviour of steel weld deposit from flux cored filler wire (FCW) can be different from that of the weld deposit from solid filler wire due to deposition and solidification of molten metal under flux cover. The variation in pulse parameters significantly influences its microstructure that primarily consists of bainite, proeutectoid ferrite and some amount of pearlite in coaxial dendritic growth in the matrix. Increase of energy input by enhancement of I_m coarsens the morphology of coaxial dendrite and transforms more ferrite in the weld deposit from flux cored steel (AWS E 70 T-5) wire (1.2 mm diameter) as shown in Fig. 6.39a, b [78]. But, at a given mean current the increase of pulse frequency or pulse duration significantly coarsens dendrite especially with respect to its growth as shown in Figs. 6.40 and 6.41 respectively possibly due to reduction of cooling rate resulting from enhancement of super heating of weld deposit. However, at a given other pulse parameters the increase of arc voltage (Fig. 6.42) and decrease of welding speed (Fig. 6.43) transforms comparatively more finely distributed ferrite in weld deposit. An appropriate control of pulse parameters along with the arc voltage and welding speed produces a comparatively finer microstructure in weld deposit with required transformation of ferrite.

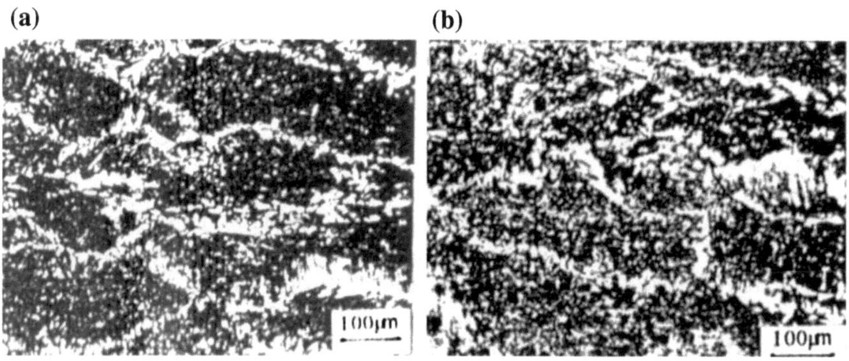

Fig. 6.39 Effect of mean current on microstructure of steel weld deposit from FC filler wire at arc voltage = 24 V, welding speed = 3 mm/s, f = 50 Hz and t_p = 6.0 ms; **a** I_m = 100 A and **b** I_m = 200 A

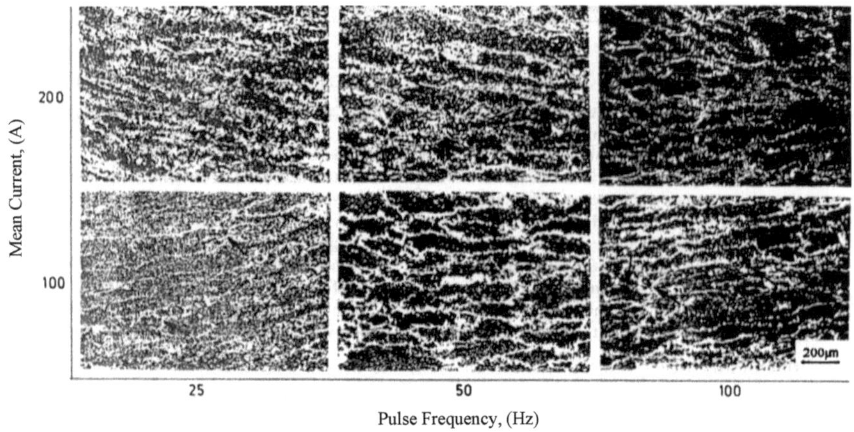

Fig. 6.40 At different mean currents effect of pulse frequency on microstructure of steel weld deposit of FCW; arc voltage = 24 V, welding speed = 3 mm/s and t_p = 6.0 ms

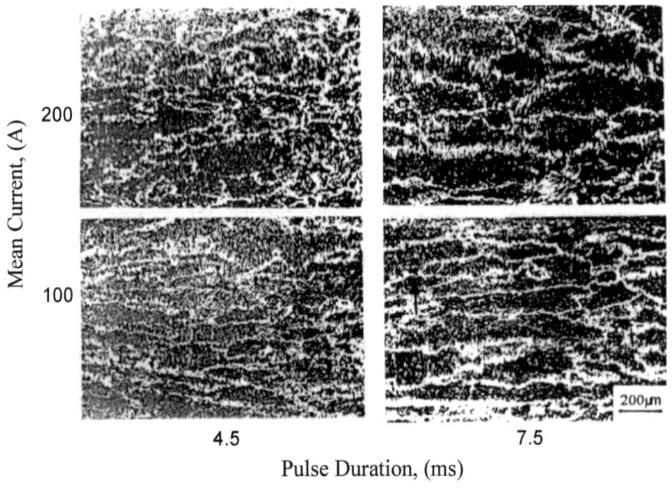

Fig. 6.41 At different mean currents effect of pulse duration on microstructure of steel weld deposit of FCW; arc voltage = 24 V, welding speed = 3 mm/s and f = 50 Hz

The variation of of weld from the FCW deposition at different welding parameters of P-GMAW process significantly changes width of HAZ and microstructure of base metal, [78, 94] adjacent to fusion line. In agreement to that discussed for Al-alloy here also at a given current an increase of pulse frequency and pulse duration relatively enhances width of HAZ (Fig. 6.44). Use of comparatively low pulse frequency and pulse duration of 25 Hz or 4.5 ms respectively at a given mean current reduces width of HAZ below that of conventional GMA weld deposited especially at a low mean current of 100 A. Reduction of arc voltage at a

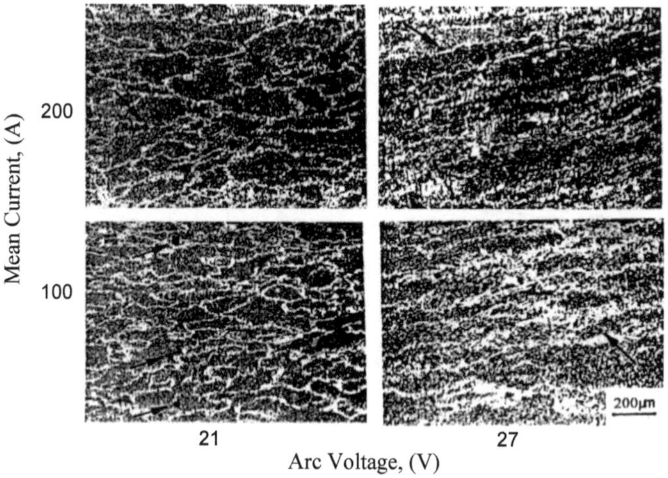

Fig. 6.42 At different mean currents effect of arc voltage on microstructure of steel weld deposit of FCW; welding speed = 3 mm/s, f = 50 Hz and t_p 6.0 ms

Fig. 6.43 At different mean currents effect of welding speed on microstructure of steel weld deposit of FCW; arc voltage = 24 V, f = 50 Hz and t_p = 6.0 ms

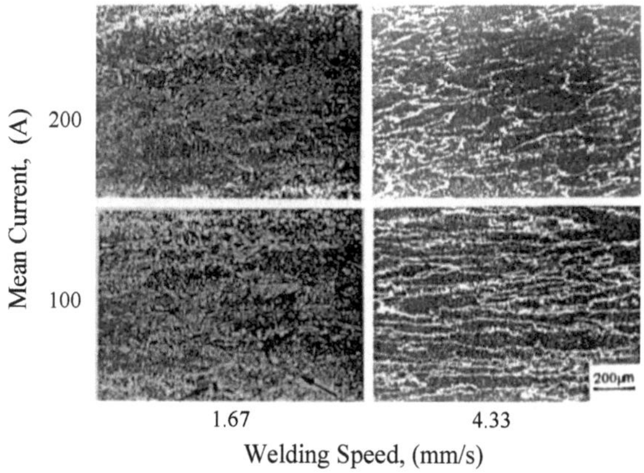

given mean current also relatively reduces width of HAZ whereas, the lowering of welding speed enhances it significantly as shown in Figs. 6.45 and 6.46 respectively. At any mean current the use of low arc voltage and high welding speed of about 21 V and 4.33 mm/s respectively gives stable weld deposition and a significantly narrower HAZ than that of continuous current (0 Hz) weld (Fig. 6.44a).

The increase of mean current and decrease of welding speed enhances width of HAZ primarily due to increase in heat input. However, at a given mean current and welding speed width of HAZ enhances with increase of pulse frequency or pulse duration. It primarily happens due to increase of super heating of weld deposit and/or heat build-up in the weld pool as a function of I_p, I_b, f and t_b. Thus, here also width of HAZ is well correlated with the summarized effect of pulse parameters

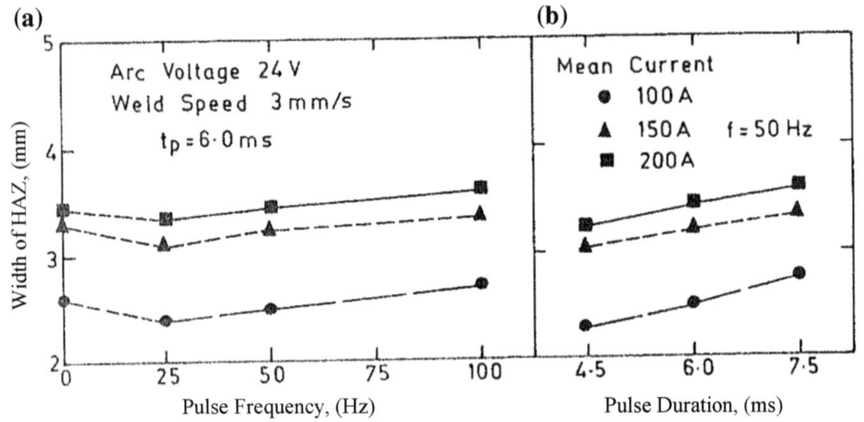

Fig. 6.44 At different mean currents effect of **a** pulse frequency and **b** pulse duration on the width of HAZ

Fig. 6.45 Effect of arc voltage on width of HAZ

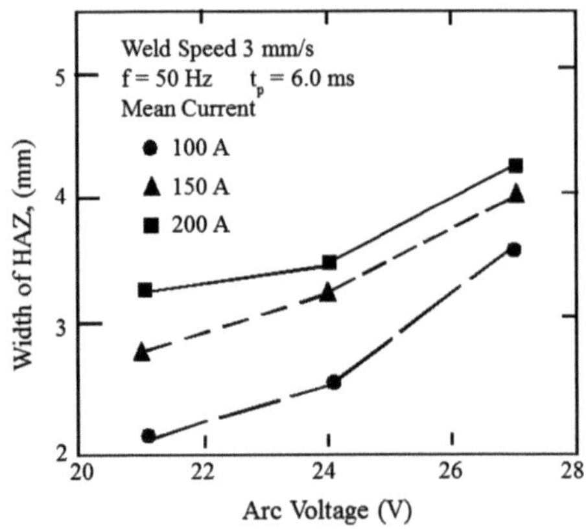

Fig. 6.46 Effect of welding speed on width of HAZ

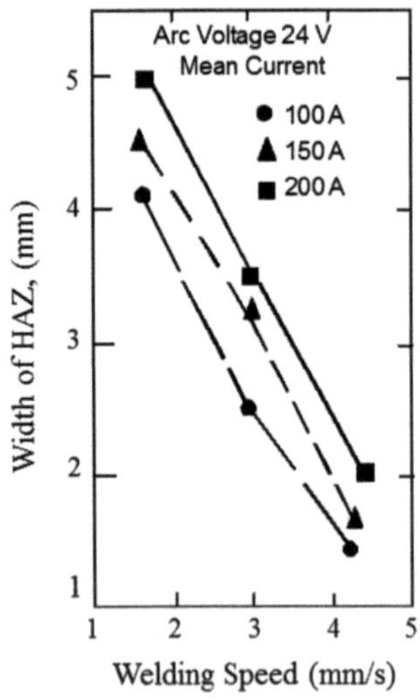

given by the factor ϕ. At given mean current width of HAZ reduces linearly (Fig. 6.47) with increase of ϕ, where arc voltage is kept constant. Arc voltage also considerably influences the I_p and I_b, and consequently the ϕ. In spite of that it cannot be considered along with the pulse frequency or duration to evaluate their influence on the width of HAZ because the arc voltage not only affects the HAZ by influencing the I_p and I_b but also by changing the arc length significantly. The expression of ϕ does not involve any functional term considering the arc length. However, the width of HAZ almost steadily reduces (Fig. 6.48) with the decrease of ϕ under relatively wide variation of arc voltage where the effect of mean current becomes practically insignificant.

Advantage of P-GMAW to achieve spray mode of metal transfer also at relatively low mean current explores the possibility of using this process in positional welding. However, at comparatively high mean current it gives enhanced deposition rate and better process economy. But, here also use of different pulse

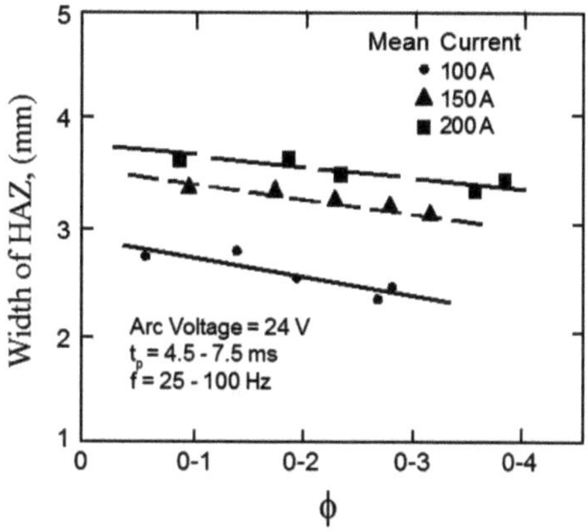

Fig. 6.47 At a given arc voltage influence of ϕ on width of HAZ

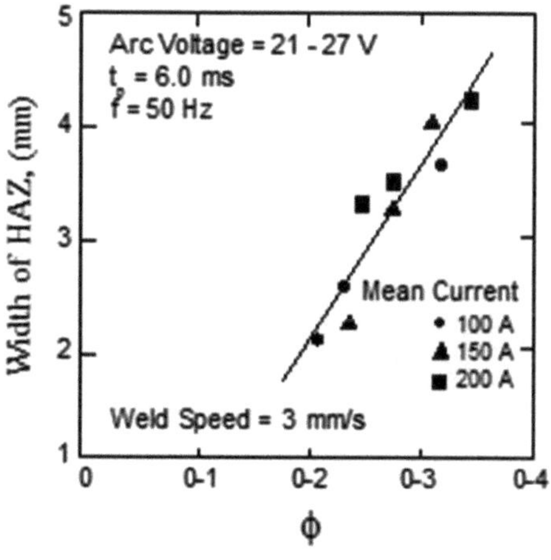

Fig. 6.48 Under varying arc voltage influence of ϕ on width of HAZ

parameters significantly affects microstructures of weld and HAZ. In vertical-up P-GMAW of 6 mm-thick structural steel plate using 1.2 mm-diameter mild steel filler wire under Ar+5% CO_2 shielding, an increase of ϕ at a given I_m appreciably refines the microstructure of weld deposit from the coaxial dendrite to a refined equiaxed one Fig. 6.49a, b where the energy input is kept constant at 2.3 kJ/mm. Such an influence becomes comparatively more significant at higher mean current. Increase of ϕ develops larger amount proeutectoid grain boundary ferrite in relatively finer-microstructure of weld metal that primarily consists of acicular ferrite, bainite, and some fine pearlite. At a given mean current, significant reduction in temperature of droplets with increase of ϕ refines microstructure of weld metal where a suitable interruption in metal deposition also plays a favorable role. This is in the line of the previous understanding of predominant role of interruption in metal deposition on microstructure refinement of weld deposit during pulse current GMAW at different pulse parameters [21, 95, 96].

In case of vertical-up P-GMAW of steel plate the prior austenite grain size of HAZ adjacent to fusion line of base metal relatively decreases with a moderate increase of mean current but significantly reduces with the increase of ϕ as shown in Fig. 6.50(i)a, b and (ii)a, b. The HAZ within about 0.5 mm from fusion line primarily consists of bainite in prior austenite grains containing grain boundary ferrite which appreciably reduces with the increase of ϕ (Fig. 6.51). The morphology of the HAZ becomes comparatively finer with increase of ϕ and distance from fusion line. The refinement of grain size of HAZ with increase of ϕ or mean current is in agreement (Fig. 4.6) to significant decrease in temperature of droplet at the time of deposition. The reduction in coarsening of morphology of HAZ with increase of distance from fusion line happens due to reduction of intensity of weld thermal cycle whereas narrowing down of coarse HAZ with increase of ϕ results from its negative influence on weld thermal cycle as discussed earlier. This phenomenon also significantly reduces the width of HAZ defined by the change of microstructure of base metal with the increase of ϕ but, at a given ϕ a moderate increase of mean current lowers the width of HAZ as shown in Fig. 6.52.

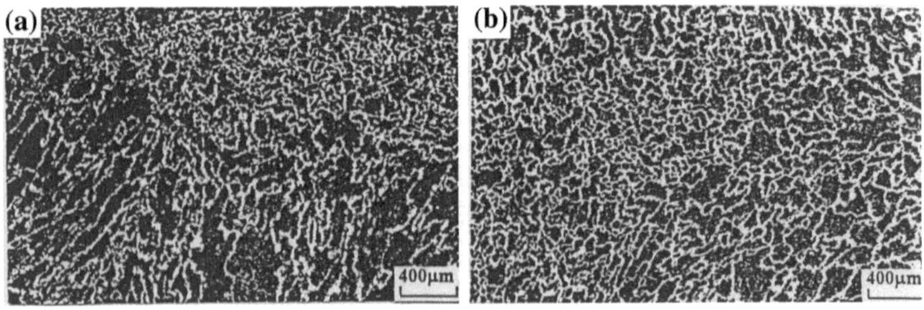

Fig. 6.49 At a given I_m = 130 A the microstructure of weld deposit at **a** ϕ = 0.07 **b** ϕ = 0.23

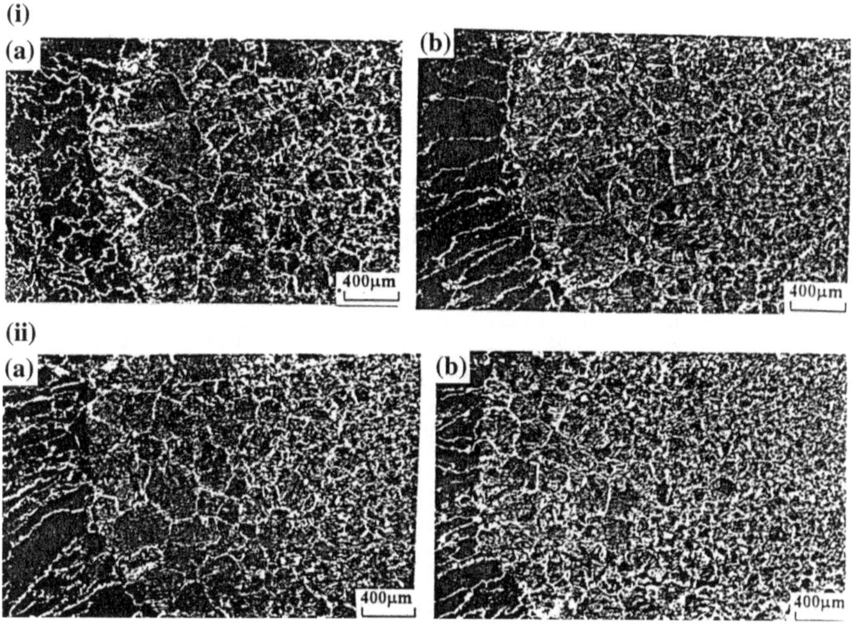

Fig. 6.50 At different mean currents influence of ϕ on microstructure of HAZ; *i* at Im = 100 A **a** $\phi = 0.056$ and **b** $\phi = 0.171$ and *ii* at $I_m = 130$ A **a** $\phi = 0.077$ and **b** $\phi = 0.229$

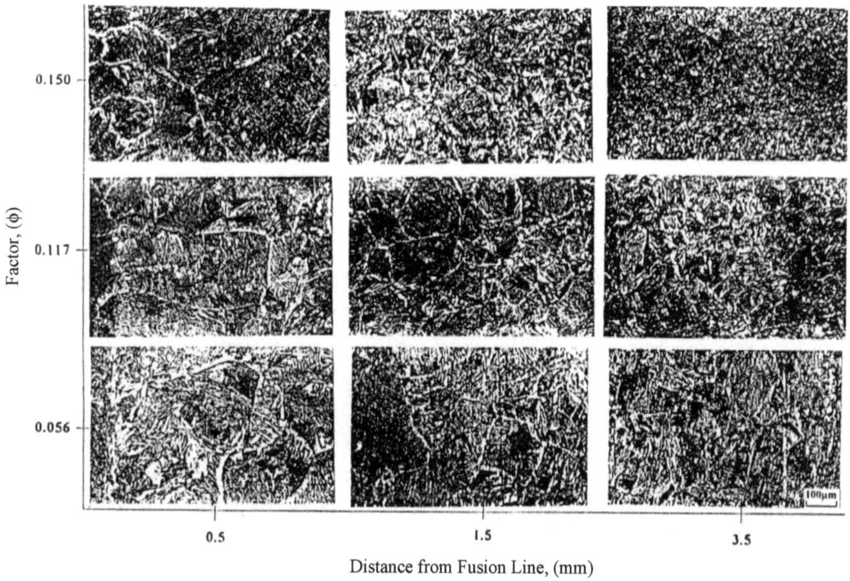

Fig. 6.51 At $I_m = 100$ A influence of ϕ on microstructure of HAZ at different distances from fusion line

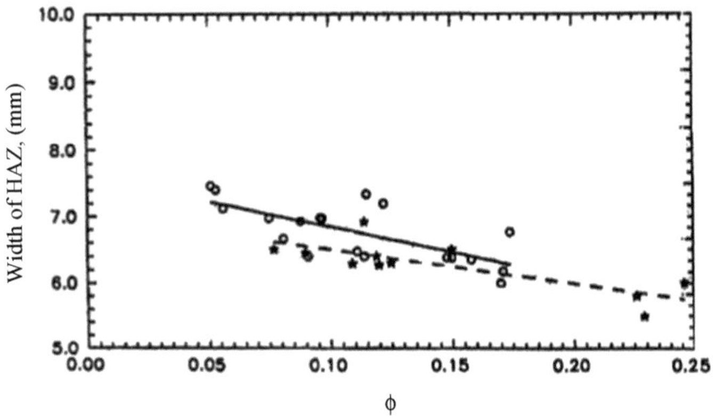

Fig. 6.52 Influence of ϕ on width of HAZ at different mean currents

6.1.4.3 Stainless Steel

The effect of P-GMAW on the microstructure of stainless steel weld and HAZ has been reported by several workers [69–71]. The microstructure of GMA weld of austenitic stainless steel basically contains co-axial dendrite where its morphology coarsens primarily due to increase of heat input (Ω) as it is typically shown in Fig. 6.53a, b for bead on plate weld deposit of 308L steel. However, in concurrence with the general observations on P-GMA welds here also coarsening of microstructure of weld deposit increases with increase of Ω but at a given Ω it appears relatively finer than that of the GMA weld [97]. In conventional arc welding the coarsening of microstructure mostly happens due to increase of energy input but during pulsed arc welding the change in microstructure is further controlled by the energy distribution in the system of pulsation through its influence on

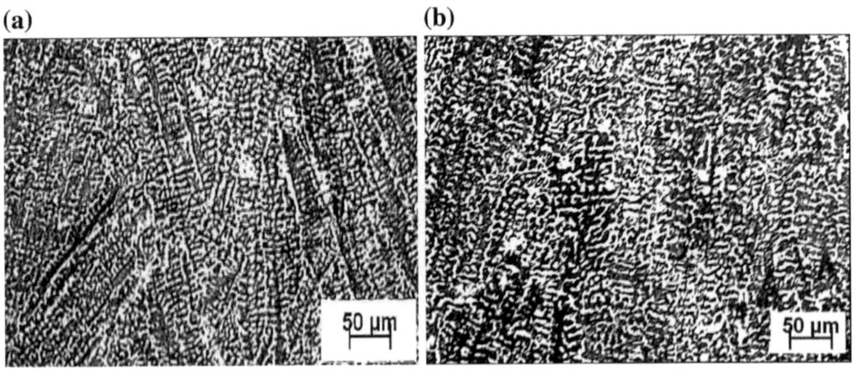

Fig. 6.53 Typical microstructure of central part of GMA weld bead deposited at different Ω of **a** 7.1 kJ/cm and **b** 9.2 kJ/cm

characteristics of arc and metal transfer. These characteristics clearly reflect in change of microstructure with a change of ϕ where the increase of ϕ from about 0.15–0.24 at a given Ω relatively coarsens microstructure due to increase of total energy input (Q_T) in the weld deposit as typically shown in Fig. 6.54. The morphological variation in coaxial growth of dendrite may have primarily happened due to thermal shock and interruption of metal deposition affecting solidification mechanism of weld deposit under pulse current as discussed earlier.

The increase of energy input quite significantly affects the grain coarsening of HAZ adjacent to fusion line (FL) as well as the width of HAZ. The typical coarsening of HAZ microstructure with the increase of heat input in GMA weld bead deposit on 304LN stainless steel plate is clearly evident in Fig. 6.55. However, in P-GMAW the lowering of total heat in depositing weld metal considerably reduces the grain coarsening and width of HAZ (Fig. 6.56) in comparison to that of the GMAW at similar Ω (Fig. 6.55b) [25, 98]. Here it is clearly understood that at given Ω variation of ϕ significantly affects Q_T of weld metal governing the weld isotherm and weld thermal cycle. Thus, it further influences grain size and width of grain growth region of HAZ as function of Ω, Q_T and ϕ. At a given Ω the significant increase of average grain size (D) of HAZ adjacent to FL with increase of ϕ can be clearly found in Fig. 6.57, where it also shows that at a given ϕ the increase of Ω also considerably enhances the D when a larger Ω effectively works at a relatively higher ϕ. The D maintains a good correlation with the Ω and ϕ. At a given ϕ the increase of Ω and vice versa significantly increases the width of grain coarsening (Z) in HAZ as shown in Fig. 6.58 by maintaining a good correlation. It is also interestingly observed (Fig. 6.59) that at a given ϕ the increase of Q_T within a given range of Ω (7.2–11.1 kJ/cm) appreciably enhances the Z. However, in all these cases of P-GMAW the Ω is primarily varied due to change of mean current where welding speed is kept constant [98]. It is further understood that at a given Ω a unit rise in Q_T increases Z by almost 20% at any ϕ

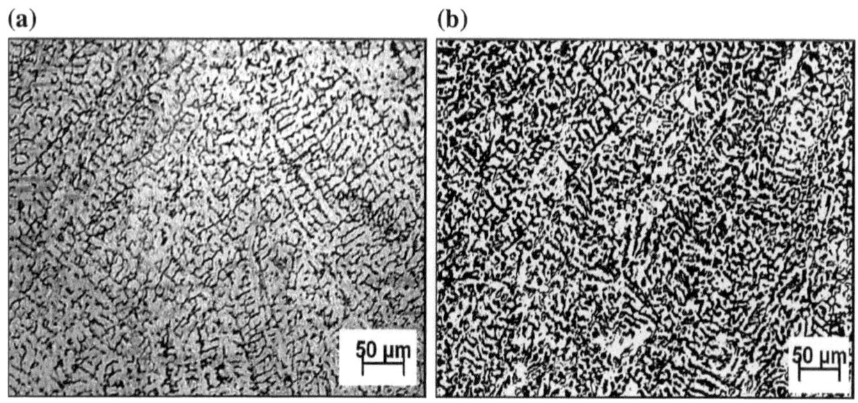

Fig. 6.54 At a given Ω of 9.66 ± 0.64 kJ/cm the variation in typical microstructure of central part of weld bead deposited at a different ϕ of **a** 0.047 **b** 0.153 and **c** 0.237

Fig. 6.55 Grain growth of HAZ adjacent to fusion line observed in GMA weld bead deposited at different Ω of **a** 7.1 kJ/cm and **b** 9.2 kJ/cm

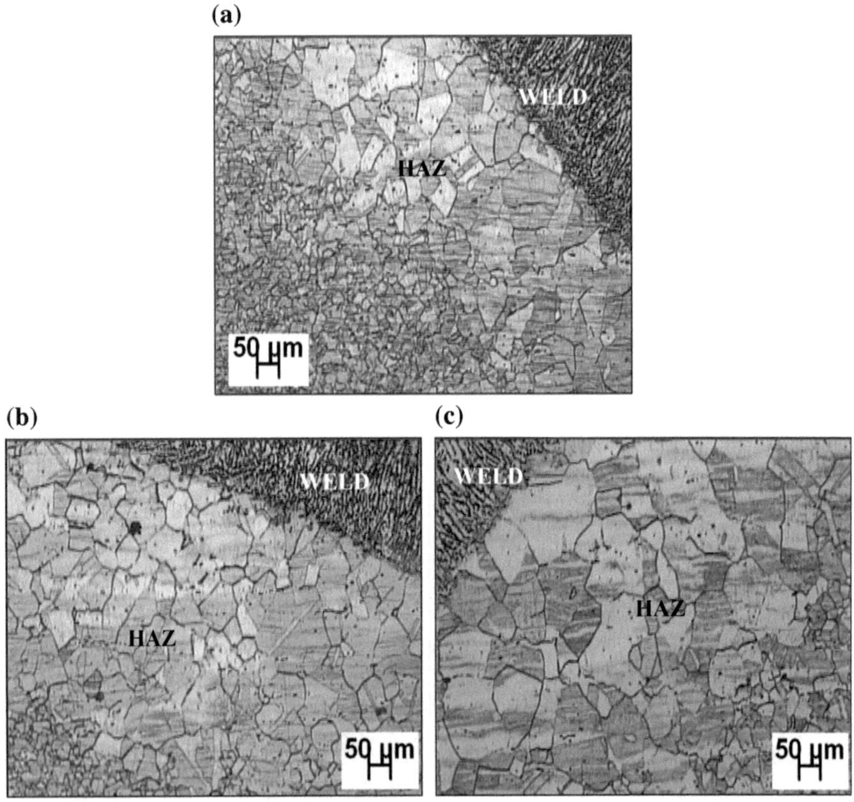

Fig. 6.56 At a given Ω of 9.66 ± 0.47 kJ/cm the variation in grain growth of HAZ adjacent to fusion line of weld bead deposited at different ϕ of **a** 0.047 **b** 0.153 and **c** 0.237

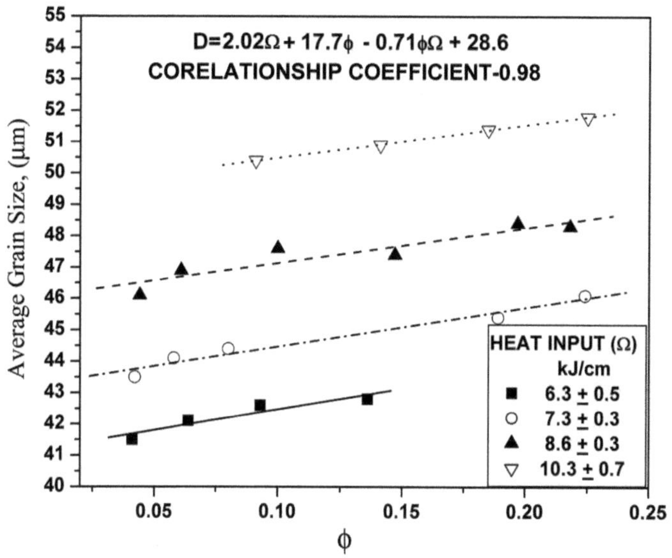

Fig. 6.57 At varied range of Ω, influence of ϕ on average grain size in HAZ adjacent to fusion line of P-GMA weld bead

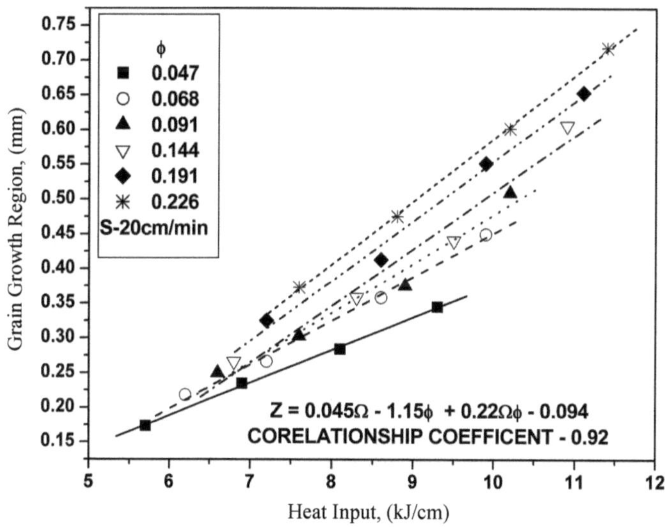

Fig. 6.58 At varied range of Ω, influence of ϕ on average grain size in HAZ adjacent to fusion line of P-GMA weld bead

(Fig. 6.59). Whereas at a given Q_T and Ω, Z increases by almost 40% with the increase of ϕ from 0.14 to 0.22. Such a dominant influence of ϕ on Z primarily indicates that thermal behaviour of weld significantly alters by varying arc stiffness

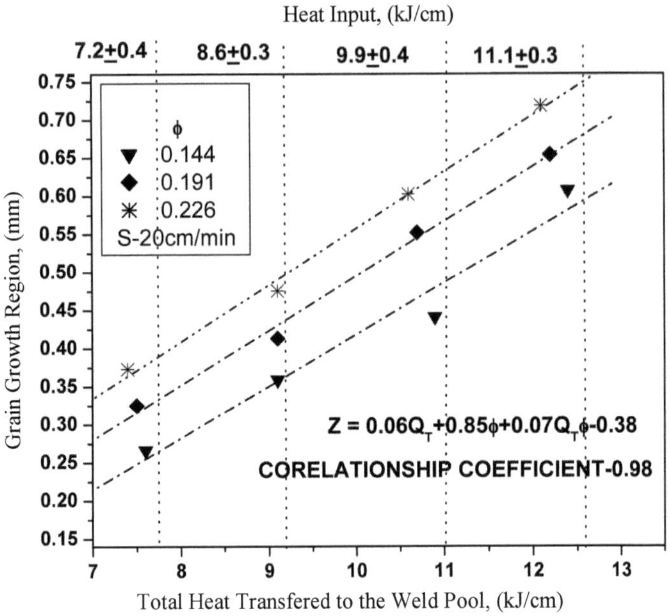

Fig. 6.59 At varied range of Ω influence of ϕ and QT on grain growth region of HAZ adjacent to fusion line at different range of ϕ of **a** 0.047–0.091 and **b** 0.144–0.226 in P-GMA weld beads

[99]. It in turn controls bulging and elongation of isotherms in transverse and weld direction of bead on plate weld deposition depending upon thermal diffusivity of base metal [26]. At given I_m, welding speed and arc voltage reduction of width of recrystallized HAZ adjacent to fusion line with increase of ϕ is in agreement to reduction in temperature of molten filler metal. It further validates the estimated thermal behaviour of the proposed model [100] in case of both the ferrous and nonferrous weld joints.

6.2 Mechanical Characteristics

Besides its good metallurgical characteristics the performance of weld joint also largely depends upon its mechanical characteristics primarily known as distortion, deformation and stresses generated in it. A proper use of P-GMAW can play a significant role in reduction of distortion/deformation and residual stresses of weld joint as a function of pulse parameter significantly influencing thermal behaviour of welding.

6.2.1 *Distortion and Shrinkage Stresses*

In case of fusion welding any undesirable change or deviation from the prescribed dimensional specification of fabricated structure or component is generally identified as distortion. It is basically caused by non uniform expansion and contraction of the regions of the fused weld and solid base metal under the heating and cooling process of welding. The differential of the expansion and contraction stresses causing locked up stresses retained in weld joint after reaching the state of apparent stability crossing through entire thermal cycle is called residual stress. In spite of having defect free weld and heat-affected zone (HAZ), the presence of adverse microstructure along with undesirable distortion and residual stresses in weld joint may badly influence the fatigue, corrosion, and stress corrosion cracking properties [27–30] of weld joint. This is especially true for high-strength metals and alloys. The properties of weld region having cast microstructure are highly sensitive to residual stresses that primarily results from shrinkage of weld metal [31, 32] depending upon weld thermal behaviour [70, 101, 102]. Thus, variation in weld thermal behavior at different pulse parameters influencing the characteristics of metal deposition often affects residual stresses of weld joint [33–35]. The variation in nature of arc and metal deposition beyond certain extent in arc welding process varies magnitude of residual stresses but, there exists a regular pattern of stress distribution in weld joint. The pattern generally follows a magnitude of maximum at weld center followed by a diminishing mode from tensile to compressive at different locations of HAZ of base metal as one goes away from fusion line. The practically instant and simultaneous occurrence of expansion and contraction in weldment especially under relatively complicated shrinkage mechanism of weld metal gives rise to a complex stress system that leads to distortion and it aggravates with subsequent weld run [36, 37]. The distortion due to shrinkage is primarily addressed as longitudinal and transverse besides angular, rotational and buckling.

6.2.1.1 Longitudinal Shrinkage Force

Longitudinal shrinkage is a dimensional reduction in a direction parallel to the length of the weld which enhances with continuous length of deposition. However, the magnitude of it along the weld length is generally found considerably lower than the transverse shrinkage. Such shrinkage can be reduced by intermittent weld wherever possible [38, 39]. The longitudinal shrinkage (Δl) as function of weld length (l) is estimated [40] as

$$\Delta l = [(F_s l)/(EA_w)] \tag{6.9}$$

where, E and A_w are elastic modulus and cross sectional area of weld respectively and longitudinal shrinkage force F_s is estimated as function of yield stress (σ_y) width of plastic zone (w_{pl}) and plate thickness (t_h). The extent of (w_{pl}) significantly depends upon heat input per unit length (q_w) of weld. The q_w in welding considerably varies with the proportionality factor (k_f) of welding process reported [41] as 61, 41 and 72 for the SMAW, GMAW and SAW respectively.

$$F_s = \sigma_y w_{pl} t_h \tag{6.10}$$

$$q_w = k_f A_w \tag{6.11}$$

Thus, the low heat input P-GMAW process can appreciably reduce the longitudinal shrinkage in weld joint through effective control of pulse parameters minimising the heat buildup in weld pool.

6.2.1.2 Transverse Shrinkage Stresss

During welding a measurement of deflection at free end of weld joint with the help of a dial gauge placed at a given distance (straining length) from weld centerline (Fig. 6.60), which may be referred as center of the weld groove, gives an estimation of distortion or bending of welded plate. After each pass of welding at certain heat input the deflection of a thick plate from its initial position gives an estimate of transverse shrinkage, transverse shrinkage stress and bending stress. They are estimated using standard mathematical expressions. The transverse shrinkage stress (σ_{tr}), that develops in weld joint, is evaluated [40] through estimation of transverse shrinkage ($\Delta_{tr(est)}$) using Eqs. 6.12 and 6.13 respectively.

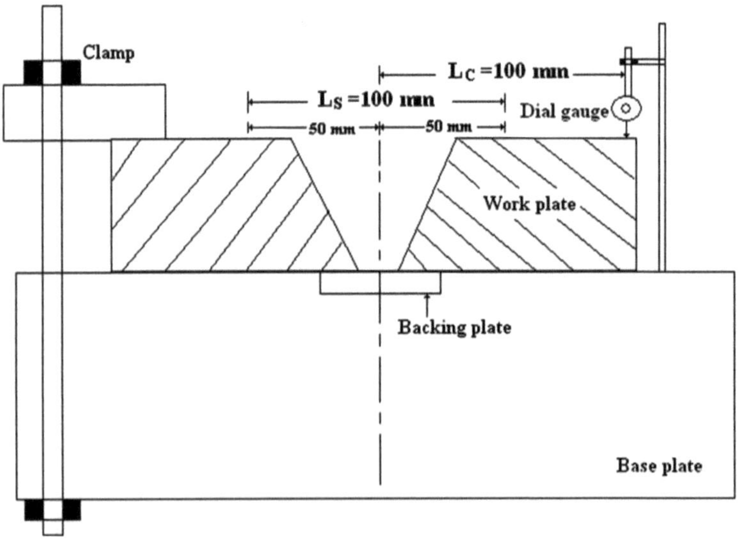

Fig. 6.60 Typical set up to study distortion and stresses in weld joint

$$\sigma_{tr} = \left[\left(R_f \Delta_{tr(est)} E \right) / L_S \right] \tag{6.12}$$

$$\Delta_{tr(est)} = (\lambda_1 K_0 A_w / t_h) + \lambda_2 b \tag{6.13}$$

where, E is the Young's modulus (210 × 10^3 Nm^{-2} for steel), L_S is the straining length (Fig. 6.60), R_f is the shape factor [42], λ_1 is linear thermal expansion. (12 × 10^{-6} K^{-1} for steel). With a change of temperature from T_0 to $((T_1-T_0)/2)$, when T_0 is initial temperature of base plate and T_1 is temperature above which the material is no longer elastic ($T_1 > T_0$) and λ_2 becomes linear thermal expansion of steel [40, 43], A_w is the cross sectional area of weld deposit (mm^2), t_h is plate thickness (mm), b is average breadth of the weld (mm) and K_0 is a constant depending up on thermal output of the welding process. Thermal characteristics of many other ferrous and nonferrous base metals are reported in various literatures [44–47] The K_0 is determined by experimentally measured transverse shrinkage ($\Delta_{tr(msd)}$).

$$K_0 = (\Delta_{tr(msd)} - 1.3\lambda_2 b)t_h / 0.78\lambda_1 A_w \tag{6.14}$$

The transverse shrinkage stress (σ_{tr}) developed in weld joints during their preparation using the plates of different thicknesses under varied thermal behaviour of different welding processes and procedures is also estimated through evaluation of transverse shrinkage (Δ_{tr}) as follows [39].

$$\sigma_{tr} = \Delta_{tr} E / L_S \tag{6.15}$$

$$\Delta_{tr} = (\mu_t 2\alpha\, q_{w)}) / C_\rho t_h \tag{6.16}$$

where, μ_t is transverse stiffness factor, α is coefficient of thermal expansion, C_ρ is specific heat per unit volume and q_w is heat input. The q_w of welding is estimated [40] in consideration of heat generated by welding arc as function of welding current (I), arc voltage (V) and welding speed (S).

$$q_w = \eta_m VI/S \tag{6.17}$$

where, η_m is the fusion efficiency estimated as a function of thermal efficiency of base metal (η_t), arc efficiency (η_h) and correction factor p.

$$\eta_m = p\, \eta_t \eta_h \tag{6.18}$$

The η_m of mild steel can be estimated [40] considering p = 1.0 and η_t = 0.368, where the η_h for SMAW and GMAW is assumed as 65–75% and 75–85% respectively. However, in P-GMAW heat input is estimated by using the mean current (I_m) Eq. 2.1 or effective current (I_{eff}) Eq. 3.19 instead of I of Eq. 6.17.

During butt welding of thick plate by multipass deposition at a practically similar order of heat input the increase of weld pass enhances the transverse shrinkage in the weld and consequently the cumulative deflection of the plates. Thus it increases the transverse shrinkage stress in weld joint. It is amply justified [48, 49] that the use of P-GMAW in place of GMAW significantly reduces the transverse shrinkage and cumulative deflection in weld joint which is finally reflected in reduction of transverse shrinkage stress in it as typically shown in Figs. 6.61, 6.62 and 6.63 respectively. This behavior is agreement to the relatively lower intensity of thermal characteristics of P-GMAW with respect to that of the GMAW process as discussed earlier. Development of a comparatively lower transverse shrinkage in weld joints of AISI:310S austenitic stainless steel prepared by double pulse GMAW process than that observed in P-GMAW and conventional GMAW processes is also reported in earlier work [50].

6.2.1.3 Bending Stress

The bending stress (σ_b) develops due to distortion in weld joint during welding is estimated [51, 52] as

Fig. 6.61 At a given heat input and weld pass effect of GMA and P-GMA welding on transverse shrinkage in 12 mm thick mild steel plate

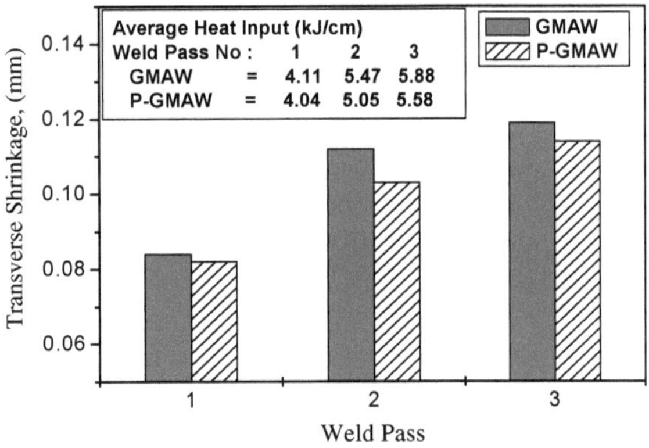

Fig. 6.62 Effect of weld passes on cumulative deflection during GMA and P-GMA welding of 12 mm thick mild steel plate

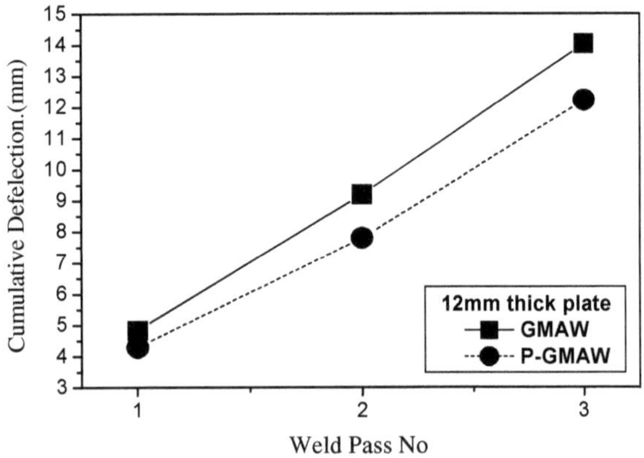

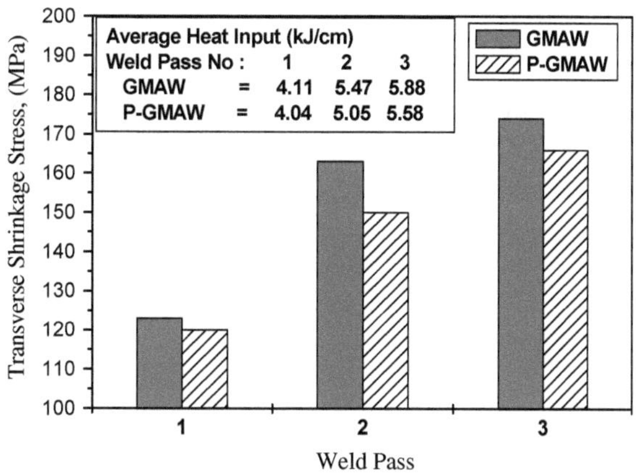

Fig. 6.63 At a given heat input and weld pass effect of GMA and P-GMA welding on transverse shrinkage stress in 12 mm thick mild steel plate

$$\sigma_b = R_f M t_h / 2I_S \tag{6.19}$$

where, M is the bending moment, L_C is the distance of measuring point of the dial gauge tip (Fig. 6.60) from the central axis of the weld joint, R_f is the shape factor of restraint intensity dependent on ratio of the thickness of weld to base plate [42] and I_S is the moment of inertia. The M and I_S are estimated as

$$M = F_d L_C \tag{6.20}$$

$$I_S = b_w t_h^3 / 12 \tag{6.21}$$

$$F_d = 3\delta E I_S / L_C^3 \tag{6.22}$$

The force (F_d) generated due to distortion at the plate of width b_w and thickness t_h is estimated through the measured deflection δ.

At similar order of heat input use of P-GMAW appreciably reduces the bending stress developed in butt weld of mild steel plate compared to that happens while using conventional GMAW process (Fig. 6.64). This is found true in the plates of different thicknesses [48]. At a given close range of heat input use of P-GMAW at relatively higher ϕ of stable welding gives rise to development of lower bending stress in weld joint of different groove design [49] as typically shown in Fig. 6.65. The development of such bending stresses in weld joint leads to retention of locked up stresses in weld commonly known as residual stresses.

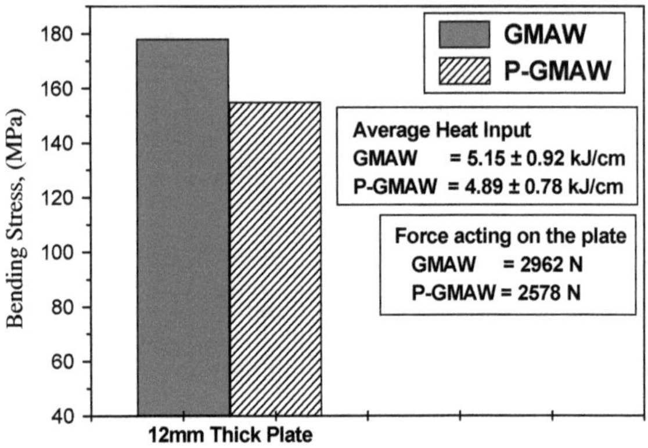

Fig. 6.64 Bending stress of weld joint of 12 mm thick mild steel plate in GMAW and P-GMAW processes

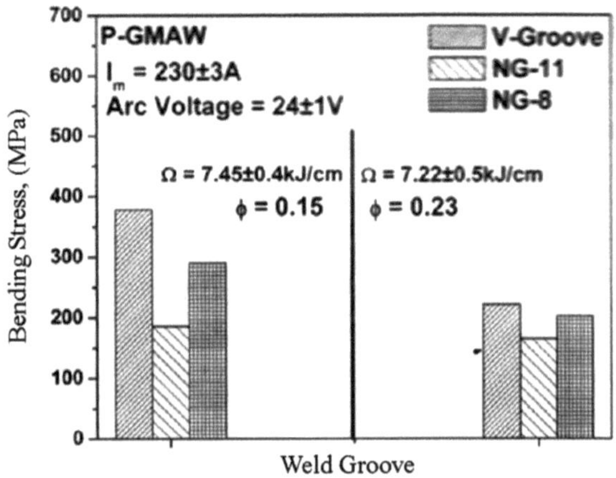

Fig. 6.65 At a given ϕ, Im and arc voltage of GMAW-P effect of different weld grooves on estimated bending stress of weld joints at different ϕ

6.2.2 *Residual Stresses*

Assuming the stresses present in the longitudinal (σ_L), transverse (σ_T) and short transverse (σ_{ST}) directions are the principal stresses the effective residual stresses (σ_R) in a weld can be estimated as

$$\sigma R = (1/\sqrt{2})[(\sigma T - \sigma L)^2 + (\sigma L - \sigma ST)^2 + (\sigma ST - \sigma T)^2]$$

The effect of residual stresses on mechanical behavior and properties of weld is well known for a long time [53, 54] influences the life and safety of weld joint. The presence of residual stress proportionately reduces the fracture toughness (K_{IC}) of the joint measured in terms of reduction in (ΔK_C) regulated by a material dependent constant [55]. It is practically impossible to avoid the development of residual stresses in fusion weld. But, a use of appropriate welding process, procedure and parameters may give access to control the residual stresses quite significantly within the acceptable level of a service condition.

Increase in number of weld pass enhances the cumulative deflection of plate weld due to accumulation of more heat with the increase of weld deposition. However, at given heat input cumulative deflection of multipass deposition (Fig. 6.62) and transverse shrinkage (Fig. 6.61) at each weld pass reduces appreciably with the use of P-GMAW compared to that observed in GMAW. It primarily happens due to relatively low severity of weld thermal cycle of the P-GMAW process. At a practically similar heat input and weld groove size the transverse shrinkage stress and bending stress develops in weld joint, estimated as a function of measured bending of the plate, relatively become low (Figs. 6.64 and 6.65) in case of P-GMA welding than GMA welding. The estimation of transverse shrinkage, transverse shrinkage stress and bending stress in mild steel plate may consider $E = 210 \times 10^3$ N/mm^2, $\mu_t = 0.5$, $\alpha = 12 \times 10^{-6}$ K^{-1} and $C_\rho = 4.9 \times 10^{-3}$ J/mm^{3}K. Thus, longitudinal and transverse residual stresses of P-GMA weld becomes significantly lower than that of GMA weld (Fig. 6.66), while measured [56] as per ASTM E837 using strain gage rosette with centre drill hole technique (Fig. 6.67). The relatively lower stress may have primarily attributed to comparatively lower force for bending of weld joint resulting from relatively smaller heat buildup arising out of interruption of weld deposition in P-GMA welding.

Lowering of welding current generally reduces residual stresses of weld centre. But, use of P-GMAW at identical condition of welding current (mean) and energy input further reduces residual stresses of a weld (Fig. 6.68) and it could be appreciably managed through appropriate control of pulse parameters (ϕ) [34]. Pulse current GMAW operated at higher ϕ beyond about 0.06 produces weld joint of comparatively lower residual stresses than conventional GMA weld. Residual stresses of weld joint depends upon variation of its various geometrical components primarily considered as areas of weld joint (WJ) and total weld pool (WP) estimated as schematically shown in Fig. 6.69 and amount of weld deposition. More precise change of residual stresses of P-GMA weld primarily happens by virtue of the

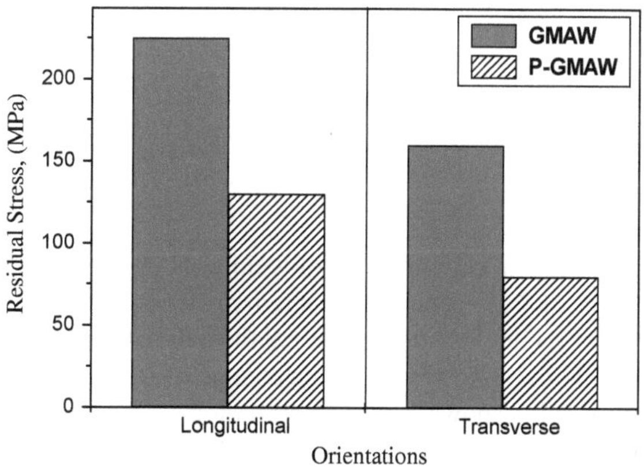

Fig. 6.66 Centre line residual stresses at different orientations of GMA and P-GMA welds

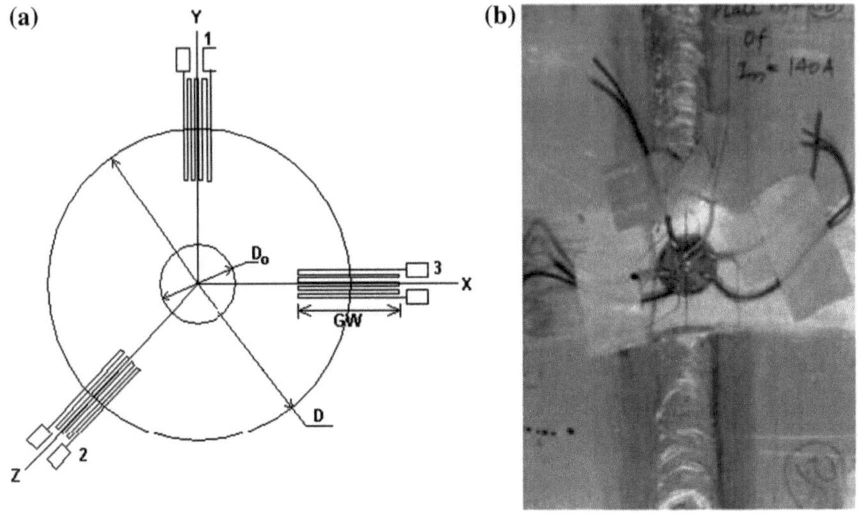

Fig. 6.67 Measurement of residual stresses of weld **a** arrangement of strain gages in a rosette having drill hole diameter Do, gage circle diameter D and strain gage length GW and **b** Placing of strain gage rosette at weld centre

Fig. 6.68 Influences of ϕ on longitudinal residual stress (σLS) at center of single pass Al–Zn–Mg alloy weld

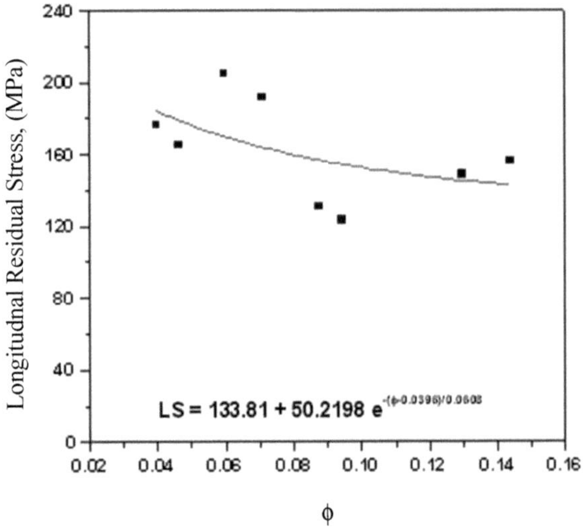

influence of ϕ on the size of weld pool (WP) (Fig. 6.70). The longitudinal residual stress of weld centre always remains comparatively higher than transverse residual stress lying close to yield strength (220–240 MPa) of the matrix (Al–Zn–Mg) alloy (Table 6.3). However, with a change of ϕ nature of variation of residual stresses of weld joint shows a close correlation to nature of variation in area of weld joint (WJ) (Fig. 6.71). It infers that residual stresses of weld joint are largely attributed to

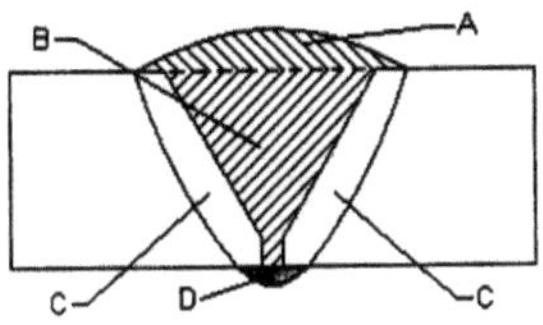

WJ (mm^2) = B + 2C and WP (mm^2) = A + B + D + 2C

Fig. 6.69 Schematic diagram of different components of weld pool in transverse section of weld joint

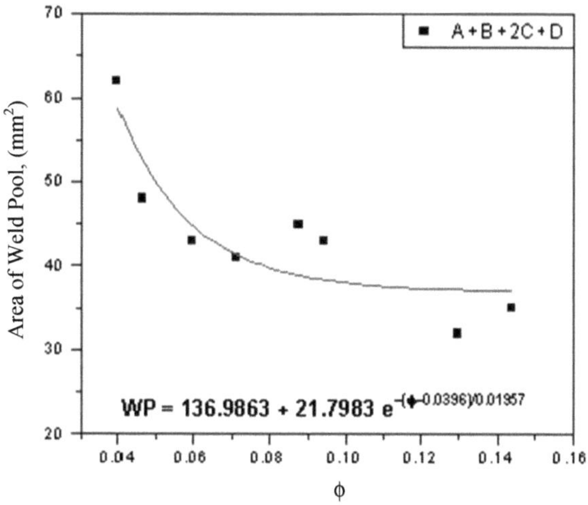

Fig. 6.70 Effect of ϕ on area of weld pool

Table 6.3 Effect of welding currents on residual stresses of conventional GMA weld of Al–Zn–Mg alloy Al–Zn–Mg alloy

Welding current (A)	Residual stresses at different orientations (MPa)	
	Longitudinal	Transverse
140	202	154
160	225	167

contraction stresses developed in weld area. Top and root reinforcements of weld deposit have practically insignificant influence on it. Contraction stresses develop during solidification is largely governed by heat content and cooling rate of weld joint comprising the weld deposit and fused base metal (2C) as shown in Fig. 6.69. Residual stresses are predominantly dictated by size of weld joint, where change in temperature of droplet at the time of deposition plays an insignificant role. Accordingly considerable amount of interest has been put forward to understand the residual stresses developed in narrow gap welding [57–61]. However, for control of residual stresses in weld joint contribution of interrupted metal deposition in pulsed current welding on modification of solidification mechanism of weld pool affecting its shrinkage stress cannot be ignored. An appropriate control of ϕ and I_m provides beneficial effect on residual stresses of weld joint due to their influence on Tde which dictates solidification behaviour of weld pool.

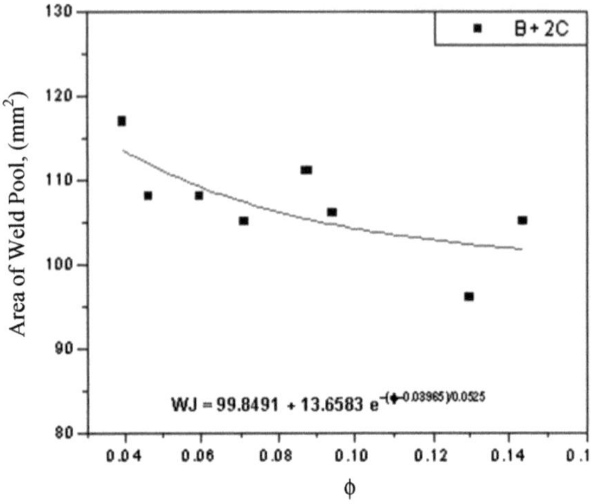

Fig. 6.71 Effect of ϕ on area of weld joint

References

1. Pearson, W.B.: Hand Book of Lattice Spacing and Structures of Metals, Pergamon Press, London, **2**: 574 (1967)
2. Ghosh, P.K., Gupta, P.C., Goyal, V.K.: Stainless steel cladding of structural steel plate using pulsed current GMAW process. Weld. J. **77**(7), 307s–312s (1998)
3. Ghosh, P.K.: The effect of dilution and heat input on the interface characteristics of stainless steel clad mild steel produced by SAW process. Tool Alloy Steels **24**(8), 255–260 (1990)
4. Delong, W.T.: Ferrite in austenitic stainless steel weld metal. Weld. J. **53**(7), 273s–286s (1974)
5. Erich, F.: Welding metallurgy of stainless steels, Springer, NY, 93 (1988)
6. Keehan, E.: Effect of microstructure on mechanical properties of high strength steel weld metals, Ph.D. Thesis, Department of Experimental Physics, Chalmers University of Technology, Sweden, ISBN:91-7291-523-4, (2004)
7. Gruhl, W., Cordier, H.: Precipitation hardening and stress corrosion in. $AlZnMg_3$. Z. Metallkunde **55**(10), 577–582 (1964)
8. Matsuda, F., Nakata, K., Shimokusu, Y.: Effect of additional element on weld solidification crack susceptibility of Al–Zn–Mg. Trans. JWRI **12**(1), 81–87 (1983)
9. Lu, Z.-J., Evns, W.J., Prker, J.D., Birley, S.: Simulation of microstructure and liquation cracking in 7017 aluminium alloy. Mater. Sc. Eng. (A) **220**(1–2), 1–7 (1996)
10. Glicksman, M.E.: Fundamentals of dendritic microstructures, aluminium transformation technology and applications, ASM, metals Park Ohio, Proceedings of Second International Symposium, Buenos Aires, Argentina, Aug. 24–26, (1981)
11. Petrov, A.V., Birman, V.I.: Solidification of weld metal in pulsed arc welding. Svar. Proiz. **15**(6), 1–3 (1968)
12. Ghosh, P.K., Gupta, P.C., Jain, N.K.: Studies on the properties of weld deposited at various pulse frequencies in MIG welding of Al–Zn–Mg alloy. Indian Weld. J. **68**(10), 550–558 (1989)
13. Threadgill P.L., Bailey, P.L.: The prospects of weld pool refinement, weld pool chemistry and metallurgy, Proceeding of International Conference, April 15–17, London, pp. 9–14 (1980)
14. Bentley, K.P.: Pulsed arc welding of thick Q7.35 plate, Weld. Metal Fab. **34**(1): 10–16 (1966)

15. Rimsky, S.T., Svetsinsky, V.G., Shejko, P.P., et al.: Consumable-electrode pulsed-arc welding of low-alloy steels in Ar+CO_2 mixture. Avtomatich. Svarka **2**, 38–41 (1993)
16. Ogle, M., Blewett, R.V.: Weld quality, inspection and acceptance specification recommended by BS 8118, part 2, Proceeding International Conference on Aluminium Weldments, 5th INALCO 92, Munich, pp. 27–29 (1992)
17. Robinson, I.B., Baysinger, F.R.: Welding aluminium alloy 7039. Weld. J. **45**, 433s–444s (1966)
18. Hussain, H.M., Ghosh, P.K., Gupta, P.C., Babu, P.N.: Fatigue crack growth properties of pulse current multipass MIG weld of Al–Zn–Mg alloy, Trans. Ind. Inst. Met. **50**(4): 275–285 (1997)
19. Goecke, S.F., Dorn, L.: Influence of least active gas additions on pulsed current MIG-welding of aluminium, Konferenz-Enizelbericht in Prod. of "Joining of aluminium and magnesium", Sebnik, pp. 19–21 (2012)
20. Lancaster, J.F.: Metallurgy of Welding, Abington Publ., 6th edn., p. 355 (1999)
21. Ghosh, P.K.: Improvement in weld metal properties by refining of microstructure using pulsed current GMAW process, Proceedings Conference Advances in Materials Processing, Dept. Met. Mater. Engg., IIT Roorkee, Nov. 9–10, pp. 108–117 (2001)
22. Mvola, B., Kah, P., Martikainen, J., Hiltunen, E.: Applications and benefits of adaptive pulsed GMAW, ISSN 1392–1207. Mechanika **19**(6), 694–701 (2013)
23. Ghosh, P.K., Yongyuth, P., Gupta, P.C., Patwardhan, A.K., Prakash, S.: Two dimensional spatial geometric solution for estimating the macro-constituents affecting the mechanical properties of multipass C-Mn steel SAW deposits, ISIJ Int., **35**(1): 63–70 (1995)
24. Ashby, M.F., Easterling, K.E.: A first report on diagrams for grain growth in welds. Acta Metall. **30**, 1969–1978 (1982)
25. Kulkarni, S.G., Ghosh, P.K., Ray, S.: Improvement of weld characteristics by variation in welding processes and parameters in joining of thick wall 304LN stainless steel pipe. ISIJ Int. **48**(11), 87–94 (2008)
26. Rajamurugan: Pulsed current gas metal arc welding of dissimilar austenitic stainless steel and HSLA steel, Ph.D. Thesis, Indian Institute of Technology Roorkee, India, (2012)
27. Tsai, Y.-L., Chang, C.-C., Chou, C.-P.: The study of the relationships between material properties and welding residual stresses, Manuf. Sci. Eng. Adv. Mater. Res., **97–101**, 703–708 (2010)
28. Tseng, K.H., Chou, C.P.: The effect of pulsed GTA welding on the residual stress of a stainless steel weldment. J. Mater. Process. Tech. **123**, 346–353 (2002)
29. Nordmark, G.E., Herbein, W.C., Dickerson, P.B., Montemarano, T.W.: Effect of weld discontinuities on fatigue of aluminum butt joints. Weld. J. **66**(6), 162s–173s (1987)
30. Toyouka, T., Tsunenari, T., Ide, R., Tange, T.: Fatigue test of residual stress induced specimens in carbon steel. Weld. J. **64**(1), 29s–36s (1985)
31. Welding Handbook, 7th edn., American Welding Society, Miami, Florida, USA, **1**, 221–244 (1981)
32. ASM Handbook, 8th edn., Welding, Brazing and Soldering, ASM, Materials Park, OH, **6**, 180–185 (1993)
33. Ghosh, P.K., Ghosh, A.K.: Effect of pulsed current GMAW on residual stresses in weld joint of high strength Al-alloy. Proceedings Seminar on Advances in Welding Technology (Weld Tech—2003), IIT Kharagpur, March 14–15, 2003
34. Ghosh, P.K., Ghosh, A.K.: Control of residual stresses affecting fatigue life of pulsed current gas metal arc weld of high strength aluminum alloy. Met. Mater. Trans. A, **35A**(8): 2439–2444 (2004)
35. Ghosh, P.K., Devakumaran, K., Goyal, V.K., Kulkarni, S.G., Ghosh, A.K.: A superior technique of using pulsed current GMAW for welding of ferrous and non-ferrous materials, Proceedings of XIV International Conference of Indian Society of Mechanical Engineers (ISME-2005), Delhi College of Engg., December 12–14, (2005)
36. Masubuchi, K.: Analysis of Welded Structures, New York, Pergamon Press, pp. 47–52 (1980)

37. Jenney, C.L., O'Brien, A.: AWS Welding Handbook, I, 9th edn., pp. 329–338 (2001)
38. Jang, C.D., Lee, C.H.: Prediction of welding deformation of ship hull blocks. Proceedings of International Workshop on Frontier Technology in Ship and Ocean Engineering, pp. 41–49 (2003)
39. Wahab, M.A., Alam, M.S., Painter, M.J., Stafford, P.E.: Experimental and numerical simulation of restraining forces in gas metal arc welded joints, welding J., **85**(2): 35s–43s (2006)
40. Radaj, Dieter: Heat Effects of Welding. Springer, Berlin (1992)
41. Haensch, A.: Schweisseigenspanungen und Formaenderungen an stabartigen Bauteilen, Berechnung und Bewertung, Berlin: VEB Verlag Technik und Duesseldorf: DVS-Verlag (1984)
42. Ueda, Y., Fukuda, K., Kim, Y.C.: Restraint stress and strain due to slit weld in rectangular plate. Trans. JWRI **7**(1), 11–16 (1978)
43. Vinokurov, V.A.: Welding Stress and Distortion. British Library, Wetherby (1977)
44. Karlsson, L.: Plane stress fields induced by moving heat sources in butt welding. J. Appl. Mech. (ASME) **99**, 231–236 (1977)
45. Friedman, E.: Thermomechanical analysis of the welding process using the finite element method. J. Press. Vessel Technol. (ASME) **97**(8), 206–213 (1975)
46. Lindh, D.V., Tocher, J.L.: Heat generation and residual stress development in resistance spot welding. Weld. J. **46**(8), 351s–360s (1967)
47. Masubuchi, K.: Analysis of welded structures. Pergamon Press, New York (1980)
48. Ghosh, P.K., Reddy, M.R., Devakumaran, K.: Distortion and transverse shrinkage stress in butt welds of steel plates under different welding procedure and parameters of GMAW and SMAW processes. Ind. Weld. J. **38**(4), 15–23 (2005)
49. Ghosh, P.K., Devkumaran, K., Pramanick, A.K.: Effect of pulse current on shrinkage stress and distortion in multipass GMA welds of different groove sizes. Weld. J. **89**(3), 43s–53s (2010)
50. Mathivanan, A., Senthilkumar, A., Devakumaran, K.: Pulsed current and dual pulse gas metal arc welding of grade AISI: 310S austenitic stainless steel. Defence Technol. **11**(3), 269–274 (2015)
51. Khurmi, R.S.: Strength of materials, S. Chand and company, India, pp. 467 (2002)
52. Maitra, G.M., Prasad, L.V.: Handbook of Mechanical Design. Tata McGrawHill Publishing Co., Ltd., New Delhi (1986)
53. Watanabe, M., Satoh, K., Minehisa, S.: The effect of residual stress on the fatigue of welded joints (in Japanese), J. Japan. Weld. Soc. **24**(8): 331–336 (1955)
54. Robelotto, R., Lambase, J.M., Toy, A.: Residual stress in welded titanium and their effects on mechanical behavior. Weld. J. **47**(7), 289s–298s (1968)
55. Lancaster, J.F.: Metallurgy of welding, 6th edn., Abington Publishing, Woodhead Publishing Ltd., Cambridge, England, p. 206 (1999)
56. Standard test method for determining residual stresses by hole drilling strain guage method, ASTM E-837**, 3.01**: 715–719 (1990)
57. Xu, J.J., Zhu, Z.Q., Chen, L.G., Ni, C.Z.: Temperature distribution and residual stresses during multipass narrow gap welding of thick plates. Mater. Sci. Technol. **22**(2), 232–236 (2006)
58. Chand, R.R., Kim, I.S., Lee, J.P., Kim, Y.S., Kim, D.G.: Numerical and experiment study of residual stress and strain in multi-pass GMA welding. J. AMME **57**(1), 31–37 (2013)
59. Liu, C., Zhang, J.X., Xue, C.B.: Numerical Investigation on residual stress distribution and evolutions during multi-pass narrow gap welding of thick-welled stainless steel pipes. J. Fusion Eng. Des. **86**, 288–295 (2011)
60. Shu, F., Lv, Y., Liu, Y., Xu, F., Sun, Z., He, P., Xu, B.: Residual stress modeling of narrow gap welded joint of aluminum alloy by cold metal transferring procedure, Constr. Building Mater., **54**: 224–235 (2014)
61. Lee, C.Y., Hwang, J.K., Bae, J.W.: Analysis of residual stress for narrow gap welding using finite element method. Int. J. Mod. Phys. B **24**, 2797 (2010)

62. Potluri, N.B., Ghosh, P.K., Gupta, P.C., Reddy, Y.S.: Studies on weld metal characteristics and their influence on tensile and fatigue properties of pulse current GMA welded Al–Zn–Mg alloy. Weld. J. **75**(2), 62s–70s (1996)
63. Ghosh, P.K., Sharma, V.: Chemical composition and microstructure in pulsed MIG welded Al–Zn–Mg alloy weldments, Materials Transactions, JIM, **32**(2): 145–150 (1991)
64. Mondolfo, L.F.: Aluminium Alloys; Structure and Properties, Butterworths, London (1979)
65. Bagryanskii, K.V., Royanov, V.A.: Svar. Proiz. **10**, 23–24 (1968)
66. Ghosh, P.K.: The influence of dilution and heat input on the characteristics of SAW stainless steel cladding on mild steel. Int. J. Joining Mater., **4**(3): 90–99 (1992)
67. DebRoy, T., David, S.A.: Physical processes in fusion welding. Rev. Mod. Phys. **67**(1), 85–112 (1995)
68. Matsuda, F., Nakata, K., Tsukamoto, K., Tohgan, S.: Combined effect of current pulsation and Zr addition on improvement of solidification cracking of Al–Zn–Mg alloy weld metal. Trans. IWRI **14**(2), 99–104 (1985)
69. Ghosh, P.K., Gupta, S.R., Gupta, P.C., Rathi, R.: Influence of pulsed current welding on the microstructure and porosity content of Al–Zn–Mg alloy weldment. Pract. Met. **27**, 613–626 (1990)
70. Ghosh, P.K., Gupta, S.R., Gupta, P.C., Rathi, R.: Pulsed MIG welding of Al–Zn–Mg alloy, materials transactions. JIM **31**(8), 723–729 (1990)
71. Hussain, H.M., Ghosh, P.K., Gupta, P.C., Potluri, N.B.: Weld characteristics of multipass pulse current MIG welded Al–Zn–Mg alloy. Int. J. Joining Mater. **9**(2), 74–79 (1997)
72. Goyal, V.K., Ghosh, P.K., Saini, J.S.: Met. Mater. Trans. A **38A**(8), 1794–1805 (2007)
73. Flemings, M.C.: Solidification Processing. McGraw Hill, New York (1974)
74. Era, T., Ueyama, T.: Spatter reduction in GMAW by current waveform control. Weld. Int. **20**(7), 496–501 (2007)
75. Rouffet, M.E., Wendt, M., Goett, G., Kozakov, R., Schoepp, H., Weltmann, K.D., Uhrlandt, D.: Spectroscopic investigation of the high-current phase of a pulsed GMAW process, J. Phys. D: Appl. Phys., **43**(43): 434003 (2010)
76. Ghosh, P.K., Gupta, P.C., Somani, R.: Influence of pulse parameters on the porosity formation in pulsed MIG weld deposit of aluminium alloy. Int. J. Joining Mater. **3**(2), 49–54 (1991)
77. Hu, J., Tsai, H.L.: Heat and mass transfer in gas metal arc welding. Part II. The metal. Int. J. Heat Mass Transf. **50**(5/6), 808–820 (2007)
78. Ghosh, P.K., Rai, B.K.: Characteristics of pulsed current bead on plate deposit in flux cored GMAW process. ISIJ Int. **36**(8), 1036–1045 (1996)
79. Ghosh, P.K., Dorn, L., Hübner, M., Goyal, V.K.: Arc characteristics and behaviour of metal transfer in pulsed current GMA welding of aluminium alloy. J. Mater. Process. Tech. **194**(4), 163–175 (2007)
80. Ghosh, P.K., Gupta, P.C.: Influence of pulsed current MIG welding on the characteristics of Al–Zn–Mg alloy weldments. Trans. Indian Inst. Met. **44**(4), 317–326 (1991)
81. Ahmed, N.: New Developments in Advanced Welding, Woodhead Publishing Ltd, Abington, pp. 9–13 (2005)
82. Randhawa, H.S.: Investigation into positional welding of structural steel using pulse current GMAW process, Ph.D. Dissertation, UOR, Roorkee (1999)
83. Ghosh, P.K., Gupta, S.R., Randhawa, H.S.: Analytical studies on characteristics of vertical up bead on plate weld deposition using pulsed current GMAW. Int. J. Joining Mater. **12**(3), 76–85 (2000)
84. Ghosh, P.K., Gupta, P.C., Breazu, M., Gupta, R.K.: The influence of some welding parameters on the properties of stainless steel strip cladding deposited by a submerged arc process. Trans. JIM **28**(7), 579–584 (1987)
85. Goyal, V.K., Ghosh, P.K., Saini, J.S.: Influence of pulse parameters on characteristics of bead-on-plate weld deposits of aluminium and its alloy in the pulsed gas metal arc welding process. Met. Mater. Trans. A **39A**(12), 3260–3275 (2008)

86. Gery, D., Long, H., Maropoulos, P.: Effects of welding speed, energy input and heat source distribution on temperature variations in butt joint welding. J. Mater. Process. Technol. **167**, 393–401 (2005)
87. Cantin, G.M.D., Francis, J.A.: Arc power and efficiency in gas tungsten arc welding of aluminium. Sci. Technol. Weld. Joining **10**(2), 200–210 (2005)
88. Goyal, V.K., Ghosh, P.K., Saini, J.S.: Analytical studies on thermal behaviour and geometry of weld pool in pulsed current gas metal arc welding. J. Mater. Process. Tech. **209**, 1318–1336 (2009)
89. Ghosh, P.K., Gupta, S.R., Randhawa, H.S.: Characteristics and criticality of pulsed current vertical up GMA weld in steel. Int. J. Joining Mater. **11**(4), 99–110 (1999)
90. Randhawa, H.S., Ghosh, P.K., Gupta, S.R.: Some basic aspects of geometrical characteristics of pulsed current vertical up GMA weld. ISIJ Int. **40**(1), 71–76 (2000)
91. Rhee, S., Asibu, E.K.: Analysis of arc pressure effect on metal transfer in gas metal arc welding. J. Appl. Phys. **70**(9), 5068–5075 (1991)
92. Joseph, A., Harwig, D., Farson, D., Richardson, R.: Measurement and calculation of arc power and heat transfer efficiency in pulsed gas metal arc welding. Sci. Technol. Weld. Joining **8**(6), 400–406 (2003)
93. Ghosh, P.K., Dorn, Lutz, Devakumaran, K., Hofmann, F.: Influence of welding parameters and shielding gas on arc characteristics and behavior of metal transfer in GMA welding of mild steel, Indian Weld. J., **41**(2): 23–33 (2008)
94. Ghosh, P.K., Rai, B.K.: Correlations of pulse parameters and bead characteristics in pulsed current flux cored GMAW process. Ind. Weld. J. **31**(4), 30–39 (1998)
95. Randhawa, H.S., Ghosh, P.K., Gupta, S.R.: Geometric characteristics of pulsed current positional GMA weld. ISIJ Int. **38**(4), 276–284 (1998)
96. Needham, J.C.: Pulse controlled consumable electrode welding arcs. Br. Weld. J. **12**(4), 191–197 (1965)
97. Radaj D.: Heat Effects of Welding', 1st edn, Springer, Berlin (1992)
98. Kulkarni, S.G.:, Effect of narrow gap welding on characteristics of weld joint of austenitic stainless steel, Ph.D. Thesis, Indian Institute of Technology Roorkee, India (2009)
99. Ghosh, P.K., Dorn, L., Kulkarni, S.G., Hofman, F.: Arc characteristics and behaviour of metal transfer in pulsed current GMA welding of stainless steel. J. Mater. Process. Technol. **209**, 1262–1274 (2009)
100. Ghosh, P.K., Goyal, V.K., Dhiman, H.K., Kumar, M.: Thermal and metal transfer behaviours in pulsed current GMA weld deposition of Al–Mg alloy. Sci. Technol. Weld. Joining **11**(2), 232–242 (2006)
101. Ghosh, P.K.: An analysis of weld characteristics as a function of pulse current MIG welding parameters. Int. J. Joining Mater. **8**(4), 157–161 (1996)
102. Ghosh, P.K., Gupta, S.R., Gupta, P.C., Rathi, R.: Fatigue characteristics of pulsed MIG welded Al–Zn–Mg alloy. J. Mat. Sci. **26**, 6161–6170 (1991)

Chapter 7
Properties of Pulse Current GMA Weld

Abstract The influence of pulse parameters on the properties of P-GMA welds of ferrous and non ferrous metals which qualifies them for various applications under the static and dynamic loading has been discussed. Control of pulse parameters affecting the various properties is addressed in terms of the factor ϕ. The hardness and tensile properties of weld joints are categorically addressed for the weld deposit and HAZ relevantly with respect to their chemistry and microstructure. The effect of pulse parameters on the properties such as toughness, fatigue, fracture mechanics and corrosion that dictate the safety and durability of weld joint in service has also been relevantly discussed in reference to the weld geometry, chemistry, microstructure and fracture behavior.

Keywords P-GMA weld joint · Ferrous and non ferrous metals · Pulse parameters · Tensile properties · Hardness · Weld deposit · HAZ · Toughness · Fatigue · Fracture mechanics · Corrosion · Weld chemistry · Microstructure · Fracture behaviour

In view of the critical role of pulse parameters on considerable change in thermal, geometrical, metallurgical and mechanical characteristics of weld joint, as discussed in earlier chapters, it is imperative to have a thorough understanding on their influence on various properties of ferrous and nonferrous welds. The required properties of weld joint widely vary with respect to its applications and safety requirement but, for general structural applications the general mechanical, fatigue and corrosion properties along with the fracture mechanics get primary consideration. The mechanical properties of weld joints are basically understood by their hardness, tensile properties and impact toughness. They primarily depend upon shape and size of grains and primary crystals of HAZ and weld respectively along with type, amount, size and distribution of phases in matrix. During P-GMA welding the variation in thermal behaviour of the process as usual significantly affects the type of phase transformation including precipitation behaviour of heat treatable alloys in weld and HAZ. But due to its ability of elaborate control of weld thermal cycle by regulation of pulse parameters affecting energy distribution in the

P.K. Ghosh, *Pulse Current Gas Metal Arc Welding*, Materials Forming, Machining and Tribology, DOI 10.1007/978-981-10-3557-9_7

process, the P-GMAW provides supplementary influence on morphology of weld and HAZ over that of GMAW, which more precisely governs the properties of weld joint.

7.1 Hardness

Hardness of weld deposit primarily depends upon its chemical composition and microstructure. However, it also reduces in presence of porosity beyond a certain extent. The chemical composition of weld deposit of dissimilar filler is largely governed by dilution whereas; its morphology is basically dictated by the solidification mechanism. Hardness of HAZ is dictated by grain growth and matrix morphology adjacent to and away from the fusion line. All such characteristics as the dilution and solidification mechanism of weld deposit and the grain growth and morphology of HAZ strongly depend upon thermal behaviour of welding. Thus the significant influence of pulse current on thermal behaviour of GMAW in various forms interestingly affects the hardness of different locations of weld joint.

7.1.1 Weld Deposit

Variation of pulse parameters significantly influences hardness of weld during welding of precipitation hardenable Al–Zn–Mg alloy using non age hardenable Al–Mg filler metal as shown in Fig. 7.1. The zinc pick-up due to dilution of weld arising out of base metal [1] fusion becomes primarily responsible for the increase of hardness of weld deposit (Fig. 7.2). The response of weld to precipitation hardening basically depends upon its zinc content. But in this regard, the influence of microstructure also cannot be ignored. The morphology of precipitation hardenable weld metal, influencing its hardness, may be primarily considered by refinement of cast structure along with the amount, size, shape and distribution of precipitates in the matrix. Influence of microstructure on hardness of the weld is

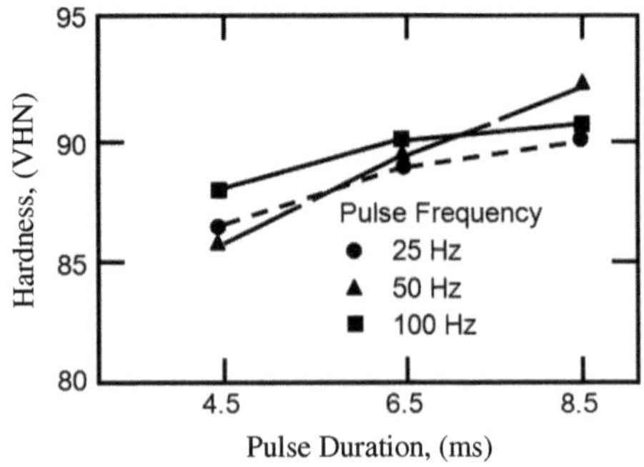

Fig. 7.1 At a given pulse frequency influence of pulse duration on hardness of weld centre

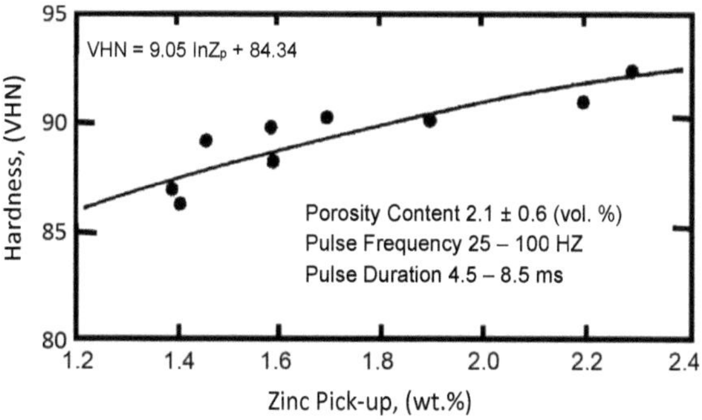

Fig. 7.2 Correlation of zinc pick-up (Zp) with hardness of weld deposit

especially realized by relatively lower hardness of conventional GMA weld in spite of its higher zinc pick-up in comparatively coarser cast structure than those of the P-GMA weld. It primarily happens due to presence of comparatively fewer coarser precipitates in the matrix of relatively coarser microstructure.

At a given mean current increase of pulse frequency or pulse duration significantly reduces hardness of weld bead on plate deposit of steel from flux cored filler wire (FCW) (Fig. 7.3), where the pulse duration and frequency are respectively kept constant [2, 3]. But, in contrast to that observed in case of welding of Al–Zn–Mg alloy using Al–Mg filler the pulsed current weld shows comparatively lower hardness than that of continuous current steel weld deposit. However, here increase of arc voltage or welding speed enhances hardness of weld deposit significantly as shown in Figs. 7.4 and 7.5 respectively, where the mean current, pulse frequency and duration are kept constant. All these behaviors are comparatively more prominently observed at higher mean current of 200 A using 1.2 mm diameter FCW. Variation in hardness of weld beads deposited by using FCW at different pulse frequency and duration, arc voltage and welding speed is in agreement to significant coarsening of their microstructure. Decrease in hardness of weld bead with increase of mean

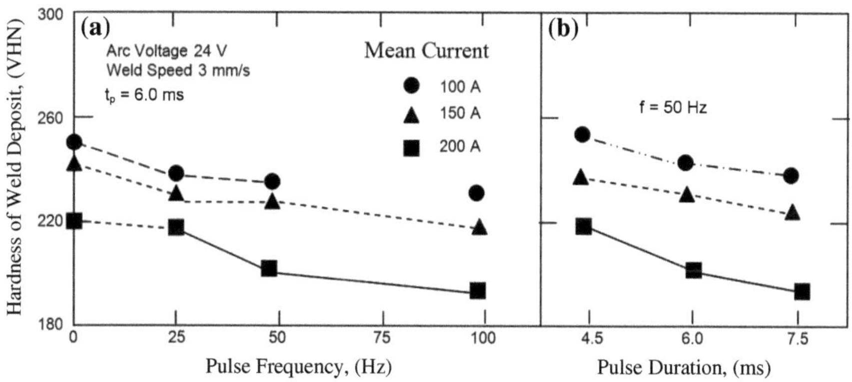

Fig. 7.3 At different mean currents effect of **a** pulse frequency and **b** pulse duration on hardness of weld bead on steel plate

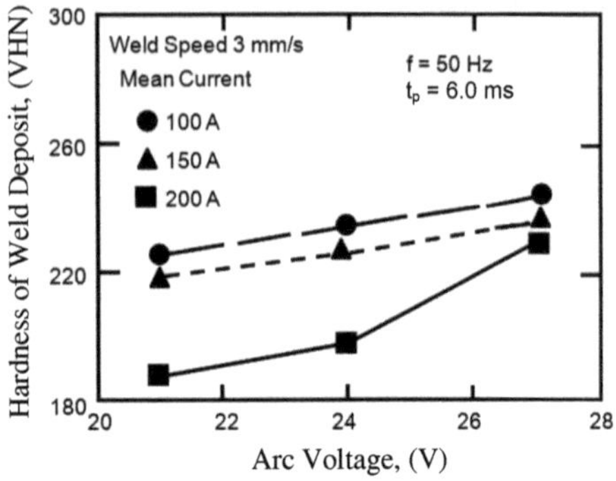

Fig. 7.4 Effect of arc voltage on hardness of weld bead on steel plate

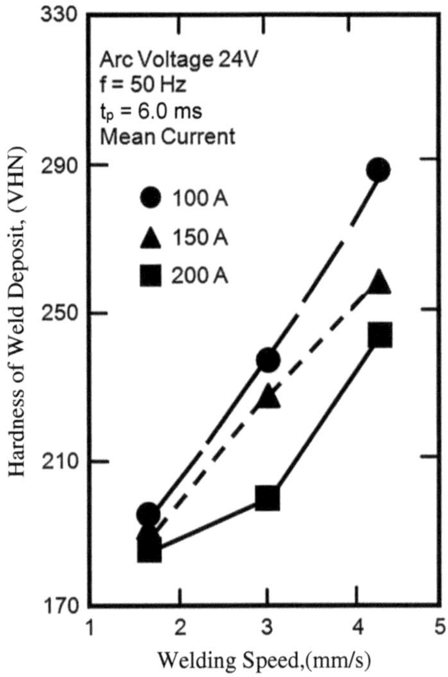

Fig. 7.5 Effect of welding speed on hardness of weld bead on steel plate

current is well understood due to increase of heat input causing coarsening of microstructure as it is also observed in continuous current (0 Hz) deposit (Fig. 7.3a). At given pulse frequency and duration increase of hardness of weld bead with increase of welding speed is well known. But, similar behaviour of hardness of weld bead with increase of arc voltage possibly attributed to change in pulse characteristics (I_p and I_b) with a change of arc voltage affecting the weld thermal cycle. This is in contrast to that generally experienced in conventional GMA welding process.

7.1.2 Heat Affected Zone

Hardness of different locations of HAZ in reference to fusion line of a given weld primarily depends upon their morphology including grain size as well as the type, amount, shape, size and distribution of metallurgical phases present depending upon weld thermal cycles at different parameters. Grain coarsening makes the matrix comparatively softer and weaker whereas, depending upon its morphology the phase transformations may compose the matrix harder and stronger than base material. During studies on non heat treatable system of weld deposition of Al–Mg alloy on aluminium the recrystallization and grain growth in HAZ close to fusion line (Fig. 7.6) reduces its hardness without being appreciably influenced by precipitation hardening or phase transformation [4]. At given mean current and pulse frequency such a reduction in hardness of HAZ with increase of pulse duration primarily happens due to increase in superheating of molten filler metal resulting from longer stay under arcing. However, the use of P-GMAW in this system at mean current beyond the transition current (220 A) of 1.6 mm diameter filler wire an appropriate control of pulse duration and pulse frequency significantly reduces the adverse influence of softening of HAZ Fig. 7.7 by suppressing the heat build-up in weld pool by effective control of interruption in solidification [4]. But, at this high mean current an increase in pulse duration at a high pulse frequency of 100 Hz again decreases the hardness of HAZ possibly by making the interruption in solidification practically invalid with respect to suppression of heat buildup in weld pool.

During welding of precipitation strengthening alloy, retaining various location of HAZ at required temperature of solutionization and ageing significantly modifies its morphological characteristics and related properties. The commonly observed adverse influence of weld thermal cycle on HAZ, as it often occurs in case of conventional GMAW process, can be favourably controlled by taking advantage of its relatively milder variation through interruption in metal deposition at appropriate pulse parameters [5, 6] of P-GMAW. In welding of Al–Zn–Mg alloy a favourable

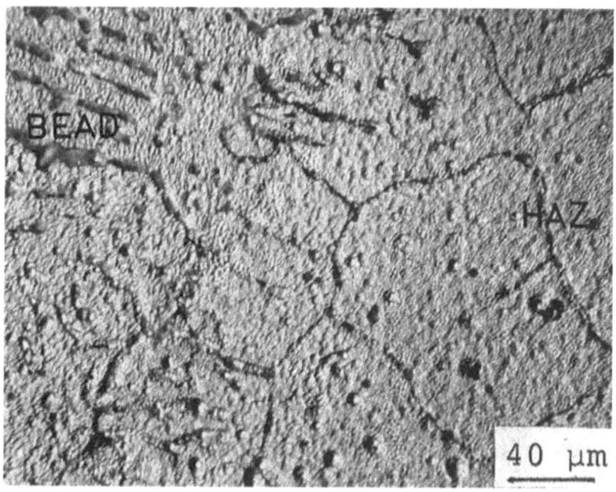

Fig. 7.6 Microstructure of HAZ showing typical recrystallization in the region adjacent to fusion line of aluminium

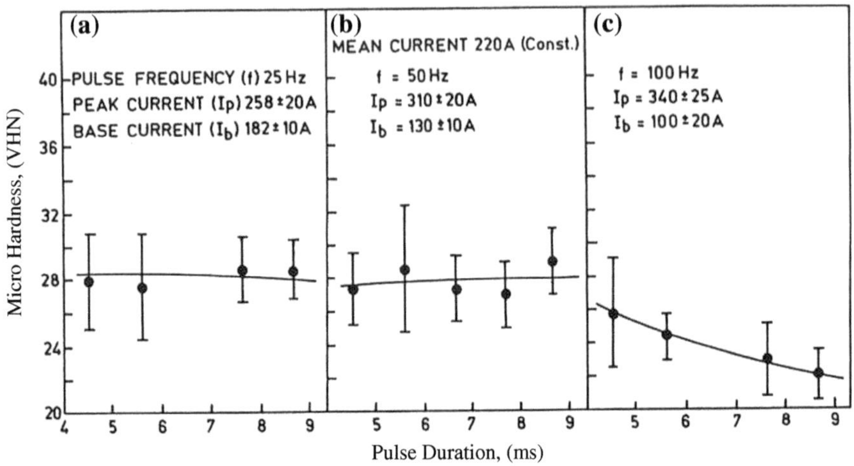

Fig. 7.7 At a given mean current 220 A influence of pulse duration on hardness of HAZ (0.5 mm from FL) observed during weld bead on plate deposition at different pulse frequencies of **a** 25, **b** 50 and **c** 100 Hz

condition of ageing contributing to the formation of comparatively finer precipitates at HAZ with the increase in pulse frequency up to 100 Hz appreciably enhances the micro hardness of HAZ at a distance 1 mm from fusion line (Fig. 7.8) in spite of appreciable grain growth.

A similar observation of considerably higher hardness of weld and HAZ than base metal (Fig. 7.9) has also been marked in case of vertical-up P-GMA welding of mild steel [7]. In spite of having similar carbon content weld metal gives comparatively higher hardness than mild steel base metal primarily due to refinement of microstructure consists of acicular ferrite, bainite and fine pearlite under

Fig. 7.8 Effect of pulse frequency on micro hardness of HAZ at a distance of 1 mm from fusion line

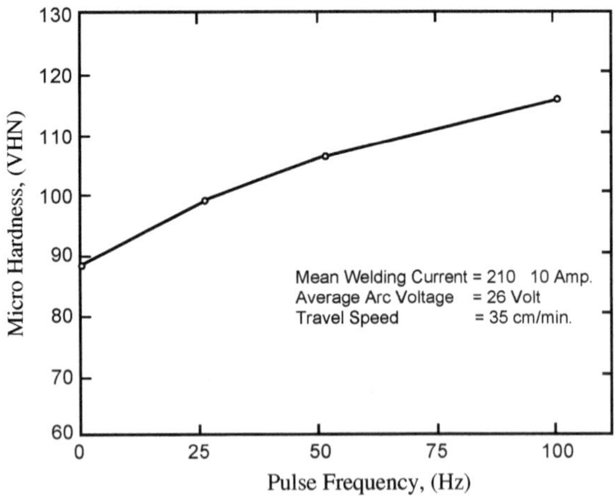

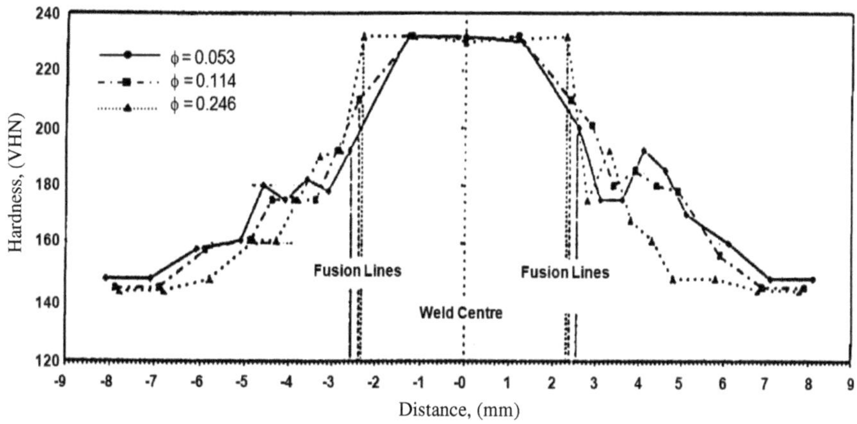

Fig. 7.9 Hardness across vertical-up steel weld produced by using pulsed current at different ϕ

P-GMAW. Even though there is significant grain growth in base metal close to fusion line, but the P-GMA welding introduces comparatively high hardness in HAZ due to favorable thermal behaviour under suitable pulse characteristics causing bainitic transformation in the matrix. However, the HAZ becomes relatively softer as one goes away from the fusion line. It occurs possibly due to lowering of the extent of austenite transformation in matrix inhibiting hard phase transformation during cooling. But, it is interesting to note that extent of HAZ having significant transformation of bainite reduces with increase of ϕ. This is in confirmation to the effect of ϕ on the reduction of width of HAZ as discussed earlier.

Increase of pulse frequency, pulse duration and arc voltage at different mean currents of weld deposition using 1.2 mm diameter flux cored wire reduces hardness of HAZ adjacent to fusion line of weld bead [5, 8] as shown in Figs. 7.10a, b and 7.11 respectively. A reduction in energy input by increase of welding speed enhances the hardness of HAZ of weld bead (Fig. 7.12) in agreement to the

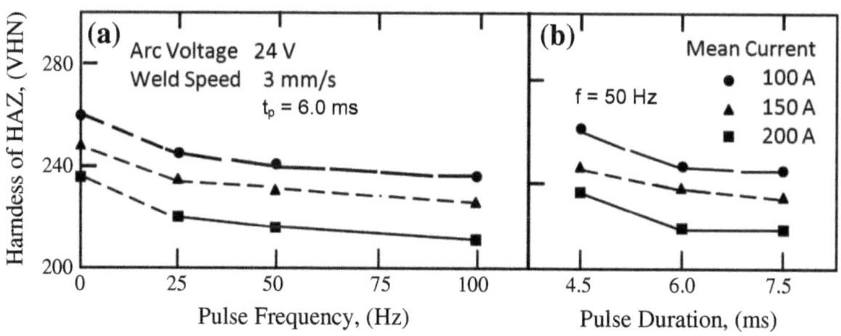

Fig. 7.10 At different mean currents effect of **a** pulse frequency and **b** pulse duration on hardness of HAZ in steel weld

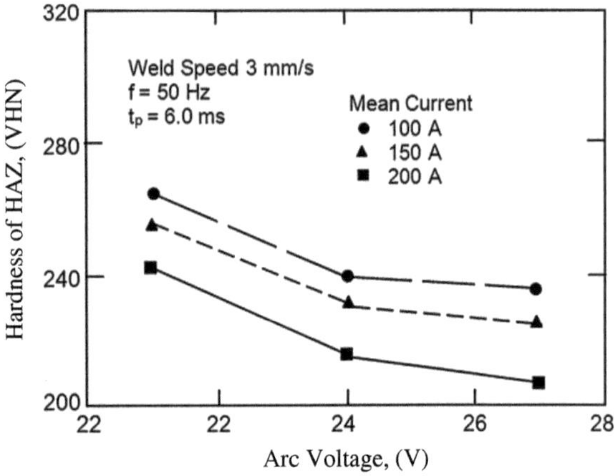

Fig. 7.11 Effect of arc voltage on hardness of HAZ in steel weld

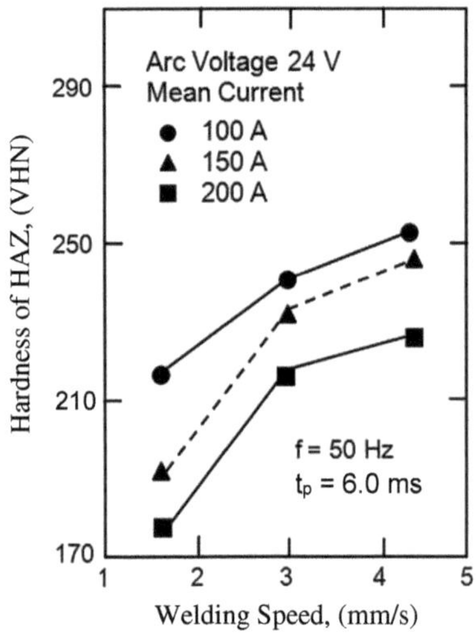

Fig. 7.12 Effect of welding speed on hardness in HAZ in steel weld

generally observed similar influence of reduction of mean current. However, at given mean current pulse current weld beads generally result comparatively softer HAZ than conventional (0 Hz) GMA weld bead (Fig. 7.10a). At given pulse parameter increase of mean current and arc voltage and decrease of welding speed reduces hardness of HAZ primarily due to coarsening of microstructure caused by increase of energy input. Whereas, at given mean current reduction in hardness of HAZ with the increase of pulse frequency or duration possibly attributed to increase in super heating and/or heat build-up in weld poolweld pool. At given mean current increase of ϕ, due to variation of pulse frequency and duration, relatively enhances hardness of HAZ almost linearly (Fig. 7.13). But, at given pulse frequency and duration enhancement of ϕ resulting from appropriate variation in I_p and I_b with a variation in arc voltage within 21–27 V reduces the hardness of HAZ significantly

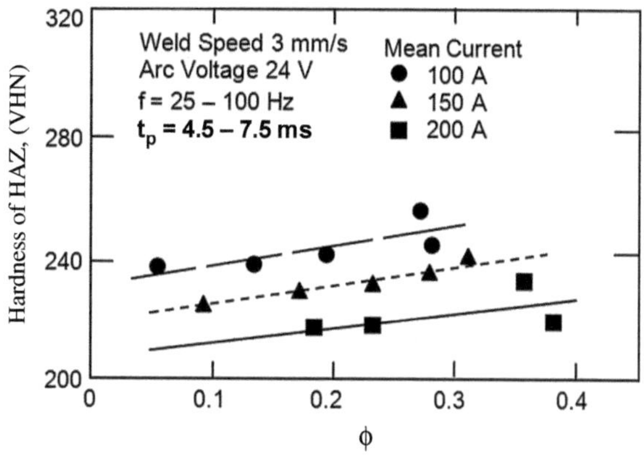

Fig. 7.13 At a given arc voltage influence of ϕ on hardness in HAZ in steel weld

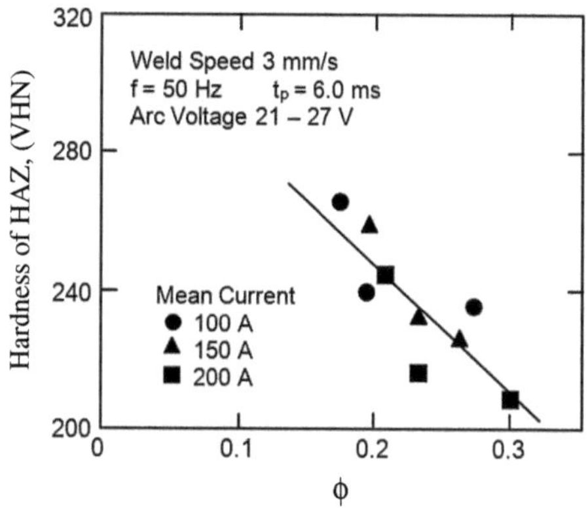

Fig. 7.14 Under varying arc voltage influence of ϕ on hardness in HAZ in steel weld

(Fig. 7.14) irrespective of the variation in I_m in the range of 100–200 A. Increase of arc voltage relatively enhances energy input and also arc length, which consequently enhances area of heating. Both of them may cause reduction of hardness of HAZ due to coarsening of microstructure. In contrast to the result shown in Fig. 7.13 where arc voltage is constant, it may be understood from Fig. 7.14 that heat distribution through arc length may predominate variation of hardness of HAZ over ϕ, causing decrease of hardness of steel in spite of lowering of heat transfer to weld deposit (Chap. 4) with the increase of ϕ.

7.2 Tensile Properties

The tensile properties of weld joint are primarily characterized by the strength and ductility of weld deposit and heat affected zone. The weakness of either of these two regions points out the deficiency of weld joint to be addressed for its improvement. The chemical composition and morphology of both the regions dictated by the weld isotherm and thermal cycle basically govern their properties. Depending upon welding of heat treatable and non heat treatable metals and alloys using similar or dissimilar filler material the dilution of weld deposit and the behaviour of phase transformation affecting the morphology of the weld and HAZ largely dictates their tensile properties.

7.2.1 Weld Deposit

Pulse current GMA welding of aluminium alloys produces radiographic quality weld having improved morphology and mechanical properties in comparison to those observed in case of their conventional GMA welding [5, 9, 10]. Single pass P-GMA weld of 10 mm thick Al–Zn–Mg (7005) alloy generally shows significantly higher tensile strength (Fig. 7.15) and elongation (Fig. 7.16) than those of the conventional GMA weld while fracturing from the Al–Mg (5183) weld deposit [1]. The mechanical properties of precipitation strengthening weld depends primarily upon amount and morphology of precipitates, which are largely governed by its chemical composition with respect to proportionate content of precipitation forming element and morphology of cast microstructure of the matrix respectively. In pulse current GMA welding of Al–Zn–Mg alloy the presence of more uniformly distributed finer precipitate in relatively refined microstructure of zinc picked Al–Mg deposit (Fig. 7.17) improves tensile properties of the weld than those observed in case of conventional GMAW. However, in this context the influence of porosity content of the matrix cannot be ignored. In spite of higher zinc pick-up and lower porosity content, the presence of comparatively coarser microstructure and bigger shrinkage

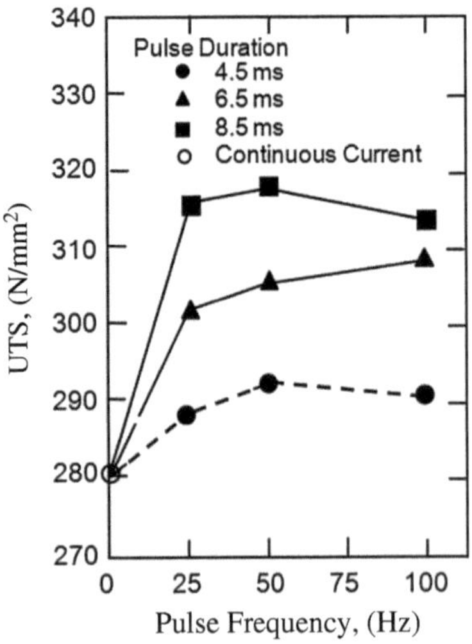

Fig. 7.15 Influence of pulse frequency and pulse duration on tensile strength of aluminium alloy weld joint

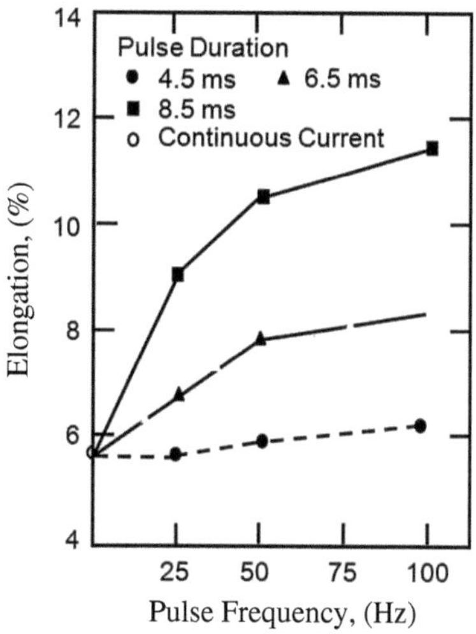

Fig. 7.16 Influence of pulse frequency and pulse duration on elongation of aluminium alloy weld deposit

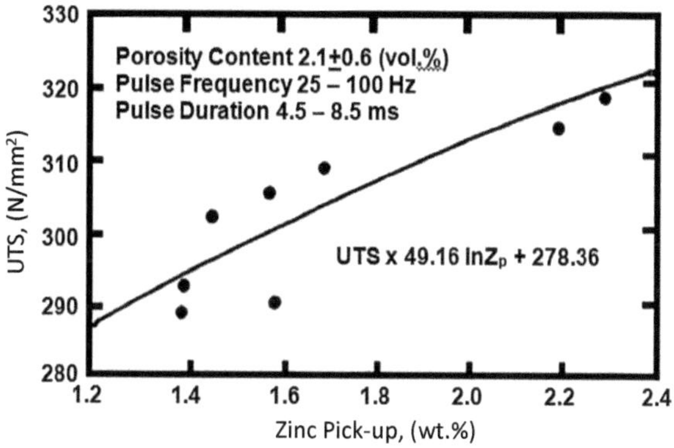

Fig. 7.17 Correlation of zinc pick-up (Zp) with ultimate tensile strength of aluminium alloy weld deposit

porosity in the matrix lowers the strength of conventional GMA weld than that of the P-GMA weld. Due to significant variation of all these phenomena of P-GMA weld with the pulse parameters they appreciably dictate the tensile properties of weld deposit. The increase of porosity lowers tensile strength of weld at a relatively faster rate at higher zinc content (Fig. 7.18) primarily due to its comparatively higher strength which makes it more sensitive to fracture initiation from pores.

In case of multipass weld joint of thick section of similar Al–Zn–Mg (T_6) alloy it is also realized that the pulse parameters significantly affects the ultimate tensile strength (σ_u), yield strength (σ_y) and elongation at fracture (e_f) when it fractures from the Al–Mg weld deposit (Fig. 7.19). But, the influence of pulse parameters on σ_y is comparatively less significant than their influence on σ_u and e_f of the weld. The tensile properties of multipass P-GMA weld also become comparatively superior to those of the conventional GMA weld as it is observed in case of the single pass weld. However, the tensile properties of multipass weld largely depend

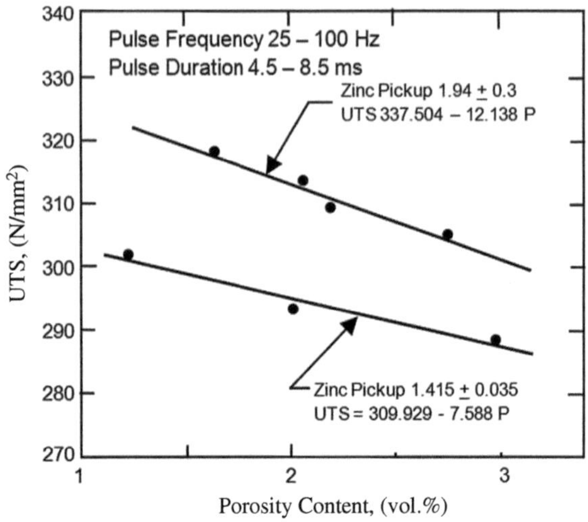

Fig. 7.18 Influence of porosity content on tensile strength of aluminium alloy weld deposit

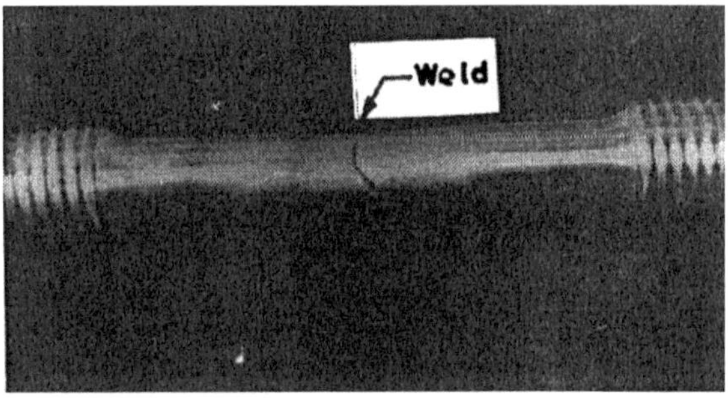

Fig. 7.19 Typical location of fracture in tensile test of aluminium alloy weld joint

also upon its reheat refined finer dendrite fraction besides the usual factors as its chemical composition, morphology of cast microstructure and porosity content. Because of appreciable influence of pulse parameters on these aspects of a multi-pass weld they significantly influence the tensile properties. It may be understood in the light of earlier discussions on single pass weld. An appreciable amount of zinc pick-up by the Al–Mg weld deposit from the Al–Zn–Mg base metal through dilution makes it susceptible to precipitation of $Mg_3Zn_3Al_2$ in the matrix as identified Table (7.1) by the presence of $(Al, Zn)_{49}Mg_{32}$ in X-ray diffraction analysis. The presence of $(Al, Zn)_{49}Mg_{32}$, which is in close approximation by nature [11] to the $Mg_3Zn_3Al_2$ has been confirmed by the inter-planer spacing (d-value), where no other interfering d-value is observed. The amount of precipitate stoichiometrically increases with the enhancement of zinc pick-up as shown in Table 7.2.

Table 7.1 X-ray diffraction phase analysis of Al–Mg weld deposit in Al–Zn–Mg base metal

Sin θ	D (Å)	(h k l)
0.761	1.0124	(400) Al
0.659	1.169	(222) Al
0.631	1.221	(311) Al
0.538	1.431	(220) Al
0.536	1.437	(941, 853) $(Al, Zn)_{49}$ Mg_{32}
0.383	2.012	(710, 550) $(Al, Zn)_{49}$ Mg_{32}
0.380	2.024	(200) Al
0.333	2.310	(611. 532) $(Al, Zn)_{49}$ Mg_{32}
0.330	2.338	(111) A1

Table 7.2 Estimated precipitate content as a function of dilution and zinc pick-up of Al–Mg weld deposit in Al–Zn–Mg base metal

Dilution (%)	Zinc pick-up (wt%)	Amount of $(Al. Zn)_{49}$ Mg_{32} precipitate (wt%)
17.5	0.64	1.0539
15.5	0.60	0.980
13	0.53	0.8563
14.5	0.56	0.9180
17.5	0.57	1.1033
15.5	0.58	0.9551
18	0.69	1.1362
19.5	0.72	1.1856
17.5	0.68	1.1198
21	0.78	1.2845

7.2.2 *Heat Affected Zone*

At a high mean current of 210 A using 1.6 mm diameter Al–Mg filler wire, increase of pulse frequency up to 100 Hz in P-GMA welding of 6 mm thick Al–Zn–Mg alloy enhances notch tensile strength and notch ductility of HAZ (Fig. 7.20) in comparison to those of the conventional GMA weld [5, 6]. The improvement in notch tensile properties of HAZ of precipitation strengthening alloy may be primarily attributed to comparatively longer retention of this zone in the range of ageing temperature (Fig. 4.16). However, prior to final fracture the crack path shows a tendency to migrate from HAZ to the weld. A similar behaviour is also observed in case of the vertical-up P-GMA welding of mild steel [7]. At given mean current increase of ϕ relatively reduces notch tensile strength of HAZ (Fig. 7.21) whereas at given ϕ increase of mean current insignificantly influences the same. Notch tensile property is a complex phenomenon of notch sensitivity, which is a function of grain size as well as type and morphology of phases present in matrix.

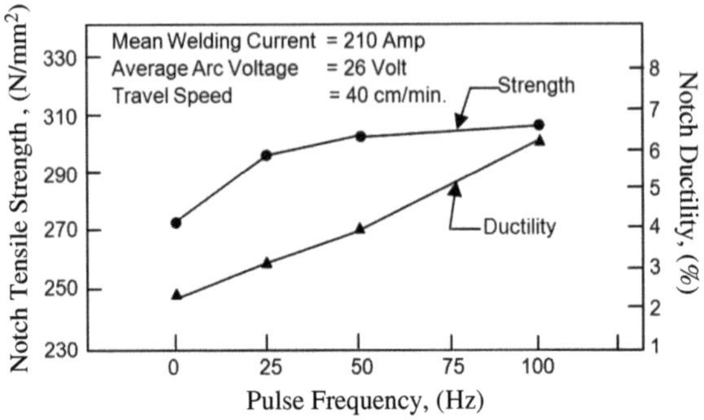

Fig. 7.20 At a given mean current influence of pulse frequency on notch tensile strength and notch ductility of HAZ 1 mm from the fusion line

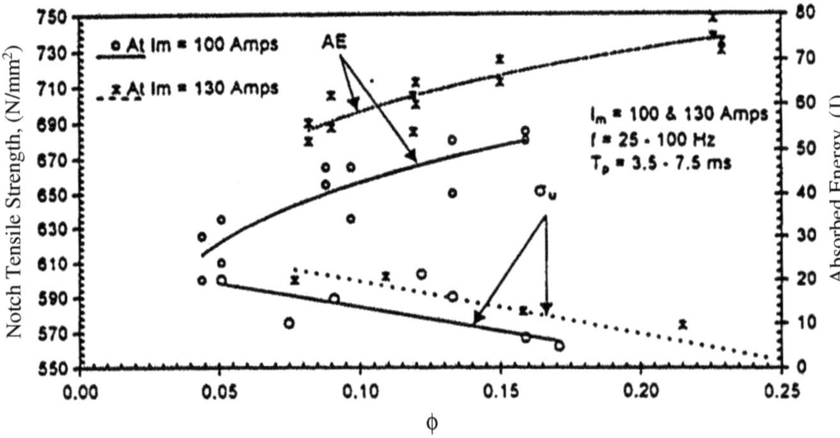

Fig. 7.21 At different mean currents influence of ϕ on notch tensile strength and Cv-toughness (at −20 °C) of HAZ

At given mean current, refinement of morphology of HAZ (Fig. 6.47) with increase of ϕ comparatively enhances hardness of HAZ (Fig. 7.6) and, thus, increases its notch sensitivity. This behavior consequently reduces the notch tensile strength of any location of matrix.

7.3 Impact Toughness

Use of P-GMAW instead of conventional GMAW in vertical-up welding of mild steel significantly improves the toughness of weld deposit. By virtue of its significant influence on weld thermal cycle to favourably influence the refinement of

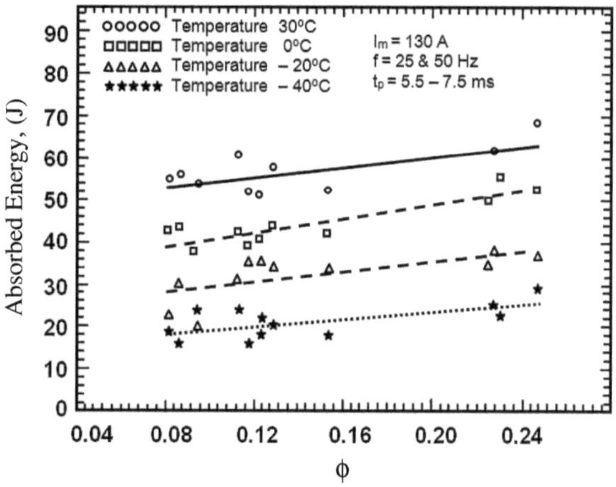

Fig. 7.22 At a given mean current of 130 A influence of ϕ on C_v-toughness of weld metal at different temperatures

microstructure and enhancement of proeutectoid ferrite of matrix, the increase of ϕ at a given I_m enhances toughness of mild steel weld deposit (Fig. 7.22). However, the toughness of weld metal reduces significantly with a decrease in test temperature from ambient to subzero level. The effective role of ϕ and I_m in governing the weld thermal cycle also dictates the toughness of HAZ. An increase of ϕ at a given I_m as well as at a given ϕ the increase of I_m significantly enhances the toughness of HAZ [7] of the vertical-up P-GMA weld of mild steel (Fig. 7.21). Increase of toughness of HAZ with increase of ϕ or I_m is predominantly attributed to reduction in coarsening of microstructure of HAZ as shown in the micrographs presented in Fig. 6.46. Reduction of grain coarsening also enhances grain boundary area and consequently increases amount of grain boundary ferrite in matrix. Thus, it also supports further the observed enhancement of toughness of HAZ with increase of ϕ or I_m. Use of high I_m as 130 A, especially at higher ϕ, with 1.2 mm diameter mild steel filler wire in P-GMA welding gives relatively higher toughness of HAZ than that occurs in case of conventional GMA welding due to more favourable state of weld thermal cycle.

7.4 Fatigue Properties

Fatigue life of weld joint fracturing from weld is primarily governed by the loading condition as well as the chemical composition, microstructure and porosity content of weld deposit affecting its tensile properties. The variation of pulse parameters significantly influences the fatigue life of weld joint fracturing from weld centre as observed [1, 12, 5.21] in both the cases of single pass or multi pass welding of Al–Zn–Mg alloy using Al–Mg filler wire (Fig. 7.23). In this regard an increase of tensile strength (σ_u), which simultaneously enhances also the yield strength, plays significant role in increasing fatigue life (Fig. 7.24) primarily due to increase of the

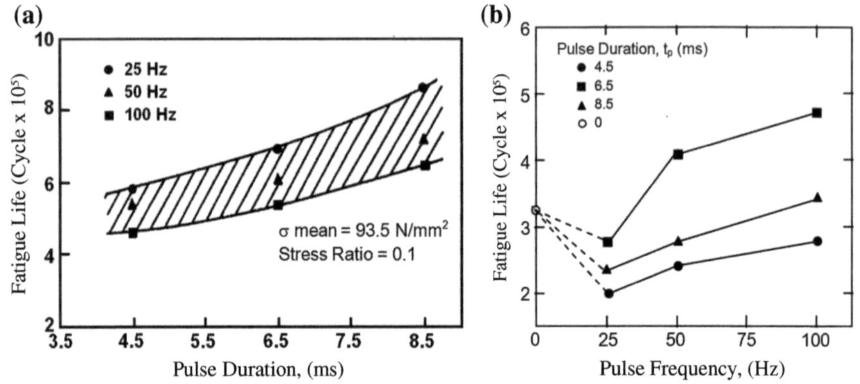

Fig. 7.23 Effect of pulse parameters on fatigue life of **a** single pass and **b** multi pass welds of Al–Zn–Mg alloy

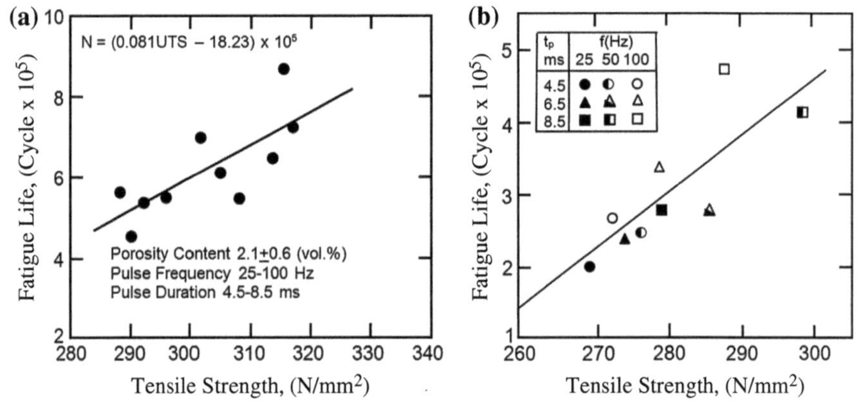

Fig. 7.24 Influence of tensile strength on fatigue life of **a** single pass and **b** multi pass welds of Al–Zn–Mg alloy

ratio of yield stress to mean stress (σ_m). An increase of porosity content (P) of weld up to a certain extent adversely influences its fatigue life (Fig. 7.25) by enhancing the possibility of appearing the pore at the surface or subsurface of the specimen inducing crack initiation (Fig. 7.26). The correlation of life N with P presented in the figure closely approximates a theoretical line reported earlier [13]. At a given porosity content a refinement of dendritic microstructure of weld also improves its fatigue life (Fig. 7.27). However, in spite of comparatively low porosity content, the fatigue life of conventional GMA weld becomes appreciably lower than that of the pulsed-current weld [13] possibly due to lower strength and coarser microstructure. Besides the microstructure and porosity content the geometry of weld also largely dictates its susceptibility to fatigue failure [13]. With a change in pulse parameters a variation in weld geometry estimated by a ratio (R_G) of the

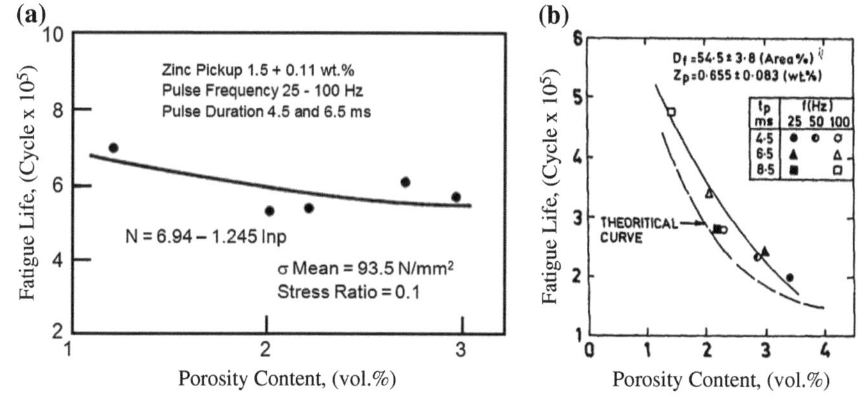

Fig. 7.25 At a given zinc pick-up and weld morphology influence of porosity content on fatigue life of **a** single pass and **b** multi pass welds of Al–Zn–Mg alloy

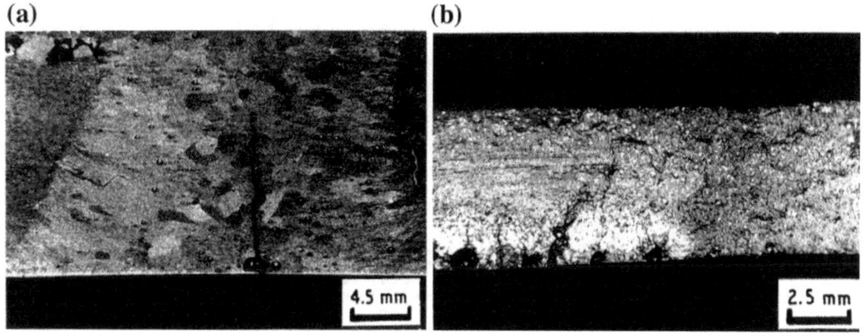

Fig. 7.26 Fatigue crack initiation from pores in aluminium alloy weld **a** micrograph and **b** fractograph of test specimen

width of the bottom (B) and the top (A) of the weld affects fatigue life of Al–Zn–Mg weld prepared by using Al–Mg filler wire. The decrease of R_G below about 0.5 significantly enhances fatigue life of the weld (Fig. 7.28) possibly by reduction of residual stresses [14] resulting from a decrease of cooling rate caused by weld having relatively low contact area with the base material that acts as a effective heat sink [15]. However, at a given R_G of the order of 0.5 the fatigue life of weld decreases significantly with increase of porosity content up to about 3.0 vol.% followed by a moderate decrease with further increase of porosity level (Fig. 7.29). Typical fractographs of the multi pass weld deposit of Al–Mg in weld joint of Al–Zn–Mg alloy having comparatively low and high fatigue life depict that the weld with a comparatively low fatigue life has a considerable amount of porosity (Fig. 7.30a) in comparison to that observed in the weld with high fatigue life (Fig. 7.30b). The fracture morphology is found comparatively finer in the weld of relatively high fatigue life in contrast to that of the weld showing comparatively low

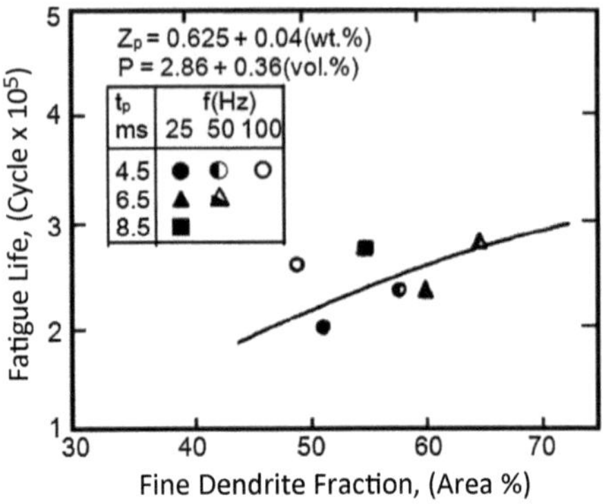

Fig. 7.27 Influence of fine dendrite fraction on fatigue life of multi pass weld of Al–Zn–Mg alloy

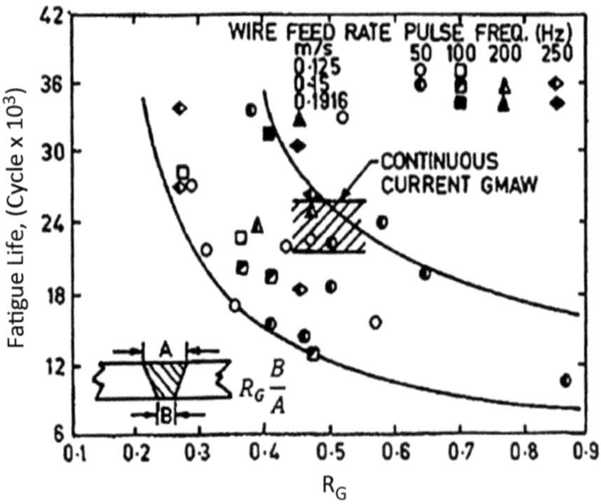

Fig. 7.28 Appropriate weld geometry improves fatigue life of pulsed current weld over the continuous current weld of aluminium alloy

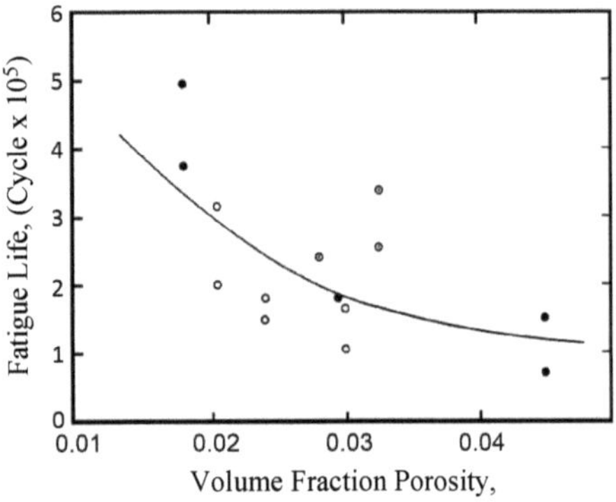

Fig. 7.29 At a given weld geometry of $R_G \approx 0.5$ influence of porosity on fatigue life of joint fractured from aluminium alloy deposit prepared at varied I_m = 150–210 A and welding speed = 27–40 cm min^{-1}

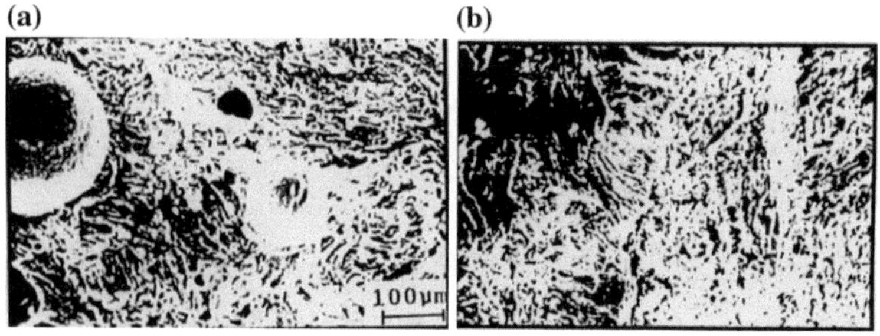

Fig. 7.30 Typical fractograph of pulsed current weld of comparatively **a** low and **b** high fatigue life

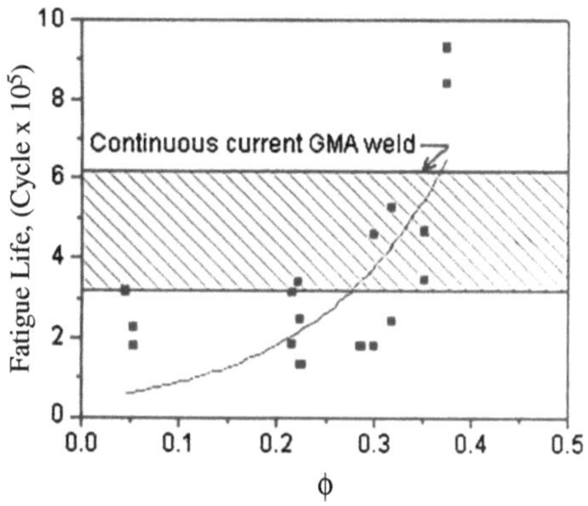

Fig. 7.31 Influence of ϕ on fatigue life of weld joint of aluminium alloy

fatigue life. This is in agreement to the occurrence of a homogeneous slip in case of finer microstructure of weld deposit resulting in enhancement of fatigue life.

Significant influence of ϕ on tensile properties and residual stresses of weld joint may significantly dictate its fatigue and stress corrosion cracking properties by influencing crack initiation and propagation. Positive influence of pulse current single pass GMAW on fatigue life of weld joint over that observed in case of conventional GMAW has been observed [13] during welding of Al–Zn–Mg alloy using Al–Mg filler wire (Fig. 7.31) where the weld joints mostly fracture from weld. A favorable change in residual stresses along with some constructive influence of strength, microstructure and porosity content/distribution improves fatigue life of weld. Variation in distribution of porosity of about 1.8 $\pm$ 0.5 vol.% in matrix causes appreciable scattering in fatigue life of weld. It happens because their presence at surface or subsurface of weld acts as a potential site for crack initiation as discussed earlier. Use of pulse current GMAW improves fatigue life of both the single and multi pass butt welds of Al–Zn–Mg alloy employing Al–Mg filler by about 30–50% over that of conventional (0 Hz) GMA weld [1, 13, 5.21] as shown in Table 7.3.

Table 7.3 Fatigue properties of conventional and pulse current GMA welds of a high strength Al–Zn–Mg alloy produced by single and multipass deposition

Type of weld joint	GMAW	Applied stresses (N/mm^2)	Stress ratio	Fatigue life (cycles)
Single pass flat butt weld	Conventional	$\sigma_{max} = 127.5$	0.5	3.5×10^5
		$\sigma_m = 93.5$	0.1	4.5×10^5
	Pulse current	$\sigma_{max} = 127.5$	0.5	5×10^5
		$\sigma_m = 93.5$	0.1	7×10^5
Multipass flat butt weld	Conventional	$\sigma_m = 80$	0.5	3.25×10^5
	Pulse current	$\sigma_m = 80$	0.5	4.75×10^5

In the context of weld geometry, it should also be noted that decrease in outer weld toe angle enhances notch formation (Fig. 5.48a–c) and favours stress concentration adversely affecting fatigue life of weld joint [7]. Due to difficulties in control of weld pool weld poolsuch an issue becomes a matter of concern especially in case of positional welding where, the use of pulse current GMAW greatly improves the situation as a function of ϕ. However, even a moderate variation of the weld toe angle of vertical-up weld of mild steel appreciably affects fatigue life of weld joint (Fig. 7.32a). The weld geometry being a function of pulse parameters, the increase of ϕ significantly enhances the fatigue life (Fig. 7.32b) due to flattening of weld reinforcement defined by increase of weld form factor (Fig. 7.32c). The increase of outer weld toe angle with the increase of ϕ is clearly revealed in Fig. 5.47. In case of conventional GMA welding at low welding current (100 A) and low-energy input the lack of fusion at weld toe (Fig. 5.49) adversely affects fatigue life of vertical-up weld. Whereas at similar level of current (I_m) the P-GMAW improves base metal fusion but, results a low outer weld toe angle (Fig. 7.33), which adversely affects fatigue life of weld joint by introducing notch effect. Fatigue life of weld joint also depends upon microstructure of HAZ adjacent to weld toe primarily dictated by thermal behavior of weld deposit at different pulse parameters. The decrease of ϕ coarsens grain size of HAZ (Fig. 5.48a) in agreement to the discussions given earlier confirming the increase of droplet temperature

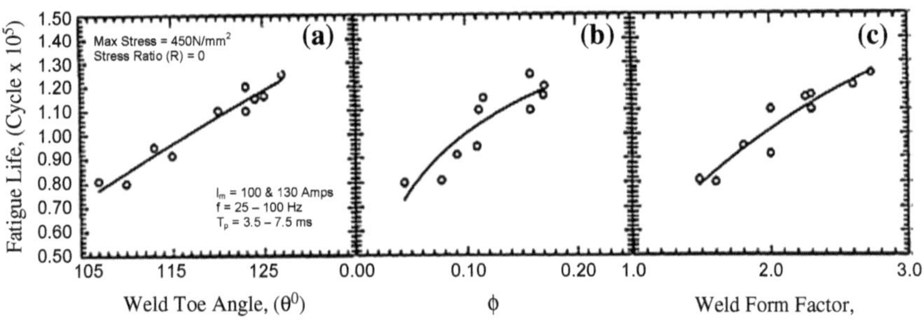

Fig. 7.32 Fatigue life of pulse current vertical-up mild steel weld joint as a function of **a** weld toe angle, **b** ϕ and **c** weld form factor

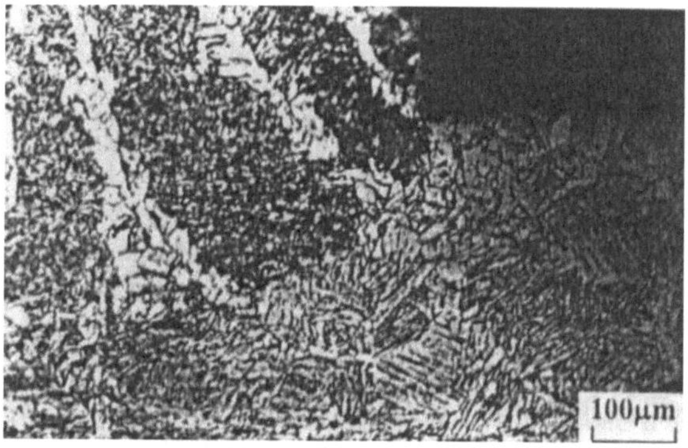

Fig. 7.33 Typical weld toe and morphology of HAZ in vertical-up pulsed-current (I_m = 100 A, f = 100 Hz, t_p = 3.5 ms, and ϕ = 0.053) GMAW of mild steel

with the decrease of ϕ. However, the morphology of HAZ at weld toe of mild steel is often found as Widmanstaten structure (Fig. 7.33) that favours improvement of fatigue life if stress concentration at weld toe is not enhanced by notch effect.

7.5 Fracture Mechanics

The strength and fracture characteristics of metal are largely governed by the presence of cracks (micro or macro) and mechanical notches in it. A true knowledge about the strength of a structural body, corroborating its safety and reliability, cannot be sufficiently worked out by using the classical theories of elasticity or plasticity alone for the stress concentration at cracks and notches. In this regard it is necessary to know the limiting strength criteria for ultimate load carrying capacity of a body leading to a catastrophic failure. The process of fracture is made up of two stages as crack initiation and crack propagation, where the crack growth may be stable or unstable depending upon limiting equilibrium state of a body. At a stable equilibrium state of crack there is no danger of immediate complete fracture, but an unstable crack growth leads to fast or catastrophic total fracture when the continuous portion of the body is no longer able to carry the applied load on it. The allowable size of an initial crack largely depends upon limiting equilibrium state evaluated by fracture mechanics. The fracture mechanics concepts provide quantitative knowledge for evaluating structural safety and reliability with respect to applied stress on a crack length developing certain stress intensity at its tip. Fracture mechanics properties of metal are generally defined by its fracture toughness under static loading. It is different from the toughness of a material, which is defined by its capacity to absorb energy during fracture. The fracture toughness essentially defines the criticality of a flaw to grow in a matrix and thus, it is primarily concerned with the geometry of a flaw and the matrix characteristics at its tip in the direction of growth.

The fracture toughness of a material is generally estimated by the plane-strain critical stress intensity factor (K_{IC}) during stable growth of a flaw, the J-integral parameters for growth initiation fracture toughness (J_{IC}), fracture resistance (J-R) curve and the crack tip opening displacement (CTOD) parameter (δ_m). Fracture toughness is commonly evaluated by standard elastic-plastic fracture toughness test method in accordance with ASTM E813 by placing a notch at desired location in standard compact tension (C-T) or three point bend (TPB) specimens. The greater the values of K_{IC}, J_{IC} and δ_m means higher the stress required for a rapid crack propagation, indicating more resistance of a body to brittle fracture and thus it ensures safety. In modern engineering and technology, the designing of structural component with improved safety and reliability requires estimation of life of a body containing detectable flaw to avoid unstable crack growth leading to its catastrophic failure in service. These parameters of fracture toughness are considered as design criterion for proper selection of material as well as soundness of its weld joint by evaluating potential danger of a flaw present in them leading to a catastrophic fracture. In designing of a component with an expected occurrence of crack it is necessary to consider certain factor of safety. Use of new material or improvement in characteristics of existing material for higher critical stress intensity factor or plane-stress fracture toughness, K_C, reduces factor of safety ($h = \sigma_u/\sigma_{max}$) on reliable operation. The value to which it can be reduced depends on the operating conditions such as the loading conditions, environmental conditions etc.

In a tough material, the J_{IC} is recognized as a conservative measurement of fracture toughness, since the material is expected to exhibit a finite amount of crack extension at higher loads [16]. At this stage the tearing modulus, allowing a stable tearing in addition to crack blunting, is proposed [17–20] to describe more correctly the critical set of conditions for fracture, following an instability criteria of $T_{applied} > T_{material}$. Thus, the tearing modulus which is describing the resistance of a material to crack expansion, instead of resistance of material to crack initiation as it is found in case of J_{IC}, may allow one to understand more about the fracture toughness of a tough material. The J-R curve data also provides an estimation of resistance to stable crack extension of a material defined by tearing modulus (T_{mod}) introduced by Paris [17] evaluated as follows.

$$T_{mod} = \frac{dJ}{da}\frac{E}{\sigma_{y\,eff}} \tag{7.1}$$

where, (dJ/da) is slope of linear mode of R-curve and E is modulus of elasticity (N/mm^2).

Under dynamic loading, the nature of crack propagation, in stable crack growth region, determining the life of a component, is defined by a correlation of crack growth rate (da/dN) with stress intensity factor range ΔK, known as Paris Law expressed [21] as

$$da/dN = C(\Delta K)^n \tag{7.2}$$

where, the n and C are the material constants, 2a is length of a flaw, N is cycle of loading and K (MPa $\sqrt{m}$) is the stress intensity factor estimated as follows.

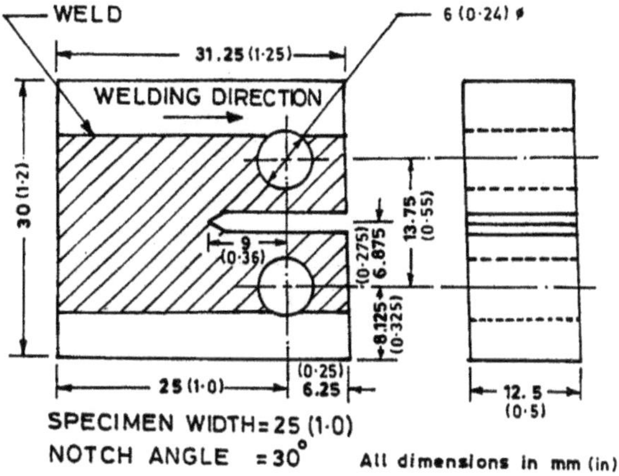

Fig. 7.34 Schematic diagram of [C-T] fracture toughness test specimen

$$K = \sigma\sqrt{\pi a} \tag{7.3}$$

As per Paris law the fatigue crack growth rate (FCGR) in any material is measured in accordance with the ASTM E 647-88 standard using a pre fatigued compact-tension (C-T) specimen (Fig. 7.34) or centre cracked tension M(T) specimen (Fig. 7.35). The FCGR is governed by both of its material dependent constants. Fatigue crack growth rate (da/dN) is general behaviour of growing cracks as a function of stress intensity factor range (ΔK) of dynamic loading as illustrated in Fig. 7.36. However, the application of Paris law has been further improved for various practices [22]. By modification of Paris law for rapid crack growth near the fracture toughness K_{IC}, it is expressed [23] as follows.

$$da/dN = [C(\Delta K)^n]/[(1 - R)\, K_{IC} - \Delta K] \tag{7.4}$$

For a better prediction of da/dN near the threshold [24] of stress intensity factor range (ΔK_{th}), the ΔK of Paris law is replaced with ($\Delta K - \Delta K_{th}$) as

$$da/dN = C(\Delta K - \Delta K_{th})^n \tag{7.5}$$

Here, ΔK_{th} is threshold stress intensity factor range below which no crack growth occurs and the crack becomes unstable at K_{IC} and R is the stress ratio. Thus, a higher value of ΔK_{th} of any material ensures its safety to fracture. The ΔK_{th} may be measured as per ASTM practices [25, 26] in case of high strength aluminium alloy it may also be broadly considered at (da/dN) of 10^{-10} m/cycle [27]. The plain strain fracture toughness, K_{IC}, as a material property, represents maximum possible rise in the value of stress intensity prior to failure. Thus, the average residual strength in a cracked element is directly related to the fracture toughness and inversely related to the square root of crack length.

The significant variation in mechanical properties of weld deposit and HAZ due to change in weld thermal cycle of different welding processes, procedures and

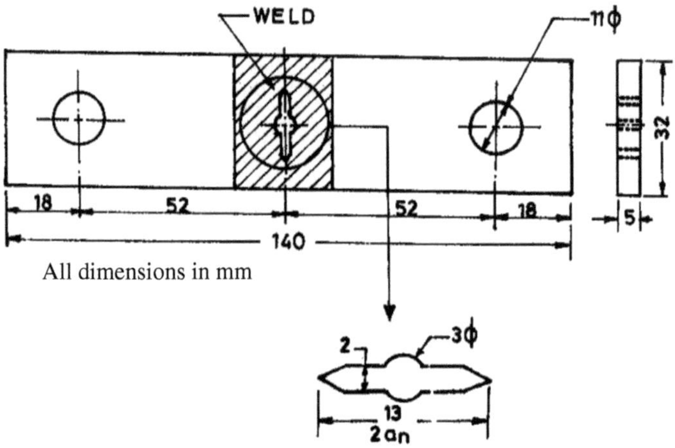

Fig. 7.35 Schematic diagram of the M(T) fatigue crack growth rate test specimen

Fig. 7.36 Typical behaviour of fatigue crack growth rate curve

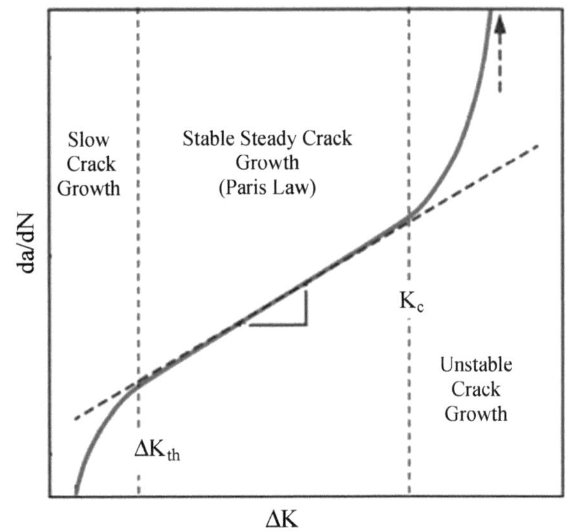

parameters appreciably dictates the initiation and growth of flaws leading to fracture of a matrix under static and dynamic loading. Depending upon their chemistry, microstructure and stresses this behaviour becomes more critical in case of high strength material. Thus, in order to understand the safety of a weld joint it is imperative to know the fracture mechanics properties of weld and HAZ, in the light of its different aspects as stated above, with respect to various conditions of weld fabrication.

Fatigue crack growth rate behavior of single pass Al–Mg–Mn weld deposit in 10 mm thick rolled plate of high strength Al-Zn-Mg alloy studied [28] under

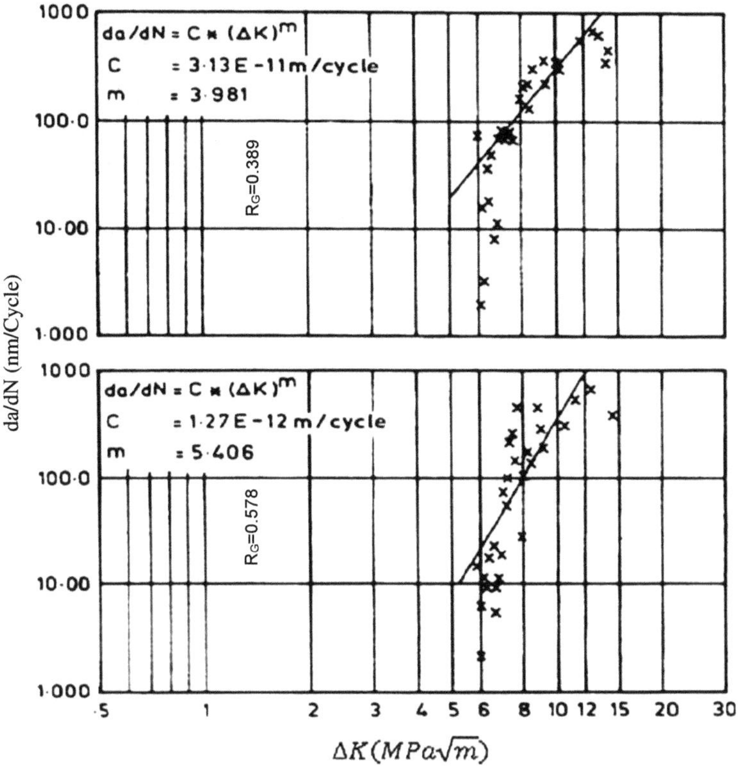

Fig. 7.37 Fatigue crack growth rate (da/dN) as a function of stress intensity factor range (ΔK) typically observed in dynamic test of weld deposit with different R_G

dynamic bending loading (Fig. 7.37) significantly depends upon weld geometry defined by the ratio (R_G) of the width of weld root to weld-top as shown in Fig. 7.38. It happens due to clear influence of R_G on the components n and C of the Eq. 7.2 as shown in Figs. 7.38 and 7.39 respectively. An appropriate control of pulse parameters keeping the R_G below 0.45 can maintain a low n than that of the conventional GMA weld that gives a slow crack growth rate (da/dN). However, an increase of R_G beyond about 0.45 considerably reduces the C. Finally the fatigue life of weld is governed by both values of n and C and in this context a proper use of P-GMAW resulting a relatively low R_G below about 0.45 improves the fatigue life of weld joint beyond that of the conventional GMA weld (Fig. 7.28). This has primarily happened due to variation in porosity content along with the size and distribution of porosity at different pulse parameters affecting the R_G. However, in this regard the influence of R_G on residual stresses in weld may not be ignored that can also affect its fatigue life.

Application of comparatively thicker sections of Al-alloys in several fields, such as pressure vessel, structural column and transport system, necessitates the use of

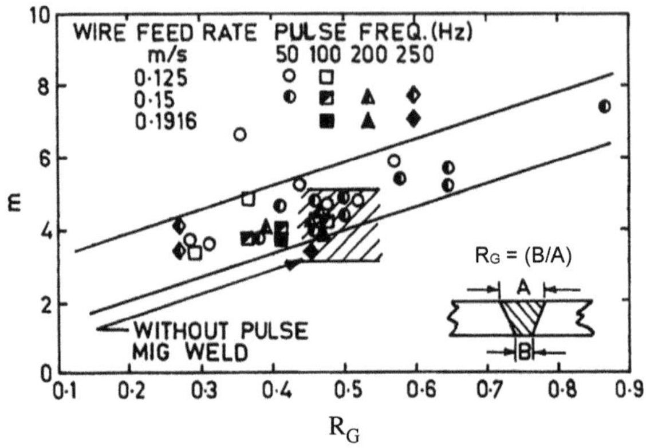

Fig. 7.38 Change in crack growth exponent m of Paris Law as a function of weld geometry R_G prepared at different wire feed rate and pulse frequency

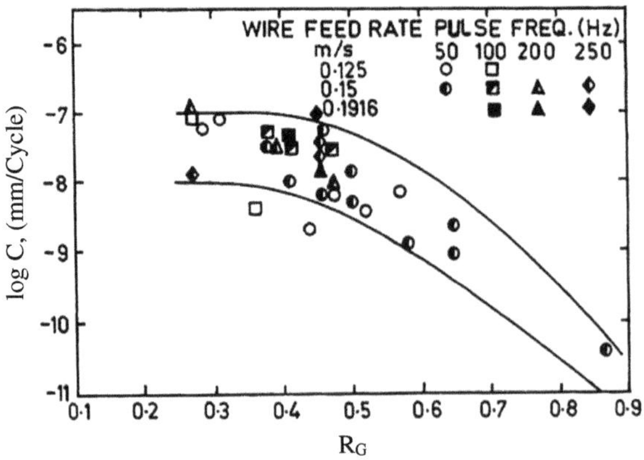

Fig. 7.39 Change in constant C of Paris Law as a function of weld geometry R_G prepared at different wire feed rate and pulse frequency

multi-pass procedure of weld fabrication. In multi-pass welding a comparatively complex nature of weld thermal cycle during filling of weld groove differently affects the chemical composition, morphology, porosity content and mechanical properties of weld from those of the single pass weld. During using the P-GMAW they affect the weld properties differently with respect to their correlations with the pulse parameters. The fracture mechanics of weld deposit as a function of the above mentioned features of the matrix and pulse parameters is duly studied in case of multipass weld (Fig. 5.36b) of thick section of high strength aluminium alloy. The correlations of fracture mechanics of weld deposit of Al–Mg alloy (5083) with various features of matrix in weld joint of Al–Zn–Mg (7005) alloy critically highlights the attractive role of P-GMAW in enhancement of weld properties. Typical macrograph (Fig. 7.40) of fractured surface of a (C-T) specimen is showing three distinctly different fracture regions as the pre-cracked (pre-fatigueing) region, stable crack growth region under static loading and forced fractured region by fatigue loading. The use of P-GMAW significantly changes the nature of J-R curve (Fig. 7.41a–c) with respect to that observed (Fig. 7.42) in case of using the conventional GMAW. During change in

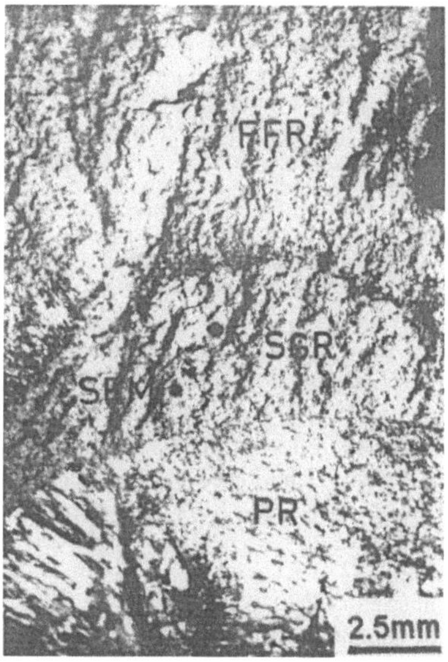

Fig. 7.40 Typical SEM fractograph of J-Integral test showing three distinctly different regions of fracture in of [C-T] specimen; Pre-cracking region (PR); Stable crack growth region (SGR); Forced fractured region (FFR)

pulse parameters of P-GMAW process character of the curve also becomes relatively different. Accordingly J_{IC} fracture toughness of the weld also significantly varies with the change in welding process and parameters, where the use of P-GMAW relatively enhances it with respect to that of the conventional (0 Hz) GMAW process (Fig. 7.43). However, the encouraging influence of P-GMAW to improve the fracture toughness of weld deposit over that of conventional one becomes more significant with a decrease of pulse duration but, the extent of its influence varies with the pulse frequency. It happens because during weld deposition from 1.6 mm diameter filler wire under argon shielding at given I_m and arc voltage of 220 A and 23 V respectively, a simultaneous variation in I_p and I_b with change in pulse frequency and duration significantly affects the matrix characteristics. As primary characteristics of multi-pass Al–Mg weld deposit in Al–Zn–Mg weld joint, an increase of zinc pick-up (Z_p), finer dendrite fraction (D_f) and tensile strength (σ_u) of weld, appreciably reduces its J_{IC} fracture toughness as shown in Figs. 7.44, 7.45, 7.46 respectively.

The change in matrix characteristics primarily as an increase of D_f and σ_u also considerably reduces the tearing modulus (T_{mod}) of multipass P-GMA weld (Figs. 7.47 and 7.48 respectively), but it always remains relatively higher than that of the conventional GMA weld [29] of the Al–Mg weld deposit in Al–Zn–Mg weld joint. The matrix of multipass deposit shows certain degree of heterogeneity with respect to its microchemistry, morphology of precipitates and amount, size and distribution of porosity. In spite of the presence of such scattering in matrix characteristics the correlation of enhancement of T_{mod} of weld deposit with the increase of porosity is largely conclusive. This is because the micro mechanism of crack propagation, in a non-homogeneous cast metal matrix like weld metal is

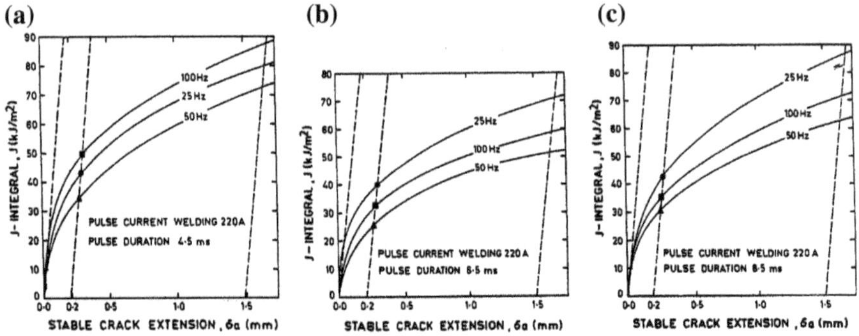

Fig. 7.41 Effect of pulse frequency at different pulse durations of; **a** 4.5, **b** 6.5 and **c** 8.5 ms on J-R curve of multipass pulse current GMA weld deposit of aluminium alloy

Fig. 7.42 J-R curve of multipass GMA weld deposit of aluminium alloy

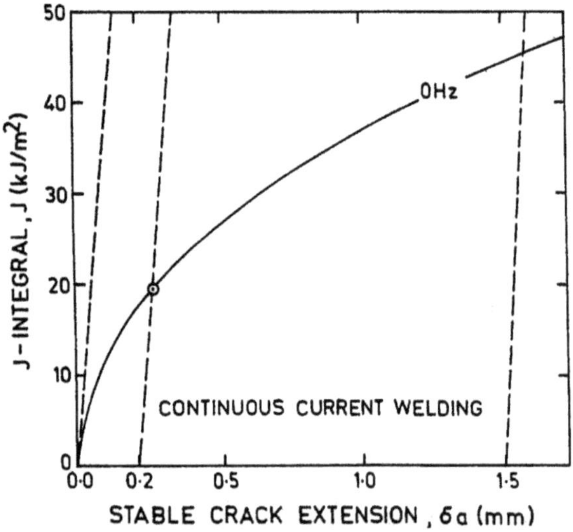

largely dictated by a competitive process of crack growth and crack blunting/arrest. The micro void nucleation, growth and coalescence favours the crack propagation and the presence of spherical pores ahead of crack tip may induce crack arrest [27], causing a delay in crack propagation depending upon their amount size and distribution in the matrix. A lower tearing modulus indicates a lower resistance of weld to stable crack extension. Typical SEM fractographs of the conventional GMA weld and the pulse current GMA welds having comparatively low and high J_{IC} are presented in Figs. 7.49 and 7.50a, b respectively. The fractographs of stable crack growth region (as typically marked in Fig. 7.40) in general show a ductile mode of fracture. However, they depict the presence of comparatively higher amount of plastic flow in the matrix when the weld is having comparatively higher J_{IC}

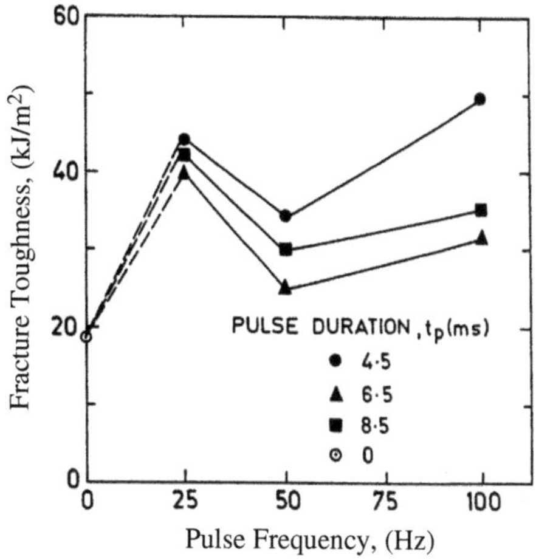

Fig. 7.43 At a given pulse duration effect of pulse frequency on J_{IC} fracture toughness of multipass pulse current GMA weld deposit of aluminium alloy

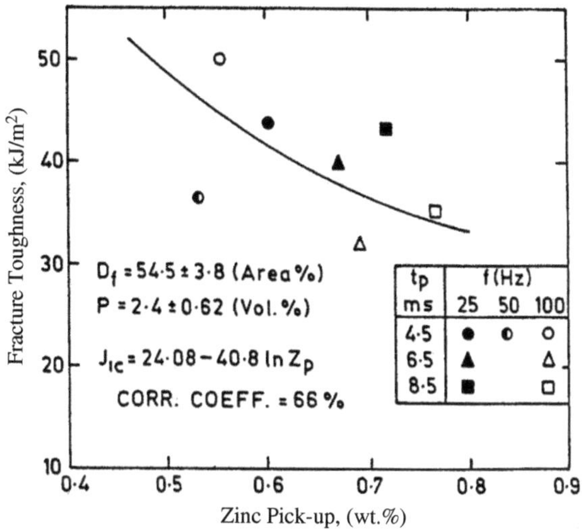

Fig. 7.44 At a given finer dendrite fraction and porosity content influence of zinc pick-up on J_{IC} fracture toughness of multipass pulse current GMA weld deposit of aluminium alloy

(Fig. 7.50b) than that observed in the welds having lower J_{IC} as revealed in Figs. 7.50a and 7.49 respectively.

A variation in pulse duration and frequency significantly influences the growth exponent m and constant c of Paris law (Fig. 7.51) of fatigue crack growth rate (da/dN) as function of ΔK in multipass P-GMA weld deposit of Al–Mg alloy (5083) in Al–Zn–Mg (7005) alloy weldment [30, 31]. At a given I_m beyond transition current of weld deposition the P-GMA weld shows (Fig. 7.52) comparatively lower (da/dN) than that of the GMA weld of same welding current. It primarily happens due to change in microstructure, zinc pick-up and porosity content of weld governed by the pulse parameters. At a given condition of the other two aspects of weld deposit, the effect of its zinc pick-up (Z_P), fine dendrite fraction (D_f) and

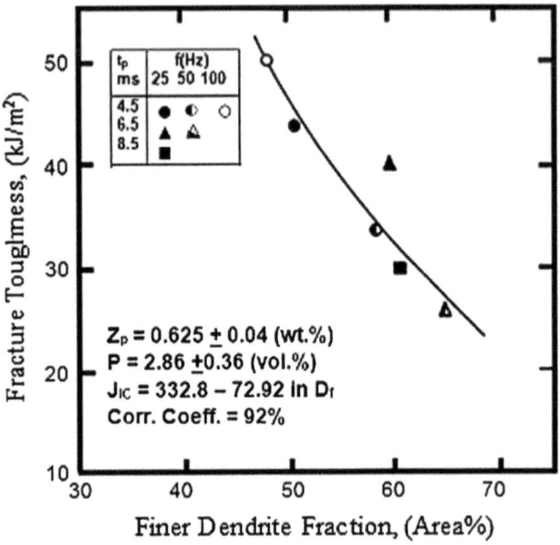

Fig. 7.45 At a given zinc pick-up and porosity content influence of finer dendrite fraction on J_{IC} fracture toughness of multipass pulse current GMA weld deposit of aluminium alloy

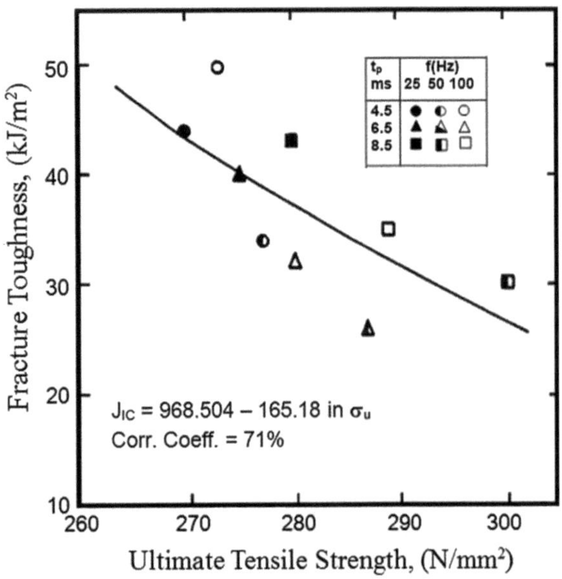

Fig. 7.46 Correlation of tensile strength and J_{IC} fracture toughness of multipass pulse current GMA weld deposit of aluminium alloy

porosity content (P) on fatigue crack growth rate (FCGR) at different ΔK depict (Figs. 7.53, 7.54, 7.55) that the increase of Z_P and P relatively reduces but, the increase of D_f significantly enhances the (da/dN). The increase of da/dN with the increase of D_f possibly happens primarily due to enhancement in number of comparatively finer precipitates in the matrix, especially at the inter dendritic region, which weakens it against crack propagation. The increase in P reduces the

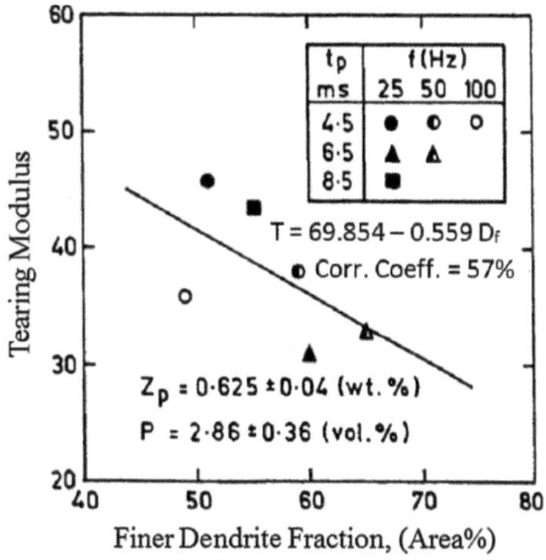

Fig. 7.47 At a given zinc pick-up and porosity content influence of finer dendrite fraction on tearing modulus of multipass pulse current GMA weld deposit of aluminium alloy

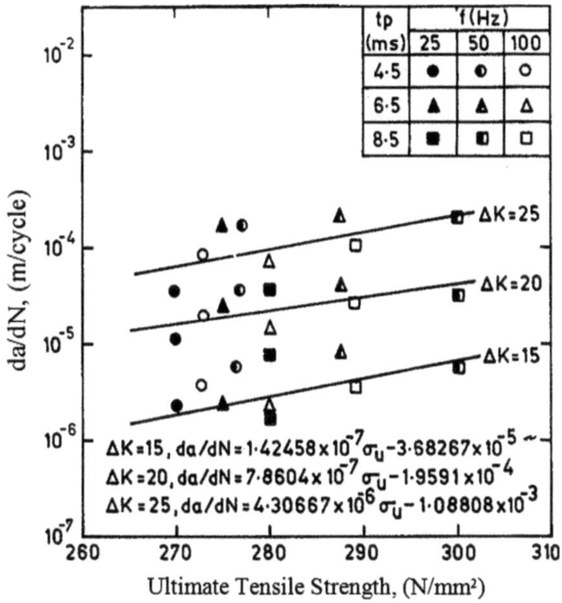

Fig. 7.48 Correlation of tensile strength and tearing modulus of multipass pulse current GMA weld deposit of aluminium alloy

da/dN largely due to interaction of the crack front with the pore acting as a potential location of crack arrest. However, at a given condition of loading the decrease in da/dN with the increase of Z_P (Fig. 7.53) may recur primarily due to increase in tensile strength of the matrix. The FCGR shows a tendency to increase with the increase of UTS of Al–Mg weld deposit (Fig. 7.56) and the increase of ΔK.

The fracture toughness J_{IC} and the δ_m at the first attainment of maximum load of single and multi-pass conventional GMA and P-GMA weld deposits of Al–Mg alloy (5083) in Al-Zn-Mg (7005) alloy weldment are compared in

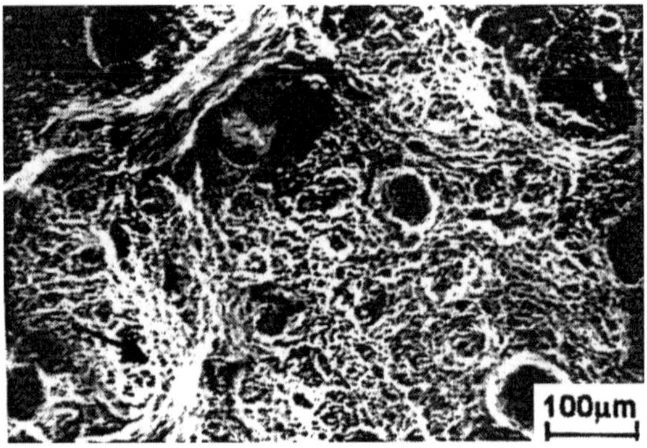

Fig. 7.49 Typical fractograph of GMA weld deposit of Al–Mg alloy showing comparatively low J_{IC} fracture toughness

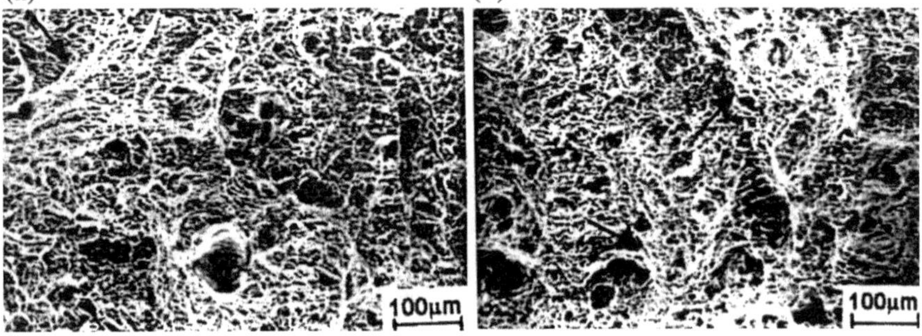

Fig. 7.50 Typical fractograph of pulse current GMA weld deposit of Al–Mg alloy showing comparatively **a** low and **b** high J_{IC} fracture toughness

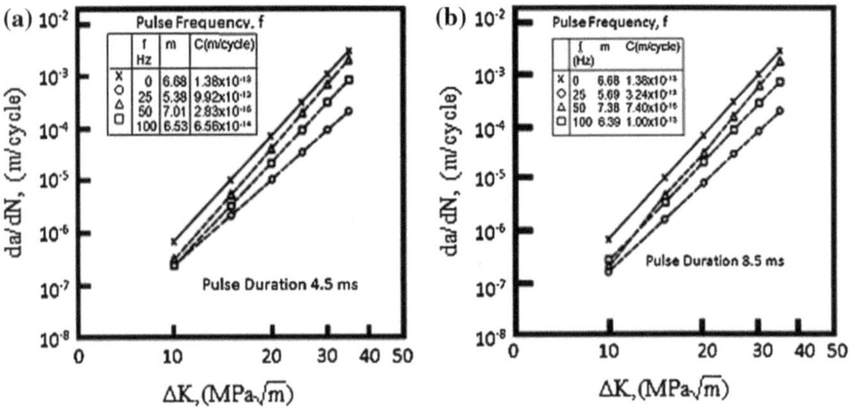

Fig. 7.51 Varied correlation of da/dN and ΔK in weld metal of Al–Mg alloy deposited at different pulse frequencies kept under a low and high pulse duration of **a** 4.5 and **b** 8.5 ms

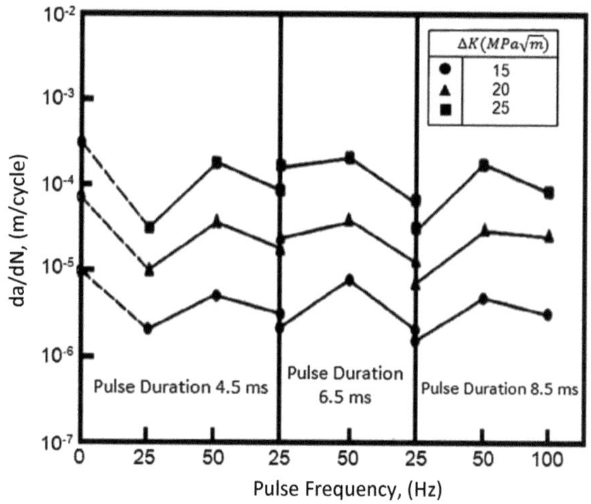

Fig. 7.52 At a given pulse duration influence of pulse frequency on FCGR in Al–Mg weld deposit

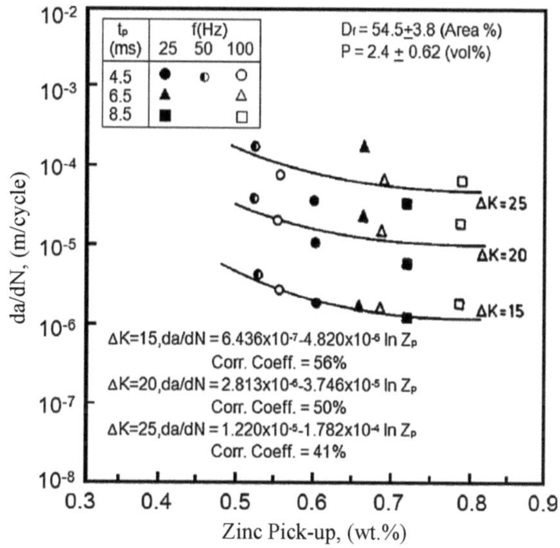

Fig. 7.53 At a given finer dendrite fraction and porosity content influence of zinc pick-up on FCGR in Al–Mg weld deposit

Table 7.4 [27, 32] It shows the superiority of P-GMA weld to that of the GMA weld. At given stress ratio ($R = \sigma_{min}/\sigma_{max}$) and ΔK, fatigue crack growth rate and material constants of Paris law of weld deposit become significantly lower in P-GMA weld in comparison to that of the GMA weld of both the single and multipass weld joints. A comparison of the ΔK_{th} of single and multipass weld deposit of the Al–Mg alloy estimated by solving the Eq. 7.4 is given in Table 7.5. The ΔK_{th} is evaluated for $m = 1$ and the relevant measured data Table (7.4) of fracture mechanics, where K_{IC} is estimated from its measured value of J_{IC} with the help of the following expression [33].

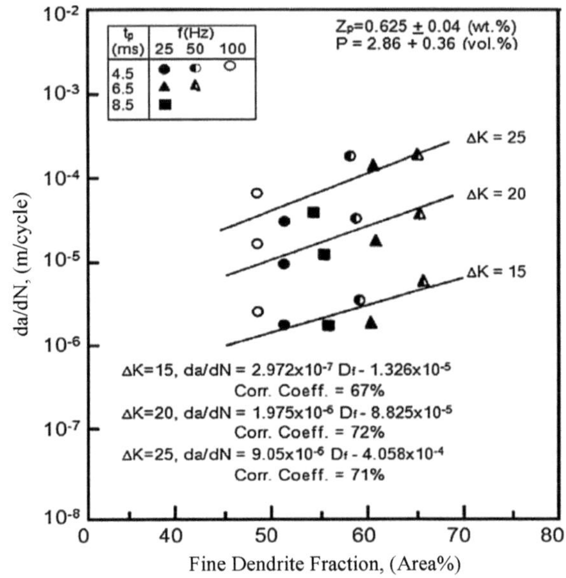

Fig. 7.54 At a given zinc pick-up and porosity content influence of finer dendrite fraction on FCGR in Al–Mg weld deposit

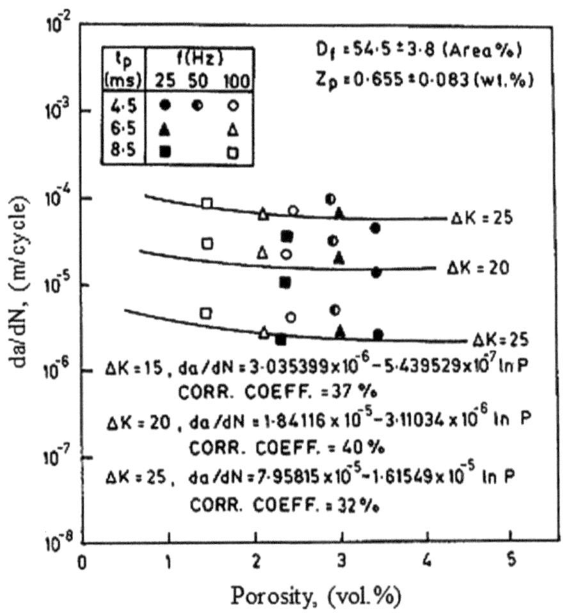

Fig. 7.55 At a given zinc pick-up and finer dendrite fraction influence of porosity content on FCGR in Al–Mg weld deposit

$$J_{IC} = [(1-\nu^2)/E]\, K_{IC}^2 \tag{7.6}$$

The Poisons ratio (ν) and modulus of elasticity (E) of Al–Mg alloy are considered [242] as 0.33 and 70.3 × 10^3 MPa respectively. The estimated ΔK_{th} of

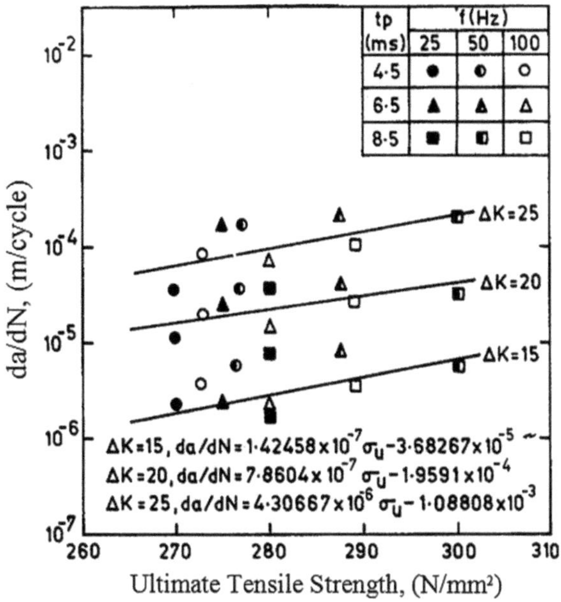

Fig. 7.56 FCGR as a function of UTS at different ΔK in Al–Mg weld deposited at different pulse parameters

Table 7.4 Fracture mechanics properties of single and multipass conventional and pulse current GMA weld deposit of 5083 Al-Mg alloy

Properties	Single pass butt weld		Multi pass butt weld	
	Conventional	Pulse current	Conventional	Pulse current
J_{IC} kJ/mm^2	13.0[a]	22.0[a]	18.0[d]	47.5 − 40.0[d]
CTOD (δ_m), mm	0.078[b]	0.113[b]	0.132[d]	0.174 − 0.164[d]
(da/dN), m/cycle	1.2×10^{-5}[a]	9.5×10^{-6}[a]	9.0×10^{-5}[c]	$7.0 \times 10^{-6} - 4.0 \times 10^{-5}$[c]
ΔK, MPa$\sqrt{m}$	20[a]	20[a]	20[c]	20[c]
Exponent, n	4.5[a]	4.3	7.1[c]	6.1 − 9.2[c]
Const., c m/cycle	2.5×10^{-8}[a]	8×10^{-8}[a]	1.4×10^{-10}[c]	$9.9 \times 10^{-10} - 3.6 \times 10^{-13}$[c]
Stress ratio, R	0.1[a]	0.1[a]	0.1[c]	0.1[c]

[a]Ref. [32]
[b]Ref. [33]
[c]Ref. [30]
[d]Ref. [29]

single pass GMA weld deposit of the 5083 alloy having a cast microstructure is in close approximation to the reported ΔK_{th} = 2.78 MPa$\sqrt{m}$ at R = 0.1 of a similar cast aluminium alloy [34]. The estimated ΔK_{th} of pulse current weld deposit of the 5083 alloy is also found to be closely in agreement to the observed [35] ΔK_{th} of the pulse GMA weld deposit of similar alloy. The Table 7.5 interestingly shows that for both the single and multipass welds the value of ΔK_{th} is significantly higher in case

Table 7.5 ΔK_{th} of the single and multipass GMA and P-GMA weld deposit of Al–Mg alloy in weld joint 7005 Al–alloy

Properties	Single pass butt weld		Multi pass butt weld	
	GMAW	P-GMAW	GMAW	P-GMAW
ΔK_{th}, MPa $\sqrt{m}$	2.2	4.3	2.5	5.5 − 3.3

of the pulse current weld than that of the conventional GMA weld. The fatigue life of a tension and bending loaded 12 mm thick bar of an engineering structure containing a semi-elliptical flaw is analyzed [35] involving both the constants and C of the Paris law (Eq. 7.2) under the fracture mechanics solutions of Newman and Raju [36]. It shows that the use of pulse current GMA welding at appropriate pulse parameter improves safety and reliability of weld joint by enhancing fatigue life of weld (Fig. 7.28) and also justifies inference of its superiority over the use of conventional GMA welding. The significant improvement in fracture toughness and fatigue crack growth characteristics of P-GMA weld primarily happens due to ability of this welding process to control more precisely the weld thermal cycle as well as the geometry, dilution, porosity content, microstructure and residual stresses of weld deposit. All these observations thoroughly justify the superiority of pulse current GMAW over the continuous current GMAW to provide better prospect of weld joint in fabrication of engineering structures and components by introducing improved safety and reliability.

7.6 Corrosion Properties

The adverse influence of thermal characteristics of arc welding on chemistry, morphology and stress distribution in weld and HAZ of a joint often reduces its corrosion resistance. It is generally difficult to avoid due to limited possibility of process control in arc welding for a significant change in weld isotherm and thermal cycle under a prescribed procedure of welding. The possibility of more precise control of behaviour of metal deposition reducing the heat build-up in pulsed current GMA weld significantly reduces the undesirable change in chemistry, morphology and stress distribution in weld and HAZ that restricts the loss of corrosion resistance of ferrous and non ferrous weld joints. In case of the welding of aluminium or aluminium alloys the use of P-GMAW appreciably reduces the grain coarsening in HAZ adjacent to weld fusion line that reduces the loss of its corrosion resistance. This is partially beneficial in case of welding of aluminium and non-hardenable aluminium alloy but, highly beneficial especially for the welding of most corrosion sensitive 7xxx high strength structural aluminium alloy. The high strength stress corrosion cracking sensitive 7xxx and 5xxx (>3% magnesium) alloys considerably enjoy the favourable influence of P-GMAW process for better performance in long term service especially at elevated temperatures (>65 °C).

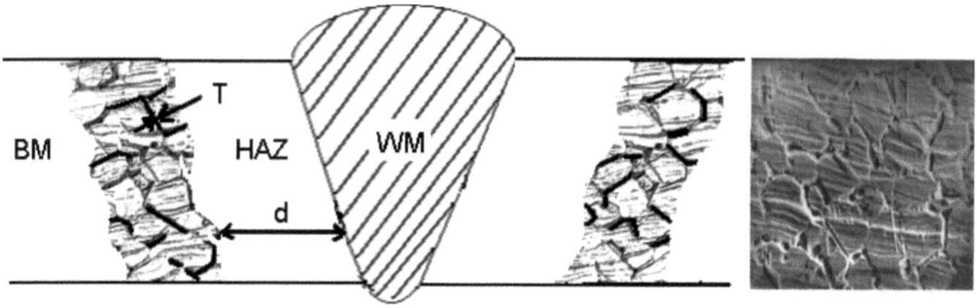

Fig. 7.57 Grain boundary thickening by intergranular corrosion attack in sensitized HAZ away from fusion line of austenitic stainless steel

Table 7.6 Improvement in corrosion resistance of HAZ in GMA and narrow gap P-GMA weld joints of 25 mm thick γ-SS plate

Welding process	Distance (d) from F. L., (mm ± σ)	1.0 ± 0.21	4.5 ± 0.2	5.0 ± 0.25	5.4 ± 0.4
GMAW	GB thickening, T (μm ± σ)	1.13 ± 0.21	2.28 ± 0.35	2.35 ± 0.67	2.83 ± 0.75
P-GMAW		0.94 ± 0.28	1.94 ± 0.47	2.05 ± 0.69	2.22 ± 0.58

Pulse current GMA (MIG) welding with relatively lower current gives excellent characteristics of stainless steel weld along with appreciably less adverse influence to the corrosion resistance of HAZ. Under the weld thermal cycle the extent of sensitization of HAZ at its different locations in reference to the fusion line of austenitic stainless steel weld can be realized by the grain boundary thickening due to chemical attack in corrosion test as shown in Fig. 7.57. The sensitization as primarily a process of formation of chromium carbide through diffusion mechanism is largely dictated by the thermal characteristics and stresses developed in HAZ during welding. The use of P-GMA welding with narrow gap welding procedure that gives relatively low severity of thermal influence combined with development of comparatively less magnitude of residual stresses significantly improves the corrosion resistance of HAZ (Table 7.6) with respect to that noted in the conventional GWA weld of 25 mm thick γ-SS plate [37]. The use of double pulse GMAW process has also been reported to reduce the susceptibility of inter-granular attack in HAZ of AISI: 310S austenitic stainless steel [38].

References

1. Potluri, N.B., Ghosh, P.K., Gupta, P.C., Reddy, Y.S.: Studies on weld metal characteristics and their influence on tensile and fatigue properties of pulse current GMA welded Al–Zn–Mg alloy. Weld. J. **75**(2), 62s–70s (1996)
2. Ghosh, P.K., Rai, B.K.: Characteristics of pulsed current bead on plate deposit in flux cored GMAW process. ISIJ Int. **36**(8), 1036–1045 (1996)

3. Ghosh, P.K., Rai, B.K.: Correlations of pulse parameters and bead characteristics in pulsed current flux cored GMAW process. Ind. Weld. J. **31**(4), 30–39 (1998)
4. Ghosh, P.K., Gupta, P.C., Somani, R.: Influence of pulse parameters on bead geometry and HAZ during bead on plate deposition by MIG welding process. Z. Metallkd. **82**(10), 756–762 (1991)
5. Gupta, P.C, Ghosh, P.K, Visa, S.: Influence of pulse frequency on the properties of HAZ in pulsed MIG welded Al–Zn–Mg alloy. In: Proceedings of International Conference on Welding Technology in developing countries present status and future needs, 1-71-77, 26–28 Sept 1988
6. Shackleton, D.N., Lucas, W.: Shielding gas mixtures for high quality mechanized GMA welding of Q&T steels. Weld. J. **53**(12), 537s–547s (1974)
7. Ghosh, P.K., Gupta, S.R., Randhawa, H.S.: Characteristics and criticality of pulsed current vertical up GMA weld in steel. Int. J. Join. Mater. **11**(4), 99–110 (1999)
8. Pintard, J.: Some experimental data on short circuit transfer, in Physics of the Welding Arc. Institute of welding, London (1966)
9. Holmes, A.W, Rogerson, J.H.: Welding problema peculiar to 7rxx Alloys. Weld. Met. Fab. **34**(9), 349 (1986)
10. Needham, J.C.: Pulse controlled consumable electrode welding arcs. Brit. Weld. J. **12**(12), 191 (1965)
11. Pearson, W.B.: Hand Book of Lattice Spacing and Structures of Metals. Pergamon Press, London, vol. 2, 574 (1967)
12. Pires, I., Quintino, L., Miranda, R.M.: Mater. Des. **28**, 1623–1631 (2007)
13. Ghosh, P.K., Gupta, S.R., Gupta, P.C., Rathi, R.: Fatigue characteristics of pulsed MIG welded Al–Zn–Mg alloy. J. Mat. Sci. **26**, 6161–6170 (1991)
14. Ghosh, P.K., Ghosh, A.K.: Control of residual stresses affecting fatigue life of pulsed current gas-metal-rc weld of high-strength aluminum alloy. Met. Mat. Trans. A **35A**, 1–6 (2004)
15. Bagryanskii, K.V., Royanov, V.A.: Svar. Proiz. **10**, 23–24 (1968)
16. Hertzberg, R.W.: Deformation and fracture mechanics of engineering materials, 2nd edn. Wiley, New York (1983)
17. Ritchie, R.O.: Why ductile fracture mechanics? Trans. ASME, J. Eng. Mater. Technol. **105** (January), 1–7 (1983)
18. Paris, P.C., Tada, H., Zahoor, A., Ernst, H.: Instability of the tearing model of elastic-plastic crack growth. In: Proceedings of Symposium on Elastic-Plastic Fracture, ASTM STP (2nd edn.), **668**(5), 5–36 (1979)
19. Hutchinson, J.W., Paris, P.C.: Stability analysis of J-controlled crack growth. Elast. Plast. Fract. ASTM STP **668**, 37–64 (1979)
20. Ernst, H., Paris, P.C., Rossow, M., Hutchinson, J.W.: Fracture mechanics. ASTM STP **677**, 581–599 (1979)
21. Paris, P.C., Gomez, M.P., Anderson, W.E.: A rational analytic theory of fatigue. Trend Eng. **13**(1), 9–14 (1961)
22. Dong, L., Haynes, R., Atluri, Satya N.: On improving the celebrated paris' power law for fatigue, by using moving least squares. CMC, Comput. Mater. Continua. **45**(1), 1–15, (2015)
23. Forman, R.G., Kearney, V.E., Engle, R.M.: Numerical analysis of crack propagation in cyclic-loaded structures. J. Basic Eng. **89**(3), 459–463 (1967)
24. Donahue, R.J., Clark, H.M., Atanmo, P., Kumble, R., McEvily, A.J.: Crack opening displacement and the rate of fatigue crack growth. Int. J. Fract. Mech. **8**(2), 209–219 (1972)
25. Donald, J.K., Bray, G.H., Bush, R.W.: An evaluation of the adjusted compliance ratio technique for determining the effective stress intensity factor. In: Panontin T.L. and Sheppard, S.D. (eds.) Fatigue and Fracture Mechanics: 29th National Symposium on Fatigue and Fracture Mechanics, ASTM STP 1332, ASTM, West Conshohocken, PA (1998)
26. Amzallag, C., Rabbe, P., Bathias, C., Benoit, D., Truchon, M.: Influence of various parameters on the determination of the fatigue crack arrest threshold. In: Hudak, Jr. S.J., Bucci, R.J. (eds.) Fatigue Crack Growth Measurement and Data Analysis, ASTM STP 738, ASTM, 29–44 (1981)

27. Forth, S.C., Herman, D.J., James, M.A.: Fatigue crack growth rate and stress-intensity factor corrections for out-of-plane crack growth. In: Daniewicz, S.R., Newman, J.C. Jr., Schwalbe, K.H. (eds.) Fatigue and Fracture Mechanics: 34th Volume, ASTM STP 1461, ASTM International, West Conshohocken, PA (2004)
28. Ghosh, P.K., Dorn, L., Issler, L.: Fatigue crack growth behaviour of pulsed current MIG weld of Al–Zn–Mg alloy. Int. J. Joining of Mater. **6**(4), 163–168 (1994)
29. Hussain, H.M., Ghosh, P.K., Gupta, P.C., Potluri, N.B.: Fracture toughness of pulse current multipass GMA weld of Al–Zn–Mg alloy. Int. J. Joining of Mater. **11**(3), 77–88 (1999)
30. Hussain, H.M., Ghosh, P.K., Gupta, P.C.: Potluri Nagesh Babu: Fatigue crack growth properties of pulse current multipass MIG weld of Al–Zn–Mg alloy. Trans. Ind. Inst. Met. **50** (4), 275–285 (1997)
31. Tiwary, M.: Studies on influence of pulse parameters on FCGR and fracture toughness behaviour of pulse current MIG welded Al–Zn–Mg alloy. Master's dissertation, Department of Mechanical & Industrial Engineering, University of Roorkee, (1995)
32. Potluri, N.B., Ghosh, P.K., Gupta, P.C., Reddy, Y.S.: Pulsed current GMA welding: a technique to improve fracture toughness of Al–Zn–Mg alloy weldments. In: Proceedings of International Welding Conference (IWC 99), Welding and Allied Technology Challenges in 21st Century, New Delhi, 732–740 15–17 Feb 1999
33. Barbagallo, S., Cerri, E.: Evaluation of the K_{IC} and J_{IC} fracture parameters in a sand cast AZ91 magnesium alloy. Eng. Fail. Anal. **11**(1), 127–140 (2004)
34. Ogle, M., Blewett, R.V.: Weld quality, inspection and acceptance specification recommended by BS 8118, part 2. In: Proceedings of International Conference on Aluminium Weldments, 5th INALCO 92, Munich, 7.1.1–7.1.31 27–29 April 1992
35. Ghosh, P.K., Dorn, L., Issler, L.: Fatigue crack growth behaviour of pulsed current MIG weld of AI–Zn–Mg alloy; Int. J. Join. Mater. **6**(4), 163 (1999)
36. Newman, J.C., Raju, I.S.: Stress-intensity factors for internal surface cracks in cylindrical pressure-vessels. J. Press. Vessel Technol. Trans. ASME **102**, 342–346 (1980)
37. Kulkarni, S.G.:, Effect of narrow gap welding on characteristics of weld joint of austenitic stainless steel. Ph.D. Thesis, Indian Institute of Technology Roorkee, India (2009)
38. Mathivanan, A., Senthilkumar, A., Devakumaran, K.: Pulsed current and dual pulse gas metal arc welding of grade AISI: 310S austenitic stainless steel. Defence Technol. **11**(3), 269–274 (2015)

Chapter 8
Prospective Use of Pulse Current GMAW Process

Abstract The utility of P-GMAW process has been discussed in reference to regulation of pulse arc in the form of short arc, chop arc and double-pulsed welding process. The unique potential features of P-GMAW process has been described in exploration of new welding procedures as extra narrow gap and deep-gap welding used in relatively thick plate and thick wall pipe welding. The use of most favourable control of weld isotherm and weld thermal cycle of P-GMAW in development of innovative products of extra thick and thin section welding is explained. The products developed improved quality which is otherwise difficult to produce by any other welding process.

Keywords Use of P-GMAW · Pulse arc regulation · Short-arc welding · Chop-arc welding · Double-pulsed welding · New welding procedure · New product development · Improved product quality

The unique characteristics of pulsed current GMAW especially as large instant heating during metal deposition with extra penetration at low heat buildup in weld pool has given new dimension to welding engineering especially by exploring its applications in several non conventional and otherwise difficult cases of welding. The P-GMAW has been emerged out as a highly prospective welding process due to its multifarious possibilities of applications through precise control of parameters which are generally not possible to consider by conventional GMAW process. Further the inherent characteristics of this process to work at high deposition rate with relatively faster travel speed producing fairly acceptable aesthetic and geometrical appearance of the weld gives highly economic performance. The ample scope of manipulation of this welding process points towards the possibilities of introducing large variation in this process in order to develop some kinds of new welding process and procedure for appropriate use in extraordinary applications. In the line of understandings of earlier chapters, it is primarily conceived as a control of energy distribution in the arc process and the nature of metal transfer governing the shape and size of weld pool weld pool along with its thermal behaviour by regulation of pulse characteristics. The extensive scope of variation in this regard

P.K. Ghosh, *Pulse Current Gas Metal Arc Welding*, Materials Forming, Machining and Tribology, DOI 10.1007/978-981-10-3557-9_8

can make use of this process in joining of extraordinarily thin and thick sections at significantly low heat. The development of such a high technically and economically viable welding process is fascinating for joining of thin ferrous and nonferrous sheets of high strength lightweight structures and components over other commonly used welding processes known as laser welding, gas tungsten arc welding (GTAW) etc. with respect to versatility of application and weld quality.

8.1 Regulated Pulsed Arc Welding Processes

The regulation of pulsed arc giving birth of new welding processes basically stems from the possibility of controlling the AC characteristics by appropriate control of pulse parameters that gives a different kind of energy distribution in the system. Thus the new process can be used for the requirements of an extra low to high heat input in welding.

8.1.1 Short-Arc Welding

The control of spattering in GMAW especially during the use of active gas or gas mixture as arc shielding always considered as a task of great technical and economical interest. The reduction of spatter loss with respect to material and welding procedure is generally achieved by process control at low energy input. The most effective approach in this regard is to control the welding current at the end of short circuiting range and at the very beginning of arc re-ignition so as to control the explosion energy of the liquid metal bridge. Eventually it is necessary to accurately control the current and voltage cycles in the three phases of welding as short circuiting, arc ignition and arcing. However, for such control of arc characteristics the power source must have precisely very fast reaction time in order to maintain the preset current and voltage cycles against the instantaneous changes of welding parameters due to any process irregularities [1, 2]. The instantaneous values of the current and voltage cycles in the three phases of short circuiting during CO_2 shielded GMAW could be meticulously controlled at desired levels using a short-arc welding through development of a transistorized inverter power source with an operating frequency of 20 kHz [3]. The power source exclusively operates by current control system instead of the waveform control [4] as tried earlier. The new short-arc welding process with right pulse shape is successfully used in CO_2 shielded GMA welding of 1–5 mm thick steel sheet with considerably reduced spatter loss at high order of process stability. The short-arc process can produce weld having easily removable fine spatter as low as about 2% in place of that exceeds 4% in case of conventional CO_2 welding process [5, 6].

8.1.2 Chop-Arc Welding

Pulse current GMAW as a well known low heat input process with ability to precisely manipulate the heat build-up in weld pool weld pool through control of arc characteristics and nature of metal transfer; it explores the possibility of using with a short-arc applicable to thin sheet welding [7]. Modifications of the short-arc welding technology was lead to development of a Chop-arc process [8–10] in Technical University of Berlin. It is applicable to welding of thin sheet in sub-millimeter range of thickness. Primary aim of the Chop-arc process is to reduce heat input during joining, in order to avoid overheating of thin sheets. It is a more economical and versatile welding process which may be used as an alternative to GTAW, laser and resistance welding. The controlled chopping of the arc results unusually low heat input as a variant of the GMAW short-arc process whereby spatter formation, porosity and distortion are minimized by providing a temporal stepwise adjustable energy input. Furthermore, as a low heating process the Chop-arc is also satisfactorily used in joining of galvanized steel sheet by brazing with Cu-base filler wires having minimum zinc evaporation.

Since ignition and stable burning of the arc requires a minimum weld current, the reduction of heat input is performed by limiting the arc burning interval through interruption of power supply (Fig. 8.1). During this off-current time a cooling of the molten zone takes place, until the wire touches the molten pool. The short circuit serves as trigger signal for switching-on the weld current and causes a droplet transfer in the way of conventional short-arc welding. The course of current starts with the current-off time, followed by an increase of short circuit current and is completed by re-ignition and burning of the arc until the next interruption takes place. The Chop-arc is a kind of hybrid process of a current controlled short-arc with a pulsed arc component assuring regular metal transfer. An appropriate use of the Chop-arc process with reduced heat input is assured through an advanced transistorized power source with manifold adaptable parameters and reproducible energy supply during the different phases of the arc. The Chop-arc process stepwise controls the energy input of the pulse form depending upon metal transfer through an electronic switch coordinating the circuit from phase to phase on detecting a trigger signal that is by a limit from duration, voltage or current. Therewith a new current set-point of the next phase is applied to the current controller. The chopping of the arc with a defined power supply switches off at the end of a phase, which is realised by a current set-point of $I_{\text{arc off}} = 0$ A.

In contrast to the conventional short-arc process, the chopping of arc not only reduces the heat input but also controls a significantly low energy input to maintain a well defined volume of molten metal giving rise to highly regular process behaviour. Such behaviour of Chop-arc process provides excellent regularity of metal transfer which can be ascribed to the controlled heat input in the beginning and end of a phase (Fig. 8.2). This is equivalent to the defined droplet size in case of the pulsed mode. A smooth control of energy input in the short circuiting phase reduces the amount and size of spatters to a great extent as well as keeps their heat

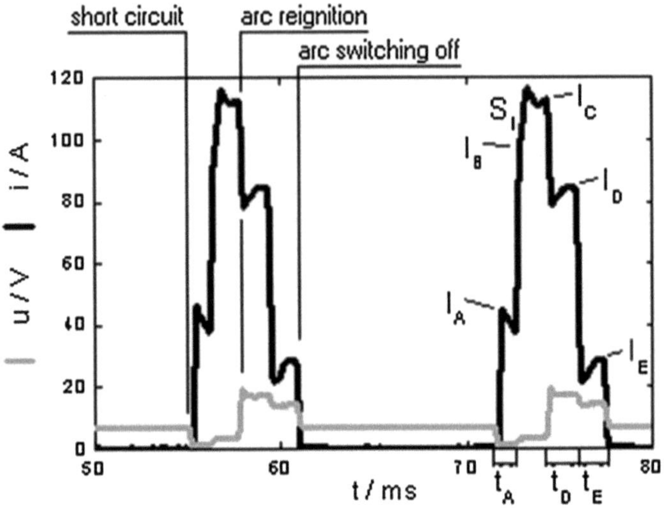

Fig. 8.1 Typical current and voltage plots of Chop-arc welding process

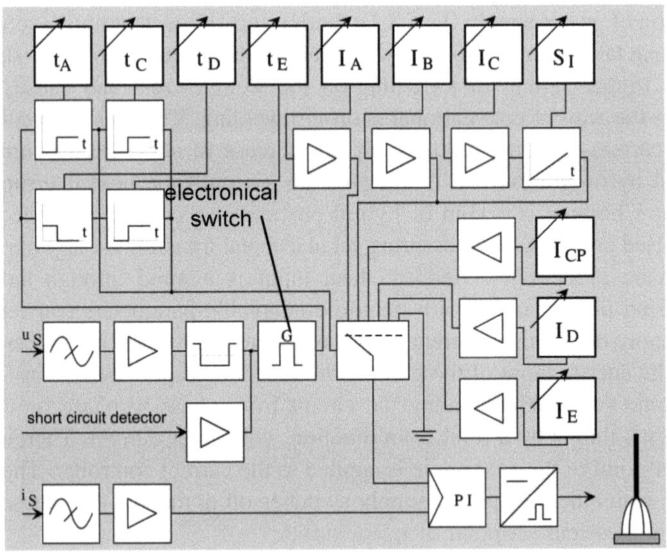

Fig. 8.2 Block diagram of Chop-arc process control

content at low level of insufficiency to adhere to the work piece. The Chop-arc process allows decoupling the link between heat inputs and welding deposition and helps the control of weld bead solidi-fication by adjusting the chopped current in arc off phase. Thus, it facilitates to limit and control the heat input more accurately enabling it to apply for welding of lower sheet thickness.

The typical welding parameters given in Table 8.1 successfully performs the Chop-arc butt welding and overlap fillet welding of 0.2 mm thick mild steel sheets (St 110–05 AM, DIN 1624/ DIN EN 10139) using Ar + 1% CO_2 + 0.5% O_2 shielding gas and 0.6 mm diameter filler wire SG 2 (DIN 8559). It produces flaw less cross section of the joint with uniform and finely rippled weld surfaces (Fig. 8.3) at a powdery spatter rate of less than 1.0%. However, a deviation from the optimised parameters in different phases of the Chop-arc process varies the spatter characteristics and cross-sectional weld shape over the full variation range of parameters. Amongst the end phase parameters the current has insignificant influence on spatter formation. However, the weld penetration increases with the increase of duration at a low current of end phase, because of direct energy transfer to the molten pool and the sheet during the arcing. Thus, the weld geometry is controlled directly by a variation of parameters of the chopping arc phase.

Table 8.1 Typical welding parameters for 0.2 mm mild steel sheets (phase "E" gives satisfactory result)

	U_m	I_m	I_A	I_B	I_C	I_D	I_E	t_A	t_C	t_D	t_E	S_I	V_S	V_D
	V	A	A	A	A	A	A	ms	ms	ms	ms	A/ms	m/min	m/min
Butt	5.5	25.0	40	85	135	90	17	0.9	3.0	1.5	2.9	50	0.5	3.2
Fillet	4.5	20.5	30	50	100	105	34	0.6	3.0	1.75	6.4	50	0.5	3.2

Fig. 8.3 Surfaces and cross sections of Chop-arc welded 0.2 mm thick steel sheet without root backing

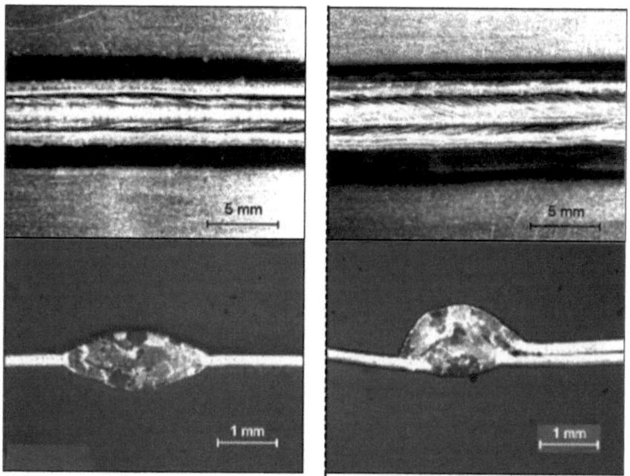

	Butt weld	Overlap fillet weld
filler	SG 2, ϕ 0,6 mm	SG 2, ϕ 0,6 mm
t_E	2,9 ms	2,4 ms
I_E	17 A	45 A
v_W	0.5 m/s	0.5 m/s
P_m	256 W	260 W
spatter rate	0.7 %	0.9 %

The low heating Chop-arc brazing process has been successfully used for lap joining of 0.7 mm thick electrolytic zinc coated (7 μm thick at 54 g/m^2) mild steel sheets using 0.8 mm diameter filler wire of Cu-base brazing alloy at optimal brazing parameters (Table 8.2). It considerably improves the arc brazing by producing practically porosity free joints at spatter rates in the range of 1.2–1.4% with appreciably low distortion and reduction of zinc evaporation preserving zinc coating on the sheets (Fig. 8.4).

A similar reduction of energy during arc process without extinguishing the arc is also developed by M/s EWM of Germany [11]. The process named as Cold-arc process is especially advantageous for brazing of Zn-coated steel sheets and dissimilar joining of Zn-coated steel sheets with aluminium sheets using Zn-Al filler wire due to its significantly low thermal characteristics. A low heat input process with controlled metal deposition using variable controlled wire speed named as the

Table 8.2 Typical brazing parameters for 0.7 mm zinc coated steel sheets (phase "E" gives satisfactory result)

	U_m	I_m	I_A	I_B	I_C	I_D	I_E	t_A	t_C	t_D	t_E	S_I	V_S	V_D
	V	A	A	A	A	A	A	ms	ms	ms	ms	A/ms	m/min	m/min
Fillet	10.0	57.0	85	70	200	115	34	2.4	4.5	4.0	10.4	50	0.5	4.6

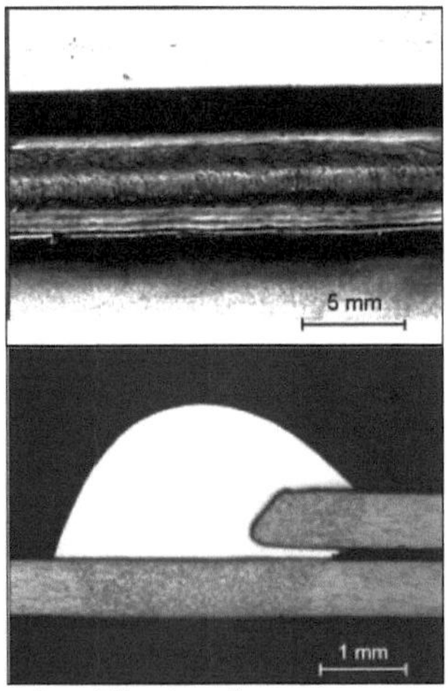

overlap fillet brazing

filler	CuSi1,8Mn1, ϕ 0,8 mm
t_E	10,4 ms
I_E	34 A
v_W	0.5 m/min
P_m	624 W
spatter rate	1,2 %

Fig. 8.4 Surface and cross section of Chop-arc brazed 0.7 mm thick zinc coated steel sheet

Cold Metal Transfer (CMT) process is developed by the M/s Fronius of Austria. It offers several advantages for welding and brazing of thin coated steel sheets as well as the sheets of coated steel and aluminium.

8.1.3 Double-Pulsed GMAW

The double pulsed (DP) GMAW works on a principle of modulation of high frequency waveform by a low frequency current pulse at both the basic stages of peak and base current duration [12–14]. It gives significant variation in thermal characteristics and mechanical stirring of the weld pool weld pool which extends the range of application of this process in various ferrous and nonferrous weld fabrications of thin to relatively thick sections. The process is also used for welding of galvanized steel sheet and brazing. It produces practically spatter free appreciably good quality weld with better economy. Many companies are offering inverter controlled synergic welding machine with double pulse functions. The DP-GMAW process improves the transverse shrinkage and mechanical and metallurgical properties of austenitic stainless steel weld joints in comparison to that of the P-GMAW process. It primarily happens due to a combined effect of pulse current and thermal pulsation (low frequency pulse). During this period, fluctuation of wire feed rate results in further increase of welding current and synchronizes decrease of arc voltage. The DP-GMAW introduces comparatively more thermal shock compared to the P-GMAW deposit and reduces the heat input and improves the properties of weld joints [15].

8.2 Exploring New Welding Procedure

It is well known that pulse current GMAW process can most conveniently perform the out of position welding and orbital welding in fabrication of engineering structures and components. The ability to produce a large variety of weld deposition at low adverse influence of thermal severity of arc welding by regulation of pulse parameters has further made the P-GMAW process also attractive for welding of thick sections and modification of weld design. This may provide further leverage to joining technology especially with respect to mechanical and metallurgical behaviours of weld joint as well as its safety and economy. The merits of P-GMAW is successfully explored and applied in narrow gap pipeline girth welds with required mechanisation that gives economical construction at improved joint properties [16–19].

The P-GMAW processes is successfully used in 3–5 mm wide narrow gap CO_2 welding of 12 mm thick steel section by deep-gap welding (DGW) procedure keeping the welding torch outside the joint while the filler wire protrudes in between the parallel weld groove [20, 21]. A mechanised DGW with direct current electrode positive (DCEP) is capable to produce practically sound weld in vertical-up position using this processes. At a given welding current the use of conventional GMAW instead of P-GMAW produces comparatively wider root and narrower weld bead. However, a variation in pulse parameters significantly affects the statistical arc stability with respect to its current-voltage relationship and weld quality. The increase of pulse frequency and decrease of off-time ratio of pulse, defined as [$R = t_b/(t_p + t_b)$ %], in the range of 150–300 Hz and 12–60% respectively improves arc stability keeping it similar to that of the conventional GMAW [21]. Reduction in gap width adversely affects the arc stability and short circuiting frequency. Due to larger current density the DGW with 1.2 mm diameter flux cored filler wire gives more spatter free stable arc welding with wider root penetration and higher admixture as compared to that results in case of using solid filler wire of same diameter.

Several merits of P-GMAW in reference to its control of arc characteristics and behavior of metal transfer give rise to highly precise intended thermal distribution in weld. Thus, it allows comfortable use of this process in narrow gap welding of thick plates which is otherwise difficult to carry out by the conventional GMAW process [22–24]. The unique ability of P-GMAW to control wide range of combination of width and penetration of weld bead with a high precision could produce extra narrow gap weld of 25 mm thick stainless steel and high strength low alloy steel [25–27] section. Such a welding with extra narrow gap could be possible by employing appropriate modification in P-GMA welding procedure with single seam deposition per layer which is otherwise not possible by conventional GMAW. Further a thorough modification in design of torch nozzle head facilitating its desired manipulation in ultra narrow weld groove opening of about 7.5–8 mm in 25 mm thick plate could produce a sound weld of dissimilar materials of austenitic stainless steel and HSLA steel plates [28, 29]. A comparative appearance of such a narrow groove weld with respect to the conventional dissimilar weld is revealed in Fig. 8.5a, b respectively. In this procedure of welding the narrow groove wall acts

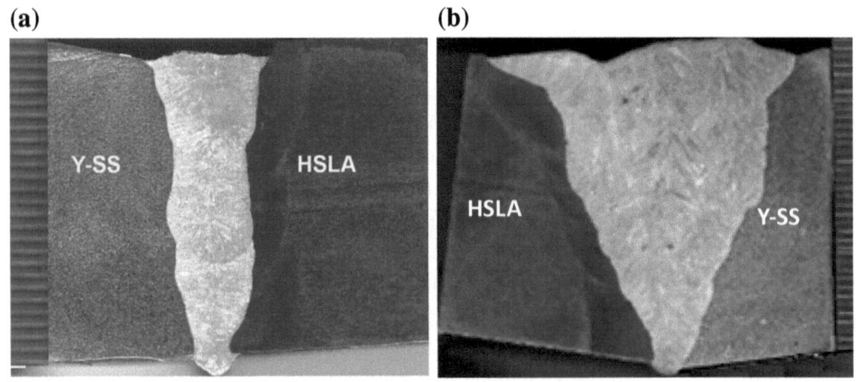

Fig. 8.5 Typical P-GMA weld of 25 mm thick HSLA steel and γ-stainless steel plates **a** ultra narrow weld groove (7.5 mm opening) and **b** conventional weld groove (28 mm opening)

as a supporting lengthwise side wall of the rectangular torch nozzle head enabling its width to narrow down up to its minimum level while keeping the flow characteristics of the shielding gas and the arc behavior inside the groove at their optimum level of acceptance to produce smooth and sound weld [28, 29]. A typical analysis of the flow characteristics of the shielding gas and the arc behavior inside the groove at 0° projection angle of the torch nozzle head used at gas flow rate of 12 l/min is shown in (Fig. 8.6). The soundness of ultra narrow groove dissimilar weld joint of austenitic stainless steel and HSLA steel plates has been typically shown in (Fig. 8.7). Use of narrow gap P-GMAW not only improves metallurgical quality of weld joint with significantly low distortion (Fig. 8.5) but also produces a further defect free weld especially with respect to its inclusion content which often arises in case of using the SMAW and SAW processes.

Fig. 8.6 Typical flow characteristics of shielding gas and arc behavior inside the ultra narrow groove (7.5 mm opening) at 0° projection angle of torch nozzle head

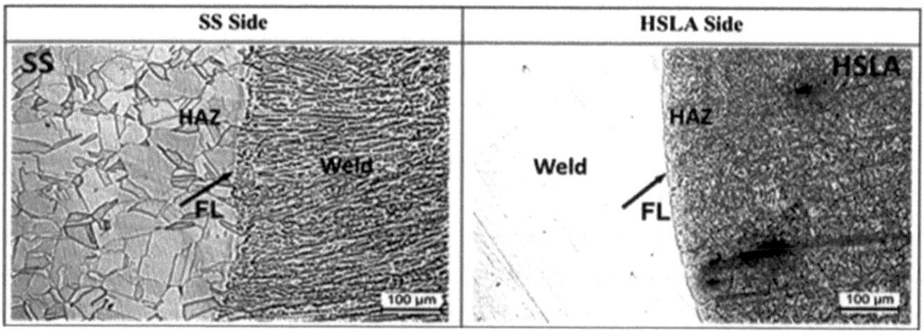

Fig. 8.7 Typical microstructure of HAZ adjacent to FL on both sides of ultra narrow groove weld joint of 25 mm thick γ-SS and HSLA steel plates

8.3 New Product and Product Quality

The unique possibility of favourable control of weld isotherm and weld thermal cycle has made the P-GMAW process suitable for application in many new challenging sectors of fabrication and product development industries that particularly involves off-shore constructions, automobile production, construction and agricultural equipment and fabrication of marine and aerospace components. In this context the successful application of this process is commonly noted in welding of tubular components, thin ($\approx$0.5 mm) sheet product and advanced ferrous and non ferrous alloys and dissimilar surfacing of steel. The merits of this welding process with respect to the thermal and transfer behavior of metal deposition improves the product quality especially in reference to weld geometry, weld chemistry, porosity, inclusion and microstructure and stress development in weld joint.

The modified spray transfer process of pulsed current GMAW addresses more efficiently the basic problems of aluminium welding primarily involving the porosity, lack of fusion, burn through and spatter with better control of weld profile and appearance. It is true in welding of both the thick and thin section of aluminium with relatively cooler weld deposition at high current that gives better acceptable weld quality than . Aluminium section of thickness 4.5–12.5 mm could be satisfactorily welded by single pass using pulse current relevantly in the range of about 25–350 A at 60% duty cycle at a comparatively higher welding speed with thicker filler wire [30–32]. Appropriate control of pulse parameters can also weld further thicker section of aluminium by multipass deposition process, which is otherwise difficult to weld due to high thermal conductivity. A control of improved penetration and bead geometry for required thermal distribution in groove wall with the help of proper designing of pulse parameters makes possible.

The pulse MIG welding process using AC current shows good gap bridging ability. The AC pulse MIG welding cyclically alternates between direct current electrode positive (DCEP) and direct current electrode negative (DCEN). It is successfully used in dissimilar welding of 1.4–1.6 mm thick C-Mn steel (SPRC440) and Al–Mg-Si alloy (6K21) sheets applying 1.2 mm diameter Al–Si alloy (4043) filler wire [33]. Control of current waves of AC pulse MIG welding in terms of the EN ratio of about 20% produces sound weld with thin (10 μm) intermetallic layer at the interface gives good tensile properties of the joint. The EN ratio is defined as

$$\mathrm{EN}(\%) = [[(\mathrm{I_{EN}} \times \mathrm{T_{EN}})/(\mathrm{I_{EP}} \times \mathrm{T_{EP}}) + (\mathrm{I_{EN}} \times \mathrm{T_{EN}})] \times 100] \qquad (8.1)$$

where, I_{EN} is average current of the section EN, (A), I_{EP} is average current of the section EP (A), T_{EN} is time of the section EN (s) and T_{EP} is time of the section EP (s).

Use of synergic MIG process with stable pulsed arc and low spatter enhances scope of production with high deposition rate in heavy section steel welding or surface overlaying with appreciable corrosion resistance. It is employed to various components in constructions of chemical, petrochemical and offshore oil and gas industries [31–33]. The P-GMAW process is satisfactorily applied for austenitic stainless steel weld overlaying especially on relatively thin C-Mn structural steel. A precise control of deposition and thermal characteristics of the overlaying through appropriate change in pulse parameters appreciably refines the microstructure of weld deposit (Figs. 8.8 and 8.9). Interestingly more refinement of microstructure giving rise to lower micro-segregation and improved corrosion properties of the overlay happens with lager deposition rate at higher mean current [34]. This may be attractive to industrial application due to better economy of production where it also gives resistance to microfissures especially at low δ-ferrite content of the matrix. An interacted layer forms at the interface (Fig. 8.10a, b) primarily due to diffusion of carbon from the mild steel substrate to the stainless steel overlay leads to hard phase transformation and microcracking dictated by the weld thermal cycle [35–37]. A control of in the system governed by the peak current as a function of pulse frequency and pulse duration at a given mean current can give controlled phase transformation at the interface.

Reducing residual stresses of weld joint due to use of pulse current GMAW [38] especially with narrow gap in welding of thick wall component is always considered beneficial for life and safety of welded structure. In case of welding of thick section stainless steel it also becomes favourable to restrict the loss of corrosion resistance of weld joint. Typical development of longitudinal residual stresses at the top and root of conventional V-groove shielded metal arc (SMA) and GMA welds and narrow groove P-GMA weld of 25 mm thick 304-LN stainless steel pipe has been shown in Fig. 8.11. It clearly shows that pulse current narrow groove welding convincingly reduces residual stresses of weld deposit to about 40% in comparison to that of common SMA weld joint and of significant order with respect to that of conventional current GMA welds.

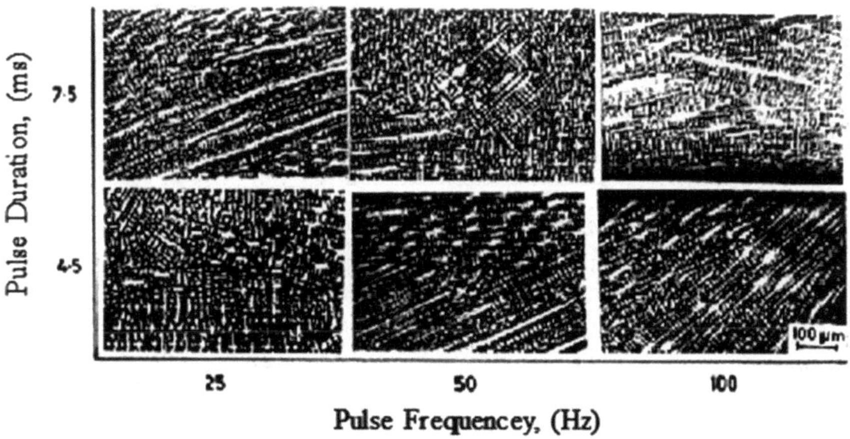

Fig. 8.8 At I_m = 160 A effect of pulse frequency and pulse duration on microstructure of austenitic stainless steel weld overlay on mild steel

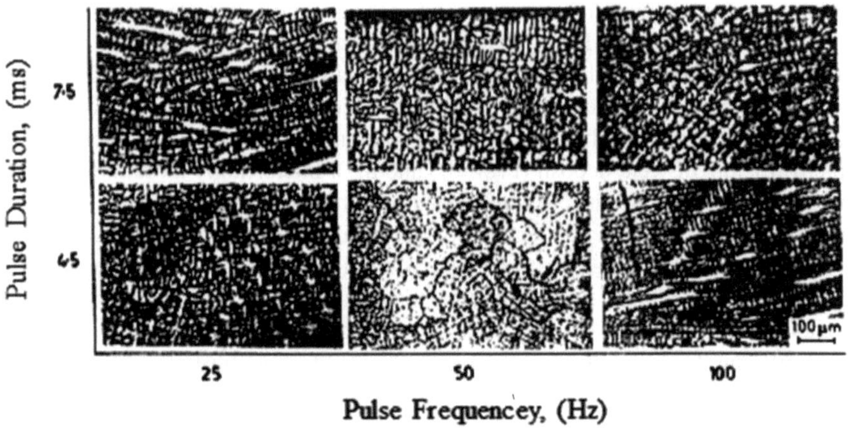

Fig. 8.9 At I_m = 200 A effect of pulse frequency and pulse duration on microstructure of austenitic stainless steel weld overlay on mild steel

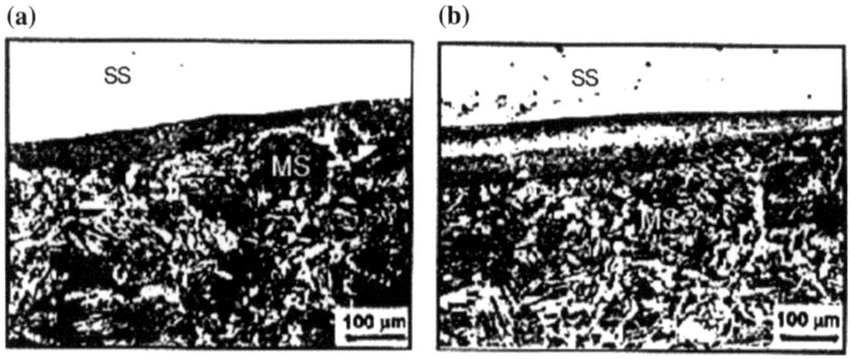

Fig. 8.10 Interacted layer at stainless steel overlay interface with mid steel produced by **a** P-GMAW and **b** conventional GMAW process

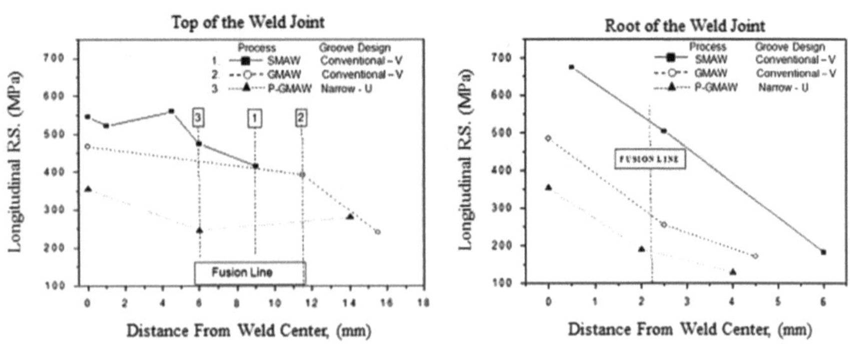

Fig. 8.11 Distribution of longitudinal residual stresses at top and root of stainless steel pipe weld

References

1. Dorn, L., Momeni, K., Rummel, Th.: Process control in the arcing and short-circuit phases during metal-arc active gas build-up welding under carbon dioxide. Schweissen und Schneiden **46**(2), 57–60 (1994)
2. Schiedermayer, M.: Improving the reliability of the inverter-based welding machines. Weld J **52**(2), 51–52 (1997)
3. Dorn, L., Chang, Y.Q., Momeni, K., Mecke, H., Rummel, Th.: Development of an advanced process control for CO_2 gas metal short arc welding. In: Proceedings 2nd International Welding Technology Symposium, IWTS'98, Istanbul, Turkey, 3–5 June, 1–9 (1998)
4. Matsuda, F., Ushio, M., Mita, T.: Wellenformsteuerungsmethode im CO_2-schweissen. Trans. of JWRI **17**(2), 11–16 (1988)
5. Gupta, S.R., Gupta, P.C.: Effect of some variables on spatter loss, Welding and Met. Fabrication November/December, 361–364 (1984)
6. Fawer, A., Winkler, R.: Schutzgase zum MAG-Schweissen—Aktueller Stand der Anwendung und Entwicklungstendenzen, DVS-ASR-Gemeinschaftstagung, Timisoara-Rumänien, 24–25 März, 21–29 (1993)
7. Dorn, L, Chang, Y,, Momeni, K.: An advanced process control to minimize spatter loss in CO_2 gas metal arc welding. In: Symposium on Joining of Materials, Tiruchirappalli, India, 20–22 Jan (2000)
8. Dorn, L., Goecke, Sven-F.: ChopArc for superior MAG-welding of thin sheets >0.2 mm. In: Proceedings 4th European conference on welding, joining and cutting. pp. 24–26 EUROJOIN 4, Cavtat-Dubrovnik, 4–26 May, 2, 91–96 (2001)
9. Dorn, L., Goecke, Sven-F.: ChopArc Process—A New GMA welding technology for very thin sheets. In: Proceedings of ESDA2002: 6th biennial conference on engineering systems design and analysis, Istanbul, Turkey, July 8–11, MAN-027 (2002)
10. Goecke, S.F., Metzke, E., Spille-Kohoff, A., Langula, M.: ChopArc—MSG arc welding for ultralightweight construction, bmb + f-gefördertes Verbundprojekt, final report, Fraunhofer IRB Verlag, ISBN 3-8167-6766-4, (2005)
11. Goecke, S.F.: Low Energy Arc Joining Process for Materials Sensitive to Heat, EWM Mündersbach, Germany, © EWM HIGHTEC WELDING GmbH 5/5 WM031801.doc1; 11.05 (2005)
12. Liu, Anhua, Tang, Xinhua, Fenggui, Lu: Weld pool profile characteristics of Al alloy in double-pulsed GMAW. Int. J. Adv. Manuf. Technol. **68**, 2015–2023 (2013)
13. Liu, Anhua, Tang, Xinhua, Fenggui, Lu: Arc profile characteristics of Al alloy in double-pulsed GMAW. Int. J. Adv. Manuf. Technol. **65**, 1–7 (2013)
14. Silva, R.H.G. Dutra, J.C. Gohr R. Jr.: Scientific and technological fundamentals for the development of the controlled short circuiting MIG/MAG welding process (CCC)—A review of the literature. Part 3 of 3: principles of controlled current short circuiting MIG/MAG systems. Weld Int. **23**(4), 251–260 (2009)
15. Mathivanan, A., Senthilkumar, A., Devakumaran, K.: Pulsed current and dual pulse gas metal arc welding of grade AISI: 310S austenitic stainless steel. Defence Technol **11**(3), 269–274 (2015)
16. Gianetto, J.A., Bowker, J.T., Dorling, D.V., Horsley, D.: Structure and Properties of X80 and X100 Pipeline Girth Welds. In: Proceedings international pipeline conference, pp. 1485–1497. Calgary, Alberta, Canada, October 4–8, Paper No. IPC2004-0316 (2004)
17. Dorllng, D.V., Loyer, A., Russell, A.N., Thompson, T.S.: Gas metal arc welding used on mainline 80 ksi pipeline in canada welding developments meet the demands from the natural gas industry for large diameter, higher strength pipe. Weld J. 55– 61 (1992)
18. Gianetto, J.A., Bowker, J.T., Dorling, D.V.: Assessment of properties and microstructure of X100 pipeline girth welds. Springer, Weld. World (2005)
19. Blackman, S.A., Dorling, D.V.: Technology advancements push pipeline welding productivity. Weld. J. **79**(8), 39–44 (2000)

20. Dorn, L., Ashram, A.Al., Rayes, M.El.: Modified narrow gap welding in vertical position. In: Proceedings international conference advances in mechanical and industrial engineering. pp. 1073–1078. University of Roorkee, India, 6–8 Feb. (1997)
21. Dorn, L., Ashram, A.Al., Rayes, M.El.: Evaluation of arc stability in pulsed current narrow-gap CO_2 welding. In: Proceedings 6th international symposium Japan welding society, 667–672. 2 Nov. (1996)
22. Kulkarni, S.G.:, Effect of narrow gap welding on characteristics of weld joint of austenitic stainless steel, Ph.D. Thesis, Indian Institute of Technology Roorkee, India, (2009)
23. Kulkarni, S.G., Ghosh, P.K., Ray, S.: Improvement of weld characteristics by variation in welding processes and parameters in joining of thick wall 304LN stainless steel pipe. ISIJ Int. **48**(11), 87–94 (2008)
24. Nakamura, T., Hiraoka, K.: Characteristics of bead formation phenomena in narrower gap welding and proposal of new welding process at ultra-narrow gap joint. Development of ultra-narrow gap GMA welding process(report 1), Quarterly Journal of the Japan Welding Society (Japan), **19**, 1, 44–53 (2001)
25. Ghosh, P.K., Agrawal, B.: Extra narrow gap gas metal arc weldingof thick high strength low alloy steel. In: 1485–1497 The 2nd South East European IIW International Congress, pp. 168–173. Pipeline welding current topic of the region, Sofia, Bulgaria, 21–24 Oct. (2010)
26. Agrawal, B.P., Ghosh, P.K.: Thermal modelling of multi pass narrow gap pulse current GMA welding by single seam per layer deposition techniques. Materials and Manufacturing Process, Tailor and Francis **25**(11), 1251–1268 (2010)
27. Ghosh, P.K., Agrawal, B.P.: Advanced technique of extra narrow groove welding of thick steel section using pulse current gas metal arc welding process. In: 64th Annual Assembly & International Conference of the International Institute of Welding, 17– 22 July, Chennai, (2011)
28. Ghosh, P.K., Anant, Ramkishor: Advance Welding Process and Procedure for Extra Narrow Gap Dissimilar Welding of Thick Section of Austenitic Stainless Steel to HSLA Steel. In: Proceedings international conference on high strength materials—challenges and applications, IIW-2015, Helsinki, Finland, June (2015)
29. Anant, Ramkishor, Ghosh, P.K.: Ultra-narrow gap welding of thick section of austenitic stainless steel to HSLA steel. J. Mater. Process. Techno. In press (2016)
30. Roehl, C., Herb, J.: New generation of pulsed MIG improves aluminium welding. Miller Electric Mfg. Co., Dec. 1, (2008)
31. Almy, D.: Pulsed MIG welding on aluminium: The details and advantages, Miller Electric Mfg. Co
32. Mita, T.: Pulse MIG welding of aluminium using constant voltage characteristic power source. Weld. Int. **16**(3), 173–182 (2002)
33. Jin Park, H., Rhee, S., Jin Kang, M., Kim, D.C.: Joining of Steel to Aluminum Alloy by AC Pulse MIG Welding, Mater Trans **50**(9), 2314–2317 (2009)
34. Ghosh, P.K., Gupta, P.C., Goyal, V.K.: Stainless steel cladding of structural steel plate using pulsed current GMAW process. Weld. J. **77**(7), 307s–312s (1998)
35. Ghosh, P.K.: The influence of dilution and heat input on the characteristics of SAW stainless steel cladding on mild steel. Int. J. for the Joining of Mater. **4**(3):90–99 (1992)
36. Ghosh, P.K.: The effect of dilution and heat input on the interface characteristics of stainless steel clad mild steel produced by SAW process". Tool and Alloy Steels **24**(8), 255–260 (1990)
37. Dorn, L., Chang, Y.Q., Momeni, K., Mecke, H., Rummel, Th.: Development of an advanced process control for CO_2 gas metal short arc welding. In: Proceedings 2nd International Welding Technology Symposium, pp. 1–9. IWTS'98, Istanbul, Turkey, 3–5 June (1998)
38. Ghosh, P.K., Ghosh, A.K.: Control of residual stresses affecting fatigue life of pulsed current gas metal arc weld of high strength aluminum alloy. Met. Mater. Trans. A **35A**,(8), 2439–2444 (2004)

Chapter 9
Advanced Power Source and Process Control

Abstract The advanced characteristic of P-GMAW power source which has widened the scope of more versatile use of this process has been discussed. The use of improved electrical efficiency of modern inverter-based power source in more precise welding at relatively less operating cost is explained. A computerized intelligent control of welding parameters through modern digital welding system that have further facilitated the use of P-GMAW process in smooth and satisfactory welding at critical applications is described. In a computerized system the possibility of selection of appropriate pulse parameters for a desired weld with the help of precise knowledge of the effects of the factor ϕ on weld characteristics has been addressed. The unique feature of modern P-GMAW power source that has significantly enhanced the commercial prospect of using this process has been presented especially also in the context of its hybridization with laser welding.

Keywords Modern power source · Pulse current · Inverter-based power source · Computerized intelligent control · Digital welding · Factor ϕ · Selection of pulse parameters · Commercial prospects

9.1 Pulse Current Power Source

Unlike to that used in case of conventional GMAW process the power source used for P-GMAW provides pulses of high current superimposed on a low background current. Pulsed arc welding power source normally consists of a 3-phase welding transformer cum rectifier unit placed in parallel with a single-phase half wave rectifier. The three-phase unit provides background current and single-phase unit supplies peak current. Both the transformer and rectifier units are mounted in a single housing having appropriate control for individual adjustments of background current and peak current. Substantial progress in applications of modern micro-electronics has prompted vast development of advanced welding power sources having highly precise rapid dynamic control over the arc and energy input. This has made the weld fabrication more versatile and economical with improved quality at

P.K. Ghosh, *Pulse Current Gas Metal Arc Welding*, Materials Forming, Machining and Tribology, DOI 10.1007/978-981-10-3557-9_9

an instant of compromise with the welder competence to a great extent. The special control functions of a modern welding power source that instantly regulates arc and behavior of metal transfer has explored possibility of successful welding of thin metal sheets of sub millimeter range especially with low risk of burn-through and warping. Whereas due to high deposition rate at low spattering it is also used in welding of relatively thicker plate of beyond 5 mm requiring especially multipass deposition with high productivity [1, 2]. In view of such developments that is attracting wide spread application of GMAW, large variety of pulse control systems are introduced in modern welding power sources produced by various manufacturers. They are for example known as CMT (Fronius), STT (Lincoln Electric), Cold Arc (EWM), SAT (ESAB) [3–5] for thin sheet welding and Force Arc (EWM), Power Arc (Migatronic), Aristo SuperPulse (ESAB) tandem welding, T.I. M.E. (Fronius) [6–8] for general purpose welding. Migatronic welding power source maker has introduced intelligent arc control function in their Sigma 400 Galaxy welding power source claiming further advantage of a low value current pulse for practically insignificant spatter used at the finishing stage of short circuiting molten drop transfer [9, 10].

A superior arc performance by its better control and improved electrical efficiency giving rise to more precise welding with less spatter and favourable thermal and mechanical characteristics produces economical product at less operating cost. The inverter-based power supplies introduced in commercial market about three decades ago could overcome most of the challenges by introducing various advanced control mechanisms in it [11]. The operational flow chart of this kind of welding power source is different from the conventional one (Fig. 9.1) which simply uses the transformer and rectifier to control the magnitude and characteristics of voltage and current in welding operation. Unlike that happens in case of the conventional welding power source here the electricity first enters instead of a transformer in to a primary rectifier in case of an inverter-based power source and changes from AC to DC and then it runs through a capacitor to smoothen out the

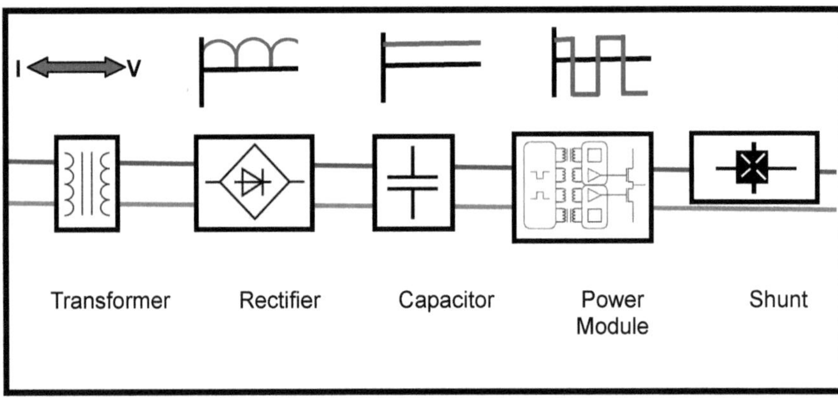

Fig. 9.1 Typical energy flow diagram of conventional transformer-based welding power source

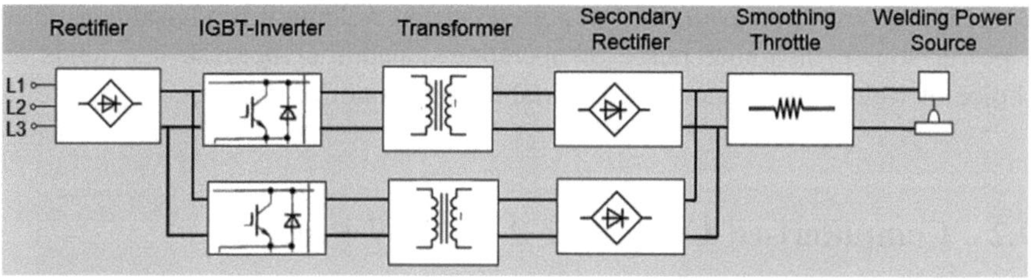

Fig. 9.2 Typical energy flow diagram of transistorized inverter-based welding power source

current. Followed by this process the current enters in to a control circuitry and transforms back to AC where the digital processing occurs (Fig. 9.2). Here the frequency of the AC current could be adjusted up to a very high level and allowed to flow through a transformer to control the ratio of amps to volts to suit the welding application. Finally the current transforms back to DC by passing through a second rectifier before putting into welding operation. The AC current in the control circuitry gets the ability to manipulate its waveform and process the information about welding condition to adapt the arc automatically based on logical judgments of proportionate function of voltage changes based on feedback collected at the weld.

Modern inverter-based welding power source functions through a computerized control circuitry that leads to development of modern digital welding system. It can intelligently control both the primary and secondary welding parameters and thus gives confidence to both the relatively new and well experienced welders to perform their job satisfactorily especially in the new challenging and critical applications by giving only the basic inputs of parameters to the machine. The intelligence of the power source is capable to take real-time decision by reading the relevant response of welding condition influencing the arc characteristics and operates fast (about 100 megabits per second) and smoothly with necessary adjustments through communication technologies. Such machines also have capacity to control arc by information feedback in the background of high memory capacity of significant number of variables. Thus, it can efficiently take care of deficiency of welder skill and effectively support the robotic welding by addressing the procedural aspects of welding through arc control by relevant adjustment of welding parameters. Some leading power source maker like Lincoln Electric offer advanced control technology for thorough customization of pulse waveform during welding for superior performance. Such system enables the power source to respond rapidly to adjust the behavior of metal transfer tailored to the wire and process. The power source often works on built-in templates for general usage of various combinations of consumable specifications and materials which substantially helps in employing pulsed GMAW process under common variation of weld fabrication. Hence appropriate selection of power source gives more confidence to welder by easy setting of highly complex process parameters of pulsed current GMAW in presence of a synergic control of the operation by automatic adjustment of the frequency and waveform. For pulsed current welding it is advisable to select a power source of higher range of operating capabilities because it extends the choice of welder for a smooth, comfortable and economic welding.

9.2 Computerised Control of P-GMAW Process

An intelligent arc control (IAC) as a function of change in welding current and arc voltage in reference to time facilitates more user friendly smooth operation of pulse current welding from the short arc to specially directed molten metal drop transfer. It provides more efficient high performance practically spatter free welding especially suitable for the out of position welding. A more specific control of weld quality by operating the GMAW process through an algorithm, which is developed as function of welding parameters and characteristics of thin sheet weld, has been successfully explored and implemented in practice [12]. The control system has been designed to operate an automated GMAW process by varying the actuator parameters of the welding robot such as its position and speed as well as other welding parameter from the power source control such as arc voltage and welding current. Here the correction of welding parameters is based on optical/visual feedback of weld characteristics received from the quality monitoring which a system that works in combination with the transient signals of the current and voltage.

In spite of the availability of all the user friendly pulsed current welding power source, welding engineers require thorough knowledge of implications of pulse parameters in control of welding performance as a function of simultaneous interaction with each other. Precise knowledge of effects of the factor ϕ as summarised influence of pulse parameters on characteristics of arc and behaviour of metal transfer provides a basic understanding to analyse the variation in weld characteristics as a function of pulse parameters of P-GMAW process. Depending upon basic analytical knowledge base of the consumable electrode arc welding process [13], understanding of the effect of ϕ may also form a basis of designing improved automation in control system of P-GMAW process. Such a development can introduce a more user friendly versatile application of P-GMAW process through selection of right pulse parameters for desired weld quality. However, simultaneous influence of pulse (peak) current (I_p), base current (I_b), pulse time (t_p) and pulse frequency (f) on each other at a given mean current (I_m) complicates the use of this concept in P-GMAW process even with a modern commercial pulse GMA welding power sources. This is primarily because most of the commercial welding power sources are basically programmed for arc control under given conditions of welding consumables but do not specifically consider its influence on weld quality. In case of the pulse current GMA welding the arc stability can be achieved in several combinations of pulse parameters while their effects on weld quality may not be the same. It happens because the energy introduced and distributed in the pulsing system varies with a change in the ratio of (I_b/I_p) and f.t_b

respectively. A good correlation of the factor ϕ with the weld characteristics as discussed in earlier Chaps. 3-7 and with the pulse parameters as (I_b/I_p) and $f.t_b$ may facilitate development of controlling algorithm for more all-round operation of the P-GMAW process for desired weld quality. The factor ϕ maintains good correlations with the parameters (I_b/I_p), $f.t_b$ and I_m as it is observed (Figs. 9.3 and 9.4) in case of P-GMA welding of Al-Zn-Mg alloy using 1.6 mm diameter AL-Mg filler wire.

$$(f_{tb}) = (\alpha\, l_n I_m - \beta)\ \ln\phi + \gamma\ \ln I_m + \delta \tag{9.1}$$

$$\left(I_b / I_p\right) = (\mu - \upsilon\, I_m)\,\phi + \eta\ I_m - \theta \tag{9.2}$$

The α, β, γ, δ, μ, υ, η and θ are empirical constants. In case of pulse current GMAW I_m is generally considered in the range beyond transition current of filler wire to ensure better weld characteristics and properties [14–16]. However, its upper limit is decided by excessive fluidity and burning of the weld. Thus, for a required ϕ giving desired weld properties it is necessary to determine [17] the values of I_b, I_p, f and t_b satisfying the preferred values of $(f.t_b)$ and (I_b/I_p) at a given

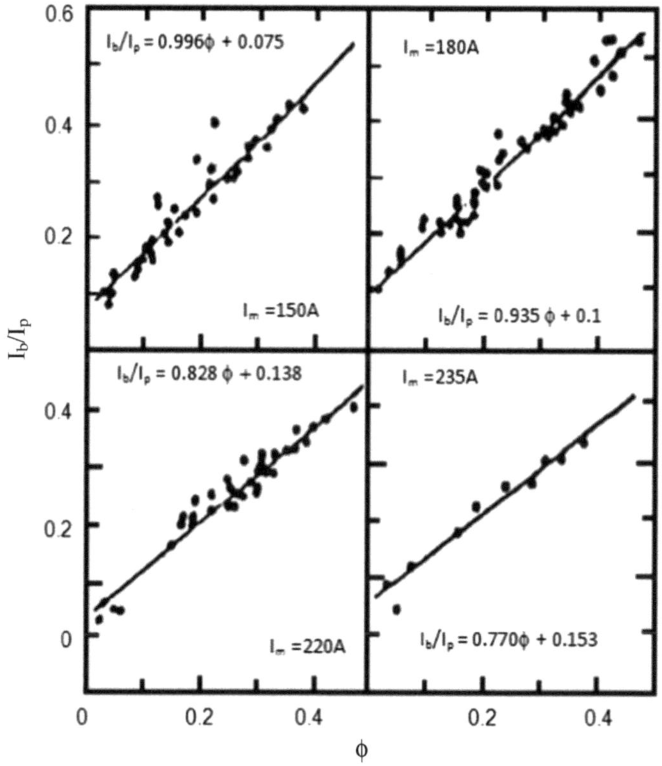

Fig. 9.3 At different I_m effect of ϕ on (I_b/I_p)

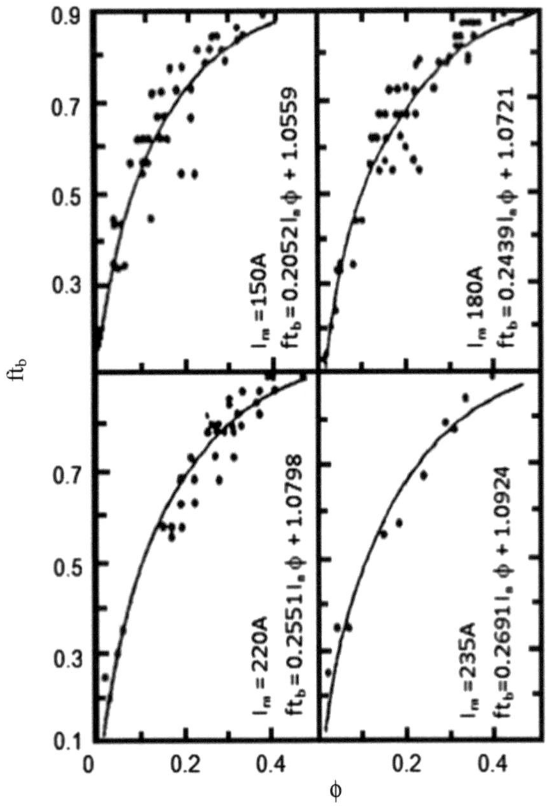

Fig. 9.4 At different I_m effect of ϕ on (ft_b)

I_m. It is possible because these parameters follow good correlations amongst them (Fig. 9.5).

$$t_b = \left[l_n \left(I_b / I_p\right) - l_n\{(\sigma' - l_n I_m + \mu')\, l_n f - (\theta\, l_n I_m + \eta')\}\right] / \left[e^{(\gamma' + \alpha' \ln Im) f} .\ (\delta' + \beta'\, l_n I_m)\right] \quad (9.3)$$

where, the σ', α', μ', θ', ν', γ', α', δ', and β are empirical constants.

Such expressions form the base to derive a methodology for selection of pulse parameters giving desired properties of pulse current GMA weld. The methodology involves solving empirical expressions correlating the properties and characteristics of the weld with the pulse parameters as discussed in earlier chapters. The methodology may follow a software based analytical programme as stated in (Fig. 9.6). This kind of analysis is able to predict the pulse parameters [18, 19] for any pulse current power source with accuracy of the order of about 80% (Fig. 9.7). However, a relatively low accuracy in estimation of t_b occurs primarily due to introducing slope in the pulse pattern of some power sources, different procedure for assessment of t_b for different power sources (Fig. 2.1) and difficulties in certain

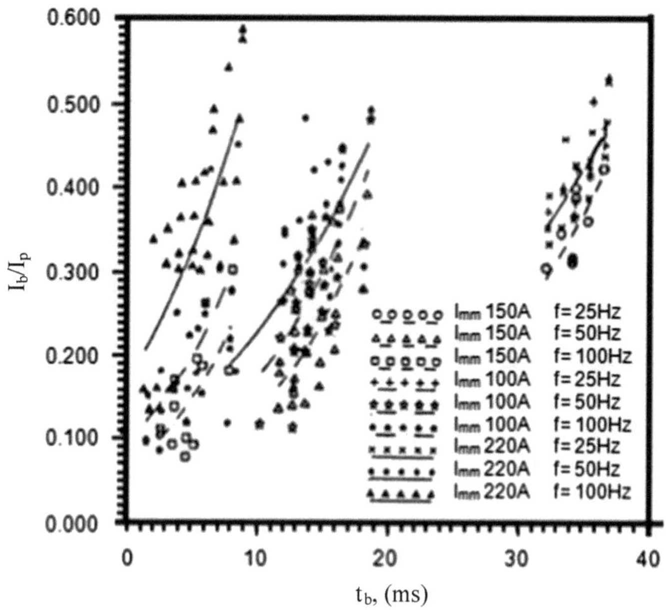

Fig. 9.5 At different I_m and ϕ effect of t_b on (I_b/I_p)

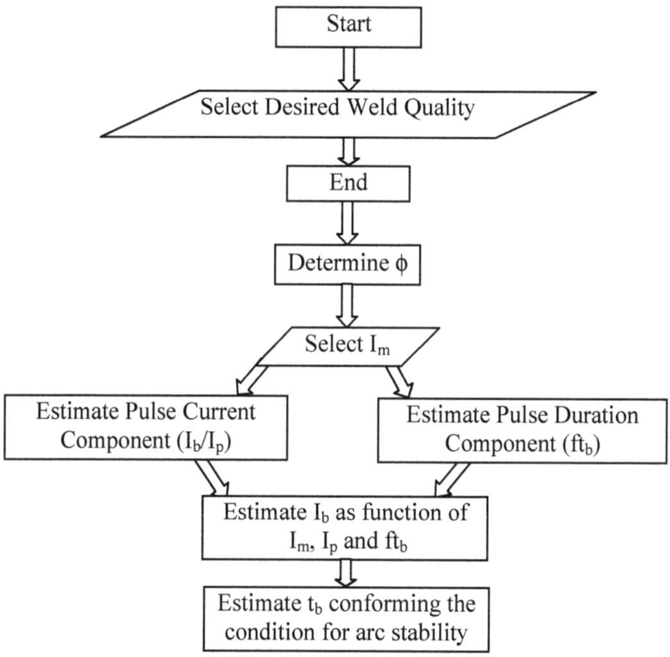

Fig. 9.6 Operating flow chart of the software

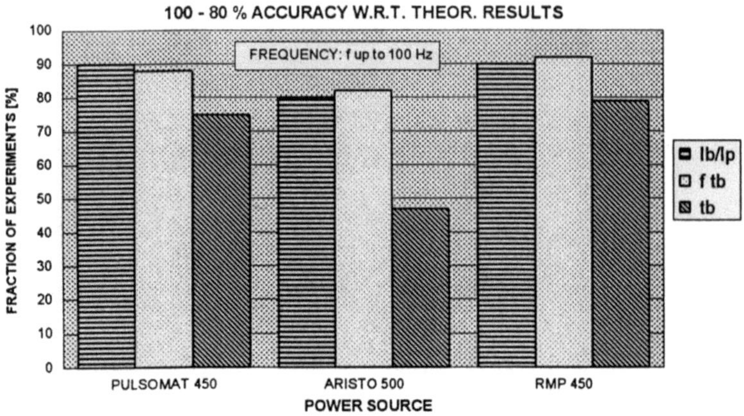

Fig. 9.7 Accuracy of experimental data with respect to theoretical results in case of different commercial power sources

cases to identify the accurate duration of t_b due to curvilinear pattern of pulse characteristics in the oscilloscope. Some typical comparative example of the estimated and measured pulse parameters checked in case of commercial pulse current power source has been shown in Tables 9.1 and 9.2. They are found practically well in agreement to each other.

Similarly the variation in accuracy of experimental to theoretical results regarding the (I_b/I_p) and $f.t_b$ with a change of pulse frequency at different power sources has been shown in Figs. 9.8 and 9.9 respectively [20, 21]. The figures show that a considerable fraction of experimental (I_b/I_p) and $f.t_b$ lies in the range of 100–80% accuracy with respect to their theoretical value under almost all pulse frequencies at different power sources. The significant universality of these empirical equations to various power sources, operating at different mode of control over pulsation, may be realized by the energy balance concept. The required energy input for a certain rate of metal deposition is almost constant for given welding conditions, which are primarily related to electrode size, material and shielding

Table 9.1 Response of power source to estimated welding parameters during deposition of weld bead on Al–Zn–Mg plate using A–Mg filler wire

Factor ϕ	Wire feed speed (m/min)	Calculated I_m (A)	Measured I_m (A)	Arc voltage (V)	I_b (A)	I_p (A)	T_b (ms)	T_p (ms)	F (Hz)
0.022	10.4	320	327	28	142	332	1.55	28.75	33
0.141	8.2	250	259	26	124	327	4.65	7.85	80
0.156	4.8	150	145	22	71	229	6.3	6.2	80
0.168	7.2	220	219	24	113	315	2.35	2.65	200
0.176	5.7	180	180	23	93	280	2.65	2.35	200
0.177	6.4	200	199	24	104	301	2.05	1.95	250

Table 9.2 Behaviour of measured values of power source in different ranges of f and I_m during operation at estimated welding parameters

ϕ Meas	I_m (A)		I_b (A)		I_p (A)		t_b (ms)		t_p (ms)		Arc voltage (V)	f (Hz)
	Cal	Mes	Cal	Mes	Cal	Mes	Cal	Mes	Cal	Mes		
0.053	149	159	29	33	230	240	6.0	5.9	9.1	9.1	21	66
0.1	149	159	49	53	282	296	9.4	9.1	7.2	7.0	20	60
0.198	161	160	86	92	311	328	10.2	10.0	4.0	4.2	21	70
0.248	150	161	101	107	311	327	10.5	10.4	3.2	3.3	22	73
0.430	149	168	139	145	293	310	12.7	12.6	1.0	0.98	21	73
0.062	205	220	50	54	300	315	12.0	12.2	21.0	21.1	20	30
0.112	202	215	74	80	353	368	16.3	16.0	15.0	14.9	22	32
0.204	208	213	117	122	394	410	17.3	17.1	7.7	7.6	22	40
0.254	208	212	135	144	399	417	14.2	14.1	4.6	4.6	23	53
0.429	195	212	184	191	391	411	14.9	14.8	1.25	1.2	21	62
0.059	248	258	66	72	318	332	3.9	3.8	10.0	9.9	20	72
0.108	248	263	90	98	371	387	6.7	6.6	8.9	8.8	25	64
0.190	242	254	135	141	425	444	7.9	7.6	5.4	5.7	23	75
0.238	248	265	155	161	439	456	8.4	8.2	4.2	4.1	26	79
0.385	253	262	210	220	452	473	9.3	9.6	2.06	1.8	23	88

Cal calculated *Mes* measured

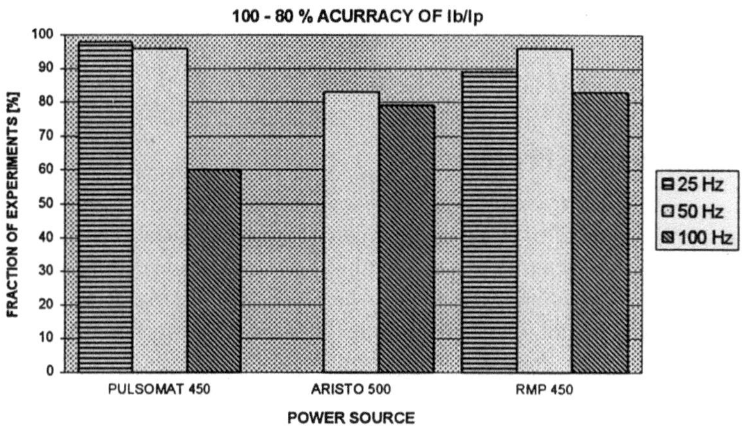

Fig. 9.8 At different pulse frequencies the accuracy of experimental (I_b/I_p) w.r.t. theoretical (I_b/I_p) observed in various power sources

(inert gas) environment. The variation in mode of pulsation with the change in pulse parameters and power sources changes only the pattern of distribution of energy input per unit time. Thus, the empirical equations of $f.t_b$, (I_b/I_p) and t_b as a function of other pulse parameters respectively (Eqs. 9.1–9.3) can be used efficiently for any set of pulse parameters and pulse current GMA welding power sources under given conditions of welding as mentioned earlier. It is also interestingly noted that the experimental data on the $f.t_b$, (I_b/I_p) and t_b are well in agreement to the theoretical

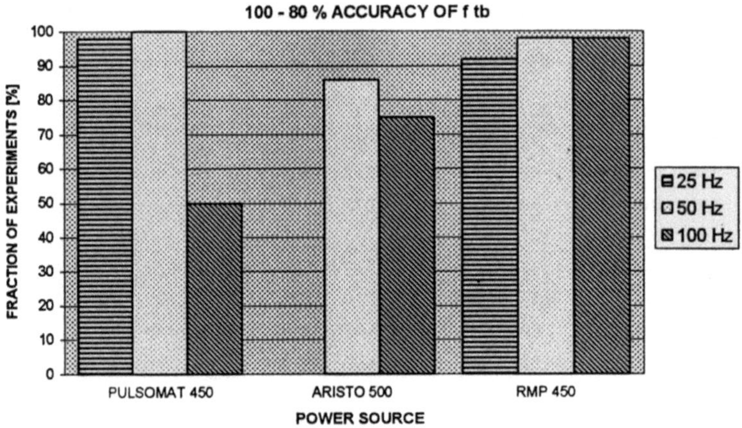

Fig. 9.9 At different pulse frequencies the accuracy of experimental ($f.t_b$) w.r.t. theoretical ($f.t_b$) observed in various power sources

results, which are evaluated with the help of the empirical correlations as function of pulse parameters without considering arc voltage that varies widely from 9–36 V. This behaviour reveals that the arc voltage is having comparatively minor influence on the empirical relationships for estimation of the pulse-off- time fraction ($f.t_b$), energy ratio (I_b/I_p) and pulse off time (t_b).

9.3 Commercial Prospect

The increasing demand of more cost effective high quality weld fabrication in cut throat competition of international market has been more competently addressed by technology driven welding engineering where the new generation welding power source has played a vital role. This has largely happened primarily by reducing the involvement of expert decision of any individual during weld fabrication. It has made the designing of weld and jig and fixturing aspects of weld preparation and execution of welding fabrication more comfortable and hassle free. The modern GMA welding power source with advanced technology mostly provides the pulse current facility that makes the welding operation more user friendly [22] to produce high quality weld at relatively cheaper cost with better long run economic return in spite of a comparatively higher capital investment on relatively costlier power source. The wide operating range of the pulsed GMAW machine allows its use in large range of applications with respect to the variety of materials, section size, weld design and location of welding [23]. It gives higher productivity due to large deposition rate at controlled thermal and mechanical effects for acceptable weld quality. The fume generation in GMAW is a matter of concern [24] especially in case of welding of certain metals including stainless steel and during the use of flux cored filler wire. The reduced severity in thermal environment with faster welding gives relatively low fume generation per unit length of welding that maintains a healthier environment with minimum welder fatigue.

The P-GMAW is conveniently used in hybrid laser welding process that facilitates more efficient production of defect free long weld joint of steel plate. The laser hybrid process is often used in ship building industry. The formation of bead humps in high-speed P-GMAW process was critically analyzed by numerical simulation based on capillary stability of the molten metal deposit and it is prevented by additional laser heat input [25]. The P-GMA welding of long steel plate requires large weld pool weld poolwith sufficient bead width to prevent capillary instability. This is achieved in presence of a large spot size at welding front with sufficient laser beam intensity. A defocused laser beam is used for melting of material in front of the weld pool weld poolto suppress the weld bead hump formation in the hybrid process.

References

1. Kola Rik, L., Rikova, M.K., Kovanda, K., Pantucek, M., Vondrou, P. : Advanced functions of a modern power source for GMAW welding of steel. Acta Polytech. **52**, 4 (2012)
2. Kolarik, L., Kolarikova, M., Kovanda, K., Pantucek, M., Vondrou, P.: Evaluation of modern power source for GMAW welding equipped with advanced arc control. MM Sci. J. July, 322–325 (2012)
3. Kola Rik, L. a kol.: GMAW svarovanı ocelovych materialu metodou force arc. TechMat 2011. Pardubice: Univerzita Pardubice, s. 184–189. ISBN 978-80-7395-431-4 (2011)
4. Lincoln electric: Surface tension transfer [online]. http://www.lincolnelectric.com/assets/enUS/Products/literature/NX220.pdf[cit. 2012–03–29] 2006
5. Esab: Swift Arc Transfer. http://products.esab.com/ESABImages/SwiftArtTransferfinal.pdf [online]. [cit. 2012–03–29] 2009
6. Fronius: Rozsirte si vedomosti. http://www.fronius.com/cps/rde/xchg/SID-77958394-23586247/froniusceskarepublika/hs.xsl/29104.htm[online]. [cit. 2004–9–13] 2009
7. Fronius: Time and time twin welding [online]. http://www.fronius.com/cps/rde/xchg/froniusinternational/hs.xsl/99163ENGHTML.htm [cit. 2012–05–29] 2011
8. Migatronic: Improved quality and higher productivity from energy-dense powerarc technology for thick-plate welding [online]. http://www.migatronic.com/default.aspx?m=4&i=115&pi=1&pr=0 [cit. 2012–03–29] 2010
9. CSN EN ISO 6947. Syarovan´ı a pˇ r´ıbuzn´e procesy – Polohy svarovan Praha: Cesky normalizaˇcn´ı institut, listopad, (2011)
10. Oddˇelen´ı v´yzkumu a v´yvoje, Migatronic A/S. Intelligent Arc Control – proces pro sniˇzov´an´ ı rozstˇriku a vnesen´eho tepla pˇri zkratov´em pˇrenosu. Svˇet svaru, **15**, 3, 12–14. ISSN 1214–4983 (2011)
11. Relyea, S.: GMAW in the digital age—Advancements in inverter technology help manufacturers achieve high arc performance, efficiency, Practical welding today, January/February, (2014)
12. Ghosh, Aritra K., Class, R.: (Daimler AG, Stuttgart), Ein robustes System zur Parameterfindung, Prozessregelung und -optimierung für automatisierte MSG Schweißprozesse, 2. VDI-Konferenz "Fügetechnik im Fahrzeugbau", Leipzig, (2012)
13. Cornu, J.: Advanced Welding Systems, Vol. 2 Consumable electrode process. IFS Publications/Springer-Verlag, Berlin, English ed., 86–96 (1988)

14. Ghosh, P.K., Gupta, S.R., Gupta, P.C., Rathi, R.: Fatigue characteristics of pulsed MIG welded Al–Zn–Mg alloy. J. Mat. Sci. **26**, 6161–6170 (1991)
15. Ghosh, P.K., Gupta, S.R., Gupta, P.C., Rathi, R.: Influence of pulsed current welding on the microstructure and porosity content of Al–Zn–Mg alloy weldment. Pract. Met. **27**, 613–626 (1990)
16. Ghosh, P.K., Gupta, S.R., Gupta, P.C., Rathi, R.: Pulsed MIG welding of Al–Zn–Mg alloy. Mater. Trans., JIM **31**(8), 723–729 (1990)
17. Ghosh, P.K.: An analysis of weld characteristics as a function of pulse current MIG welding parameters. Int. J. Joining Mat. **8**(4), 157–161 (1996)
18. Weber, Jeff: Pulsed GMAW a plus for mid western metalworker. Weld. J. **61**(11), 51–52 (1982)
19. Ghosh, P.K., Dorn, L., Hübner, M., Goyal, V.K.: Arc characteristics and behaviour of metal transfer in pulsed current GMA welding of aluminium alloy. J. Mater. Process Tech. **194**(4), 163–175 (2007)
20. Dorn, L., Ghosh, P.K., Goecke, S., Possibility of aluminium MIG welding by modification of pulse parameters using their correlationships valid for different power sources. In: Procd. IIW Asia Pacific Welding Congress, Auckland, New Zealand, 4–9 February, 897–903 (1996)
21. Ghosh, P.K., Dorn, L., Goecke, S.F.: Universality of correlationships among pulse parameters for different MIG welding power sources. Int. J. Joining Mater. **13**(2), 40–47 (2001)
22. Mvola, B., Kah, P., Martikainen, J., Hiltunen, E. : Applications and benefits of adaptive pulsed GMAW, ISSN 1392—1207. Mechanika **19**, 6, 694–701 (2013)
23. Dorling, D. V. and Taylor, D.: Developments in tandem pulsed gas metal arc welding for grade 690 pipelines, Canadian Welding Association Journal, 8–10, (2005)
24. Chae, H., Kim, C., Kim, J., Rhee, S.: Fume generation behaviors in short circuit mode during gas metal arc welding and flux cored arc welding. Mater. Trans. **47**(7), 1859–1863 (2006)
25. Cho, M.H., Farson, D.F.: Simulation Study of a Hybrid Process for the Prevention of Weld Bead Hump Formation. Weld. J. **86**(9), 253s–262s (2007)

Erratum to: Pulse Current Gas Metal Arc Welding

Prakriti Kumar Ghosh

Erratum to:
P.K. Ghosh, *Pulse Current Gas Metal Arc Welding*, Materials Forming, Machining and Tribology, DOI 10.1007/978-981-10-3557-9

The original version of the book was inadvertently published without the preface in Frontmatter. The book has been updated with the change.

The updated online version of this book can be found at
http://dx.doi.org/10.1007/978-981-10-3557-9

P.K. Ghosh, *Pulse Current Gas Metal Arc Welding*, Materials Forming, Machining and Tribology, DOI 10.1007/978-981-10-3557-9_10

Index

P.K. Ghosh, *Pulse Current Gas Metal Arc Welding*, Materials Forming, Machining and Tribology, DOI 10.1007/978-981-10-3557-9

MIX
Papier aus verantwortungsvollen Quellen
Paper from responsible sources
FSC® C105338

If you have any concerns about our products,
you can contact us on
ProductSafety@springernature.com

In case Publisher is established outside the EU,
the EU authorized representative is:
Springer Nature Customer Service Center GmbH
Europaplatz 3, 69115 Heidelberg, Germany

Printed by Libri Plureos GmbH
in Hamburg, Germany